Foams

Foams

J. J. Bikerman

Springer-Verlag New York . Heidelberg . Berlin

1973

Library of Congress Cataloging in Publication Data

Bikerman, Jacob Joseph, 1898-
Foams

(Applied physics and engineering, v. 10)
Includes bibliographical references.
1. Foam.   I. Title.
QD549.B5 1973   541'.34514   72-94700
ISBN 0-387-06108-8

Printed in the United States of America

ISBN 0-387-06108-8   Springer-Verlag New York . Heidelberg . Berlin
ISBN 3-540-06108-8   Springer-Verlag Berlin . Heidelberg . New York

Preface

The book *Foams: Theory and Industrial Applications*, written by the undersigned and three collaborators and published in 1953, is still the only monograph on liquid foam in the English language. Naturally the science of foams had advanced in the intervening years so that a practically new book had to be prepared to give justice to the present state of our knowledge. This monograph has only one author and does not deal with solid foams, fire-fighting foams, and flotation, on which information is available elsewhere. The other applications of foam and its fundamental properties are reviewed at length and, whenever possible, attempts are made to reach the truth through a maze of conflicting evidence.

February 1973   J. J. BIKERMAN

Contents

viii   *Contents*

Foams

CHAPTER 1

GENERAL. FOAM FILMS

§1. *Definitions.* The foams with which this books deals are agglomerations of gas bubbles separated from each other by thin liquid films. The term froth in the usual parlance often implies low stability but is not often employed in the scientific literature. Dictionaries list also the word *spume*, nearly synonymous with foam and froth, but it is not used by scientists.

Foams are colloidal systems. In the classification based on the state of aggregation of the two phases, foams commonly belong to the first group. The list is:

1. Gases dispersed in liquids (foams, gas emulsions)
2. Liquids dispersed in gases (mists, fogs, liquid aerosols)
3. Gases dispersed in solids (solid foams)
4. Solids dispersed in gases (fumes, smokes, solid aerosols)
5. Liquids dispersed in liquids (emulsions)
6. Liquids dispersed in solids (some gels)
7. Solids dispersed in liquids (suspensions, sols)
8. Solids dispersed in solids

The third class in the list, that is, solid foams, at present has great commercial importance; everyone is familiar with foam rubber, polyurethane foams, etc. Those interested in these systems are served by special monographs, such as that by Benning.[1] In all true foams, whether the films are liquid or solid, each bubble is closed, that is, has no gas-filled channels connecting it with the neighboring bubbles (also known as cells in solid foams). When such connections exist, a sponge is present. In sponges both phases (that is, gas and solid) are continuous; whereas in foams the gas forms a discontinuous or disperse phase, and the continuous phase (or the dispersion medium) is liquid or solid. Bread usually is mainly a sponge and partly a solid foam.

Dispersions of a gas in a liquid, in which the thickness of the interstitial liquid layers is at least commensurate with the bubble diameter, are commonly designated as *gas emulsions*; the term "gas dispersions" also is met with, and some scientists also in English speaking countries use the German word *Kugelschaum* (= sphere foam). The boundary between

1

foams (sometimes denoted in Germany as *Polyederschaum*) and gas emulsions naturally is not quantitatively defined. If 1 cm³ of foam or gas emulsion contains φ cm³ of liquid and $1 - \varphi$ cm³ of gas, it is customary to speak of foams when φ is less than 0.1 and of gas emulsions when φ exceeds 0.9; the region in between is a "gray zone." The $(1 - \varphi)$ of a gas emulsion frequently is referred to as its *hold-up*, φ is known as wetness, and the ratio of foam volume to the volume of liquid in foam, that is $1/\varphi$, is the expansion factor or expansion ratio.

§2. **Gas emulsions.**   Several important operations, both in the laboratory and in industry, are performed in gas emulsions. For instance, when moist air is bubbled through sulfuric acid to dry the former, a gas emulsion is present, and the process is an example of mass transfer from one phase to the other. When hot air is injected into a cool liquid, heat transfer takes place in the gas emulsion produced. These effects are considered in books on chemical engineering (see, for instance, Valentin[2]), and are outside the scope of the present monograph. Some insight into the formation of gas emulsions is given in Chapter 2 because this information helps in the understanding of foam formation. The most basic facts of the colloid behavior of these systems will be found in this section and in §4.

Gas emulsions are unstable in the gravitational field. If a gas, whose density and viscosity are, respectively, ρ_1 (g/cm³) and η_1 (g/cm·sec), is present as a spherical bubble (of radius r) in a liquid of density ρ_2 and viscosity η_2, then the bubble rises with the velocity:

$$u = \frac{2gr^2}{9\eta_2}(\rho_2 - \rho_1)\frac{3\eta_1 + 3\eta_2}{3\eta_1 + 2\eta_2} \text{ cm/sec} \qquad (1.1)$$

g (cm/sec²) is the acceleration due to gravity. This is the equation of Rybczynski and Hadamar. Since, in general, $\rho_2 \gg \rho_1$ and $\eta_2 \gg \eta_1$, this equation can be simplified to

$$u = \frac{gr^2\rho_2}{3\eta_2}. \qquad (1.2)$$

For water at room temperature, $\rho_2 = 1$ g/cm³ and $\eta_2 = 0.01$ g/cm sec. Hence, u is approximately $3 \times 10^4 r^2$ cm/sec. If r is $1\,\mu$ ($= 10^{-4}$ cm), u is near $3\,\mu$/sec or 1 cm/hr; and if r is $10\,\mu$, u is about $300\,\mu$/sec. It is clear that a gas emulsion of visible bubbles in water or another liquid of similar ρ_2 and η_2 would lose its disperse phase within minutes or hours.

The experimental values of the rate u of ascent usually are smaller than the theory predicts and in many instances satisfy the equation

$$u = \frac{2gr^2\rho}{9\eta_2}, \qquad (1.3)$$

valid for solid spheres in a liquid, better than formula (1.2); ρ here means the difference between the densities of the two phases. Several reasons have been advanced for the observed deviation from the theory, and two of these reasons are of fundamental importance to the science of foams.

One is the Marangoni effect discussed in more detail in §11. Unless the liquid is unusually pure, it is likely to contain surface-active ingredients, that is, substances that lower the surface tension γ of the solution and, consequently, accumulate in the surface layer (§135). When a bubble rises, the liquid around it flows from the upper to the lower pole of the "sphere" and thus tends to depress the local concentration of the surface-active impurity at the zenith and to augment it at the nadir. Hence, the tension of the gas-liquid boundary tends to be higher near the top than near the bottom of the bubble. This difference in γ values causes flow of the inter-facial layer from the bottom toward the top, i.e., in the direction opposite to that associated with the bubble rise. As a result, this rise is retarded.

§3. *Capillary pressure.* The other reason derives from Laplace's law of capillary pressure, which is invoked many times in this book. If a gas-liquid interface is curved and the two principal radii of curvature at a point are R_1 and R_2, then the pressure in one phase is greater than the pressure in the other by the amount:

$$P_c = \gamma\left(\frac{1}{R_1}+\frac{1}{R_2}\right);\qquad(1.4)$$

P_c is the capillary pressure (g/cm sec^2) at this point. Both radii are perpendicular to the surface as the radii of a circle are perpendicular to the circumference. The two planes in which the radii swing are perpendicular to each other. If both radii are on one side of the interface, as in a sphere, both are considered positive and P_c also is positive, according to Equation 1.4. This means that the pressure in a sphere is greater than outside; in general, the pressure steps up by P_c whenever an interface is crossed from the convex to the concave side. A comparison with a blown-up rubber balloon may help in remembering this rule.

In many instances the signs of R_1 and R_2 are different. One of the two principal radii of curvature of a saddle is above the horse and swings from the head to the tail; and the other is below the horse and swings from one flank to the other. If R_1 happens to be equal to $-R_2$, then the pressure on both sides of the doubly curved interface is identical.

Hydrostatic pressure, if present, has to be added to the capillary pressure to predict the shape of a liquid surface. If a bubble is suspended in a liquid and the hydrostatic pressure at the top of the bubble is p_0, then it is $p_0+g\rho_1 z$ in the bubble next to its bottom; the height of the bubble is z cm. At the same bottom pole but in the liquid, hydrostatic pressure is $p_0+g\rho_2 z$. Thus the difference between the hydrostatic pressures inside and outside the bubble is zero at the upper pole and $g(\rho_2-\rho_1)z$ at the lower pole. In equilibrium, this difference must be compensated by the difference in capillary pressures. If the two radii of curvature of the top point of the bubble are each equal to R_a (they must be identical in an axially symmetric body) and the two radii of the bottom surface are each R_b, then the equation

$$2\gamma\left(\frac{1}{R_a}-\frac{1}{R_b}\right) = g(\rho_2-\rho_1)z \qquad (1.5)$$

must be valid. Since ρ_2 is greater than ρ_1, also $1/R_a$ must be greater than $1/R_b$, that is, the radius of curvature increases from the top to the bottom. Thus a bubble cannot be truly spherical in the gravitational field, and Equation 1.1 derived for spheres cannot be exact.

It is clear that the difference between R_a and R_b is greater the greater z. In other words, the bigger the bubble the more it deviates from the spherical shape. When the bubble is not stationary, its shape is affected also by viscous traction and inertia, and a large rising bubble may be more like a medusa than a ball.

§4. *Gas emulsions.*   As long as gas emulsions exist, their properties are, or ought to be, similar to those of liquid-inliquid emulsions. A difference between the two types of dispersion is caused by the fact that liquids are almost incompressible while gases are readily compressed. A bubble at the bottom of a 10-m tall tank filled with water is at a pressure of about 2 atmospheres and, consequently, has only half the volume of a bubble of identical mass but situated near the air-water surface. Hence, a rising bubble expands. The capillary pressure of §3 may have an even greater effect. The pressure in a spherical bubble of radius R is $p_0+(2\gamma/R)$. For water at room temperature, $\gamma = 72$ g/sec^2. The capillary pressure is about 1.4×10^6 dynes/cm^2 when $R = 1$ μ that is, markedly greater than the atmospheric pressure which usually is near 10^6 dynes/cm^2 (or 1 bar). The gas pressure in this bubble is 2.4 bars or 2.4 times as great as the outside pressure (near the top of the liquid column).

The compression of gas in the bubble should be taken into account when the ratio of the gas mass m_1 to the mass m_2 of liquid is computed from the volume ratio φ defined in §1. Obviously, $m_1/m_2 = \rho_1(1-\varphi)/\rho_2\varphi$, but ρ_1 is the density of the compressed gas which may be several times that of the gas at atmospheric pressure.

It cannot be doubted that the viscosity of a gas emulsion composed of very small bubbles is nearly equal to that of an ordinary emulsion having an identical value of φ; naturally the dispersion medium also must be identical in the two systems. The restriction to very small bubbles is necessary because large bubbles are too easily deformed and the extent of deformation depends on the velocity gradient (or shear stress); (see §§2 and 3).

Gas bubbles move in an electrostatic field, that is, exhibit electrophoresis, in the same manner as do emulsion droplets. If the bubbles are "protected," that is, coated with a film, their electrokinetic potential depends on the nature of that film; when the film is a protein, the potential varies with the pH of the liquid as if the protein only, and no gas, were present.

The coagulation of gas emulsions (i.e., formation of a few large from many small bubbles) is complicated by the continual rise of the bubbles and has not been well studied.

Foam Films

§5. Film shape. Foams are agglomerations of bubbles, but they are also structures composed of thin liquid films. It is convenient to discuss some properties of these films before reviewing the characteristics of the final systems.

To prepare a foam film it is sufficient to immerse a wire loop into a soap solution and to withdraw the loop into the air; a film spanning the loop usually remains. Plateau[3], the blind physicist who was the first investigator of these films, sometimes encountered difficulties in that the film slipped off the loop and rejoined the aqueous phase. To avoid such mishaps, Boys[4] recommended employing tinned iron wires, about 1 mm in diameter, or similar aluminum filaments. They should be washed in a soap solution before use.

An important effect was observed by Plateau in a slightly more complicated arrangement. Two horizontal wire rings are placed one on the other, and the groove between them is filled with a soap solution (Fig. 1.1). Then the upper ring is cautiously lifted to a position exemplified by Fig. 1.2. A roughly cylindrical vertical film (*l* in Fig. 1.2) remains between the rings. When the distance between them is further increased, the film acquires the shape of an hourglass (Fig. 1.3); if the upper ring is raised further, the "waist" of the hourglass becomes thinner and, finally, the film snaps and gives rise to two films spread across the two rings.

Why is the nearly cylindrical film unstable? Equation 1.4 is the answer. The films usually are so thin that gravitation may be disregarded and the shape is determined by capillary forces alone. The same air, under identical pressure, is present both outside and inside the cylinder. Hence the pressure difference across the film must be zero. This is possible only as long as the sum $(1/R_1)+(1/R_2)$ is equal to zero at every point of the film. Obviously, a cylinder does not satisfy this condition. The radius swinging in a plane perpendicular to the drawing (Fig. 1.2) is equal to the radius of the cylinder, but the radius swinging in the plane of the drawing is infinitely great (because the generatrix of a cylinder is a straight line); thus the above sum cannot be zero.

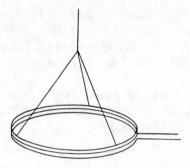

Fig. 1.1. A soap film between two rings in contact.

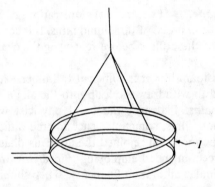

Fig. 1.2. A nearly cylindrical soap film (*1*) between two neighboring rings.

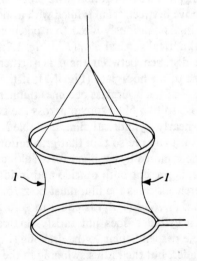

Fig. 1.3. A catenoid soap film between two distant rings. The film cannot remain a cylinder.

There is only one three-dimensional figure for which the equation

$$\frac{1}{R_1} + \frac{1}{R_2} = 0 \tag{1.6}$$

is valid. It is called *catenoid*. Its profile is a catenary curve; thus the two curves marked *1* in Fig. 1.3 are catenaries. This is the shape which a flexible cord assumes when it is attached to two supports situated on the same level in a gravitational field. The two radii of curvature are on the opposite sides of the film and, consequently, have opposite signs, as indicated in §3. If a point of the film is in the distance x cm from the vertical axis of the system and the film makes an (acute) angle α with the horizontal at this point, then the inner radius of curvature there is $x/\sin \alpha$. The outside radii of curvature are those of curves *1* of Fig. 1.3 and swing in the plane of the paper. It follows from the equation for catenoids that the above-mentioned transformation of an hourglass-like film into two horizontal films takes place when the distance between the rings is two-thirds of the ring diameter. This conclusion was confirmed in Plateau's experiments. For a recent paper on catenoids see Dickey[5].

§6.   If, contrary to the catenoid of §5, a film (or an assembly of films) encloses a volume, the gas pressure in the latter may be different from the pressure outside but the difference P_c will be constant everywhere and satisfy Equation 1.4. **Example 1:** a wire frame consisting of 12 cube edges is dipped for a moment into a solution capable of foaming. When it emerges again, the six films connecting the wires are flat and form a cube. Since all radii of curvature are infinitely great, the pressures inside and outside are identical. **Example 2:** two wire rings stiffened with a connecting wire (*1* in Fig. 1.4) are dipped and lifted back. Three films remain on the

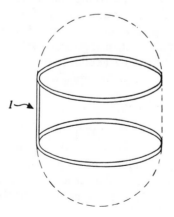

Fig. 1.4.   A soap film partly cylindrical and partly spherical. The radius of the spheres is twice the radius of the cylinder. *1* is the wire connecting two solid rings.

frame. One is cylindrical and vertical, and the other two are more or less horizontal and rest one on each of the rings. If the external radius of each ring is r, one of the radii of curvature of the cylindrical surface also is equal to r, while the other is infinity, as pointed out in §5. Thus $P_c = \gamma/r$. The P_c across the other films also must be equal to γ/r. This means that these films are not plane; they are spherical segments and concave toward the center of the system, as indicated by the dotted lines; each of their radii of curvature is equal to $2r$. When $R_1 = R_2 = 2r$, then $\gamma[(1/R_1)+(1/R_2)] = \gamma/r$, as it should be.

Several surfaces are known for which the sum of the inverse radii of curvature is constant everywhere. A plane and a cylinder are referred to above. Plateau discovered a nodoid and an unduloid which look like corrugated cylinders. None of these surfaces alone can surround a volume. On the other hand, a sphere has a constant value of $(1/R_1)+(1/R_2)$ (because the radius of the sphere is equal to both R_1 and R_2) and also encloses a volume. This is the reason for bubbles being spherical in the absence of gravitation and motion.

§7. *Plateau borders.*   Here the effect of gravitation on films is reviewed.

When the film is vertical, as in Fig. 1.2, the weight suspended on its upper end is obviously greater than that hanging near its lower end. The weight difference is gmz, if g is the acceleration due to gravity, m the mass of the film for 1 cm height (minus the mass of an equal volume of air), and z is the vertical distance of the selected point above the lower wire ring. If, at every level, this weight is balanced by the surface tension force $2(w+\delta)\gamma$ [w is the width and δ the thickness of the film so that $2(w+\delta)$ is its perimeter], then surface tension γ must increase with z.[6] A liquid consisting of only one chemical compound cannot have more than one value of surface tension (at a given temperature); thus no pure liquid can give rise to stable lamellae. When a surface-active solute is present, its concentration can be greater at smaller z than at greater z values, and $d\gamma/dz$ can be a positive quantity as required by equilibrium considerations. Generally speaking, $d\gamma/dz$ is very small. If in the equation

$$gmz = 2(w+\delta)\gamma \qquad (1.7)$$

$\delta = 10^{-4}$cm, $w = 1$ cm, and $m = 10^{-4}$g/cm, the ratio γ/z (which in this instance is identical with $d\gamma/d\dot{z}$) is near 0.05 g/cm·sec². Most data on surface tension have precision limits of 0.1 g/sec².

Equation 1.7 indicates how tall a vertical film can be at the most. The mass m is equal to $w\delta(\rho_2-\rho_1)$; ρ_1 and ρ_2 are the densities of gas and liquid. Hence for each height z the relation

$$z = \frac{2(w+\delta)\gamma}{g(\rho_2-\rho_1)\delta w} \qquad (1.8)$$

must be valid. Since usually δ is much smaller than w and ρ_1 is much smaller than ρ_2, the equation

$$z = \frac{2\gamma}{g\rho_2\delta} \tag{1.9}$$

is sufficiently exact. For aqueous soap solutions the greatest γ can be about 70 g/sec^2 (at room temperature) and $\rho_2 = 1$ g/cm^3. Thus the greatest vertical extent of a film must be about $140/980\delta \approx 0.15/\delta$. The tallest films described in the literature[7] were over 600 cm tall. Their average thickness was not recorded, but presumably it was near 2.5×10^{-4} cm.

These films were produced by withdrawing a large brass ring from a saturated solution of a surfactant consisting of several sodium alkyl sulfonates $C_nH_{2n+1}SO_3Na$, n being 12 to 18. Films of indefinite length (many kilometers) are produced daily, but they are solid (at least after the withdrawal), whereas the aqueous sulfonate films were liquid. The material collected after the bursting of the films could be used again for making another film.

Equation 1.7 implies that inertia forces are negligibly small, that is, the film drains at a very small rate; this assumption should be tested in every instance (see §97).

When the film is horizontal, γ commonly is constant over its whole extent, but complications arise in the region of contact between the film and its rigid support. Usually the support is perfectly wetted by the film. In the instance of Fig. 1.5, which illustrates a lamella closing a vertical cylindrical tube, this means that the gas-liquid interface is vertical (i.e. parallel to the gas-solid interface) along the three-phase line in which the gas, the liquid, and the solid meet. In this manner the two film surfaces, which are plane in the middle of the tube, are strongly curved along the periphery of the film.

The upper of these surfaces still may be in equilibrium; in fact, it may have the shape of the surface of bulk liquid partly filling a beaker. Fig. 1.6 is a magnified profile of the upper left corner of the lamella; mm is the level of the plane portion of the upper interface. A point a, on the curved

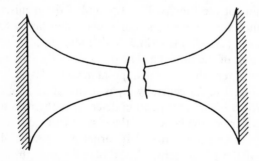

Fig. 1.5. A foam lamella in a cylindrical tube.

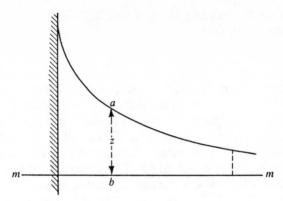

Fig. 1.6. The upper region of contact between the lamella and the wall. Point a is z cm above point b lying in the plane mm of the liquid surface far from the wall.

meniscus, is z cm above mm; hence the hydrostatic pressure in the liquid at b is by $g\rho_2 z$ greater than at a. Denote the gas pressure at a with p_0. If the capillary pressure at this point is P_c, the pressure in the liquid there is $p_0 - P_c$, and the pressure at b is $p_0 - P_c + g\rho_2 z$. Above the plane area of the lamella and next to it, the gas pressure is $p_0 + g\rho_1 z$. As there the lamella is plane, also the pressure in the liquid underneath the plane boundary is $p_0 + g\rho_1 z$. In equilibrium, the equation $p_0 - P_c + g\rho_2 z = p_0 + g\rho_1 z$, that is,

$$P_c = g(\rho_2 - \rho_1)z \qquad (1.10)$$

or

$$\frac{1}{R_1} + \frac{1}{R_2} = \frac{g\rho z}{\gamma} \qquad (1.11)$$

must be satisfied; ρ again is the difference between the two densities. Equation 1.11 or an analogous equation is met with very often when the shape of a liquid surface is determined by both gravitation and capillarity.

Both radii of curvature (R_1 and R_2) are perpendicular to the interface. If r is the radius of the tube of Fig. 1.5, and if the meniscus at point a makes an (acute) angle α with the horizontal, then (as in §5) one of the radii of curvature (say, R_1) is $r/\sin \alpha$. This radius swings in a plane perpendicular to that of the drawing. The other, i.e., R_2, is the radius of the circle which, for a very short length at point a, coincides with the meniscus curve seen in Fig. 1.6. As generally R_1 is many times as great as R_2, $P_c \approx \gamma/R_2$, and $R_2 \approx \gamma/g\rho z$. This relation determines the meniscus profile of the upper boundary of the lamella if the system is in equilibrium.

Fig. 1.7 represents the bottom left corner of the same film (Fig. 1.5). The pressure of the gas underneath the plane liquid surface and in the liquid right above this surface is equal; let its value be p_1. In equilibrium, the pressure in the liquid at a also would be p_1. Hence the hydrostatic pressure in the liquid at b would amount to $p_1 + g\rho_2 z$. Capillary pressure is $-P_c$ whenever an interface is crossed from the concave to the convex side. Thus the pressure in the gas at b would be $p_1 + g\rho_2 z + P_c$. On the other hand, this pressure must be equal to $p_1 + g\rho_1 z$. The resulting equation, $P_c = g(\rho_1 - \rho_2)z$ cannot be correct, since P_c is positive and $\rho_1 - \rho_2$ is negative. It must be concluded that no hydrostatic equilibrium exists in the film and that the pressure in point a is smaller than that in the central part of the lamella.

Those parts of the film that have curved surfaces are called Plateau's borders, as Plateau[8] was their discoverer. Because pressure in these borders is less than in the plane parts, liquid streams from the center to the periphery of every lamella. This flow is an important cause of foam drainage and foam rupture.

§8. Bursting.   Thin lamellae are inherently unstable because their surface area is too big for their volume. The volume of a film of an area (on one side) of 100 cm^2 and a thickness 10^{-4} cm is 10^{-2} cm^3. The radius of a spherical drop of this volume is about 0.1336 cm and the surface of the drop is 0.2244 cm^2. Transformation of the lamella into a sphere would lower the area by $(200.0000 - 0.2244)$ cm^2. If surface tension is 50 g/sec^2, the amount of energy recovered is almost 10^4 ergs. The energy liberated per gram (if $\rho_2 = 1$ g/cm^3) is 10^6 ergs.

If m is the mass of a particle and u its velocity, the kinetic energy is $\frac{1}{2}mu^2$. The m of the above lamella is 10^{-2} g. Thus, if the energy 10^4 ergs is completely transformed into kinetic energy, u will be approximately 1.4×10^3 cm/sec. This means that, when the above film bursts, the velocity of the liquid may be as high as 14 m/sec.

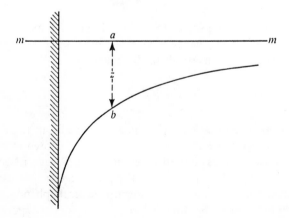

Fig. 1.7.   The lower region of contact between the lamella and the wall. Point a is situated in the plane mm of the liquid surface far from the wall and z cm above point b in the interface. The pressure in the film cannot be uniform.

A little more sophisticated is the calculation given by Dupré[9]. At the moment when the radius of the hole in the bursting film is r cm, the surface area, which has vanished, is practically equal to $2\pi r^2$ so that the available energy is $2\pi r^2\gamma$ ergs. The mass originally present in what now is a hole was $\pi r^2\delta\rho_2$; δ is the film thickness. If the surface energy is completely transformed in the kinetic energy of this mass, then

$$2\pi r^2\gamma = 0.5\pi r^2\delta\rho_2 u^2. \tag{1.12}$$

Hence,

$$u = \left(\frac{4\gamma}{\delta\rho_2}\right)^{0.5}; \tag{1.13}$$

the linear rate of expansion of the hole is independent of r and consequently is constant in time. With $\gamma = 50$ g/sec^2, $\delta = 2\times 10^{-3}$ cm, and $\rho_2 = 1$ g/cm^3, u is about 316 cm/sec.

A different result is obtained in the following (unpublished) treatment. As rapid photography shows, the liquid which initially occupied the future hole, forms a torus-like "welt" along the rim of the hole. It is acted upon by the surface tension of the two (as yet undisturbed) surfaces. Thus the force is $2\pi r \cdot 2\gamma = 4\pi r\gamma$; $2\pi r$ is the perimeter of the hole. A force $4\pi r\gamma$ acting on the mass $\pi r^2\delta\rho_2$ imparts an acceleration equal to $4\pi r\gamma/\pi r^2\delta\rho_2$. Hence,

$$\frac{du}{dt} = \frac{4\gamma}{\delta\rho_2 r}. \tag{1.14}$$

But $du/dt = d^2r/dt^2$, t being time. The resulting equation can be integrated to

$$u = \left(\frac{8\gamma}{\delta\rho_2}\right)^{0.5}\left(\ln\frac{r}{r_0}\right)^{0.5}, \tag{1.15}$$

r_0 being the radius of the small hole which initiated bursting. According to Equation 1.15, the advance of the rim is more rapid the bigger the hole.

Two other derivations result in the equation:

$$u = \left(\frac{2\gamma}{\delta\rho_2}\right)^{0.5}. \tag{1.16}$$

In one[10] it is assumed that u is independent of r; and in the other[11] the surface energy liberated when r increases by dr is supposed to be spent on accelerating only the mass initially present in the dr-wide ribbon.

§9.   None of the equations of the preceding section can be completely correct because they all disregard several phenomena which undoubtedly exist. Probably the neglect of the gas phase, pointed out in reference[12], is the least excusable. A liquid rapidly moving in air must involve some air mass in its motion, and the kinetic energy of the air may well be commensurate with that of the thin liquid film. Then, the viscous forces in the liquid must retard its every flow so that a part of the surface energy is directly transformed into heat rather than into any kinetic energy of a macroscopic mass. An experimental confirmation of this concept has recently been provided[13].

High-speed cinematography supplied valuable information on the process of bursting. If a surfactant film is pierced by, for instance, an electric spark, the hole formed is approximately a circle, and this circle gradually invades the rest of the lamella; in other words, the lamella does not shatter. In those instances in which spontaneous film rupture could be observed, it also proceeded by a gradual enlargement of a hole, which usually appeared near the top of the film[14]. Thus the model leading to Equations 1.13 or 1.15 seems to be essentially correct.

The rate, u, of bursting is of the order of magnitude predicted by Dupré. For instance, the tall hollow cylinders of §7 ruptured at a rate of 20 to 30 m/sec. The rate of expansion of a hole was almost constant for every lamella and again equal to a few meters per second[14]. The constancy of u cannot be used as an argument against the validity of Equation 1.15 because the secondary effects mentioned in the beginning of this section also increase in intensity when velocity increases. Both the speed of air surrounding the contracting film and the friction losses in the lamella increase with u and thus may keep the resultant velocity nearly constant, as the rate of ascent of a bubble in a liquid may be constant because the liquid resistance is proportional to velocity.

In many instances it was more difficult to measure the thickness δ of the lamella than the rate u of its rupture. Still, Ranz[15] concluded that the experimental u was only by about 10% smaller than predicted by Equation 1.13. The absolute velocities were between 3 and 7 m/sec.

McEntee[16] could determine u as a function of δ. The u exhibited a maximum (equal to about 40 m/sec) at about $\delta = 200$Å and was regularly smaller at greater thicknesses, as is predicted by the formulae of §8. The u at $\delta = 10\ \mu$ was only 2 m/sec.

§10. Causes of film stability. It is easy to see why films burst, but an explanation is needed for the fact that they can be produced at all and that some of them can be preserved for minutes, hours, or even years.

The most obvious cause of stability was pointed out by Plateau[17]. He called it *viscosité propre des couches superficielles*; the modern term is surface viscosity. The essential idea is that each film is stratified, has a sandwich-like structure. The inner layer has the viscosity of the liquid in bulk but the two exterior layers (adjacent to the gas phase) are much more viscous. It may be added at present, that these layers frequently have no Newtonian viscosity at all. They possess a yield point, that is, they do not flow under very small stresses and, in this respect, are analogous to solids. If the yield stress (g/cm·sec²), that is, the minimum stress necessary for flow, is high, the surface layer may be treated as a solid. Suppose that it consists of soap or a polymer of high molecular weight. When the internal liquid has drained out or evaporated, a twin sheet of solid soap or solid plastic remains and, naturally, can exist for an indefinite time.

No complete review of surface viscosity can be offered here; see Joly[18]. Only some information relevant to film stability is given in the following.

The earliest measurements of surface viscosity were performed by Plateau and later shown to be erroneous. Plateau suspended a needle in the surface of the liquid to be tested. When the suspension wire was twisted, the needle rotated in the horizontal plane. The speed of rotation was said to be smaller than expected whenever the viscosity (η_s) of the surface layers was greater than the bulk viscosity (η) of the underlying solution. The dimension of bulk viscosity is, of course, g/cm·sec, and the dimension of η_s is g/sec. Marangoni[19] pointed out the fallacy of Plateau's method. When the needle moves, it compresses the surface layer in front of it and leaves a relatively clean (swept) surface behind. Foaming agents very often lower the surface tension of the liquid, and this lowering often is greater the higher the concentration. Consequently, the surface tension (γ_1) of the surface in front of the moving needle is lower than that (γ_0) of the nearly pure solvent in the back of it. In this manner, the advance of the needle is retarded by a force equal to $\gamma_0 - \gamma_1$ g/sec^2 for every centimeter of needle length. This resistance simulates an enhanced viscosity and exaggerates the value of η_s. Mechanical effects of the difference between surface tensions at different points of a liquid surface now are commonly denoted by the term *Marangoni effect* (see §§2 and 11).

To measure η_s without interference from this effect, thin circular disks are used instead of a needle. A suspended horizontal disk is forced to rotate (say, 2 cm) below the surface and then in the surface of the liquid. If the resistance encountered is greater in the second arrangement, a measurable η_s is present. An even better comparison is possible if the disk is turned in the surface of an uncontaminated liquid, some foaming agent is added to the latter, and the disk is turned again. As before, if the resistance is greater during the second experiment, η_s exists and can be measured. This comparison is valid because small amounts of common foaming agents have only a negligible effect on the bulk viscosity.

In another method the surface of a liquid is separated into two areas by a bar in which a narrow slit is left open. One of the two areas is contaminated with a foaming agent. Because of the Marangoni effect the surface impurity flows through the slit and invades the initially pure area. This flow takes more time the greater η_s.

In Plateau's time there were only two classes of foaming agents, namely soaps (chiefly from olive oil) and saponins. The η_s of many soap solutions is small and cannot account for the stability of their films, but saponins possess not only a marked η_s but also a yield point (see §134).

When a bubble of an aqueous saponin solution is blown at the end of a tube and the air is rapidly withdrawn, the bubble in its contraction does not maintain its spherical shape (as a soap bubble does) but forms wrinkles; this is a certain indication of the fact that the state of the surface is not determined by surface tension and bulk viscosity alone. It is well-nigh impossible to doubt that saponin foam films are long-lived because the saponin "crusts" in them are almost solid. Additional data on the mechanical properties of foam films will be found in Chapter 7.

§11. Marangoni effect.　　　　No special viscosity and no rigidity can be detected in the surface layer of many solutions capable of measurable foaming; dilute solutions of butanol in water would be a suitable example. In these instances the Marangoni effect first mentioned in §2 is generally believed to be the main cause of film stability[19,20]. Before a film starts bursting, its surface usually has to be indented. Indentation means local increase of surface area. As the diffusion of foaming agent (such as soap or butanol) from the bulk of solution to the surface requires time, the concentration of the agent over the extended surface is smaller than that over the undisturbed area. Hence the surface tension of the indented spot exceeds that of the surrounding surface and the greater γ tends to "close the wound." When the surface crust is pushed toward the indented spot, it drags with it, because of viscosity, some of the liquid filling the middle volume of the sandwich; thus that spot in the film which was made thin by indentation becomes thicker again[21].

The above explanation is based on the concept that the diffusion of the surface-active solute (or foaming agent) is so slow that the Marangoni effect has time to operate, that is, the different tensions in the surface are maintained long enough to cause some flow of the liquid. This idea is confirmed by direct measurements of "dynamic surface tension," i.e., of the surface tension of solutions as a function of the time after the creation of a new surface. For instance[22], the surface tension of a 0.001% aqueous solution of 1-decanol reaches its equilibrium value in about 40 sec. The lifetime of many soap films is shorter than this.

Because of their historical interest, some data published by Rayleigh[23] are shown in Table 1. In it the γ of surfaces 0.01 sec old is compared with that of equilibrium surfaces; the figures are recalculated by the present author. An "infusion of horse chestnut" is impure saponin. The table shows that the tension of a fresh surface is so much greater than that in an equilibrium system, that the Marangoni effect may cause violent movements in the surface layer.

Marangoni called the above difference in tensions "elastic surface force." Gibbs[6] pointed out that the surface tension of a thin film may be lowered almost permanently during an extension of the film because the total amount of the solute may be depleted by adsorption. If the lamella is sufficiently thin and sufficiently large, the diffusion from the adjacent

Table 1. Surface Tension of 0.01-Sec-Old and of Equilibrium Surfaces

Liquid	Surface Tension (g/sec²)	
	Fresh	Aged
Water	72	72
Sodium oleate 1:40	56	25
1:80	59	25
1:400	76	25
1:4000	76	53
Infusion of horse chestnut	73	49
Water	72	72

liquid volumes into it may be neglected so that the total amount M of the solute is practically constant when the film area (on one side) is expanded from A to $A + A_1$. Before the extension, this M is $c_0 \delta A + 2\Gamma_0 A$; c_0 is the concentration of the middle layer (see §10), δ is the lamella thickness, and Γ_0 is the amount adsorbed per unit area. After the extension, the concentration in the middle layer is $M - 2 \Gamma_1(A + A_1)/\delta(A + A_1)$, i.e., smaller than c_0; Γ_1, the adsorption after stretching, is a little smaller than Γ_0, but usually this difference may be disregarded. Since the solute is surface-active, γ generally rises when the above concentration decreases (see §135); thus another kind of surface elasticity is possible. If A is the film area, the derivative $d\gamma/dA$ may serve as the quantitative expression of surface elasticity.

§12. Many observations on, and measurements of, the Marangoni effect can be found in the literature. Here only three papers, all dealing with typical foam films or bubbles, can be reviewed.

A convenient way of measuring surface tension utilizes capillary pull; see, for instance, reference[22]. A thin vertical slide is suspended at the end of a balance beam, and the bottom of the slide is made to touch a horizontal liquid surface. If the breadth and thickness of the slide are w and δ, and if the liquid forms a contact angle θ with the material of the slide, then the pull exerted by the liquid on the slide is $2(w + \delta)\gamma \cos \theta$; the buoyancy force, directed upward, is $w\delta zg\rho_2$; z is the immersion depth of the slide, ρ_2 the specific gravity of the liquid, and g the acceleration due to gravity. The difference between the weight W of the slide in contact with the liquid and the weight W_0 of the same slide in air is $W - W_0 = 2(w + \delta)\gamma \cos \theta - w\delta zg\rho_2$. This equation permits determination of the product $\gamma \cos \theta$. When the experimenter is interested in the value of γ, a slide is selected such that $\cos \theta = 1$, and γ is readily calculated.

Such a slide, marked *1*, is shown in Fig. 1.8; *2* is a balance beam. A light wire frame *3* is partly immersed in liquid *L*. When *3* is gradually withdrawn from the liquid, the total interface A increases and the γ value indicated by the capillary pull on slide *l* also changes. Experiments of this type have been performed[24], for instance, on dilute solutions of 1-dodecanol, sodium dodecyl sulfate, and sodium dodecyl sulfate (0.25 wt %) plus 1-dodecanol (0.001 %). Apparently, spreading of the surfactant over the whole interface was about as rapid as the withdrawal of the frame. Consequently, the surface tension at the slide increased during the withdrawal and the value of $A \cdot d\gamma/dA$ was from 5 to 25 g/sec^2. However, the increase of γ was temporary only; the initial lower values usually were restored in about 10 sec. This time interval is not very different from the 40 sec mentioned in §11.

In another set of experiments[25], two frames similar to *3* in Fig. 1.8 were employed. One was pushed into the liquid at the same rate, as the other, simultaneously, was withdrawn. Naturally the film was thicker in the latter frame, since less time was available for drainage. On the other hand, a thicker vertical film in equilibrium will require a greater surface tension to keep it suspended; see Gibbs' treatment indicated in §7. If this treatment is followed, it appears that the film being withdrawn has a greater γ than that which is being lowered. A kind of surface elasticity is simulated in this manner. Its relation to the Marangoni effect is difficult to define.

In the third arrangement[26], a bubble is formed at the tip of a capillary fixed so near to the surface that the "dome" of the bubble rises above the horizontal expanse of the liquid. The air pressure in the bubble is periodi-

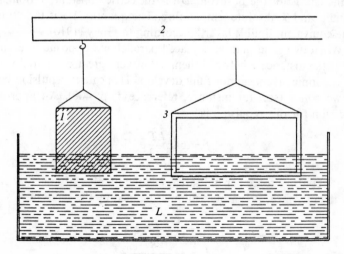

Fig. 1.8. Measurement of the Marangoni effect. *l* is a vertical slide suspended on balance beam *2*; the pull on it is a measure of the surface tension of liquid *L*. Raising or lowering wire frame *3* alters the adsorbed amount of the surfactant and thus this surface tension.

cally varied (30 to 150 times per second), and the work required to expand
and contract the bubble area is measured; it is spent on overcoming the
Marangoni elasticity. The method was applied to solutions of hexanoic,
octanoic, and decanoic acids in 0.07 N hydrochloric acid, and to aqueous
solutions of typical surfactants (sodium decyl sulfate, sodium dodecyl
sulfate). The importance of the Marangoni effect for foam stability is
reviewed in Chapter 10.

   §13.   A third cause of the relative stability of many foam lamellae
was found[12,21,27] in the mutual repulsion of electric double layers. When
a solid is brought into contact with a liquid, redistribution of electric
charges immediately takes place so that the solid carries a (positive or
negative) charge, and the liquid, an equal and opposite amount of electricity. Let the solid be negative. Then the liquid contains an excess of cations.
This excess is not present in an infinitely thin layer adjacent to the interface,
as would be concluded from classic (continuum) electrostatics, but rather
is spread over a space so that the major part of the positive charge occupies
a layer $1/\alpha$ cm thick. The value of $\alpha(\text{cm}^{-1})$ is a function of the electrolyte
concentration. When the potential difference Ψ between the liquid-solid
interface and the central regions of the liquid is small (say, less than 0.1
volt) and the concentration c of the ions is less than, say, 10^{21} per cm^3,
then

$$\alpha = 2\varepsilon \left(\frac{2\pi c}{DkT} \right)^{0.5};\qquad (1.17)$$

ε is the charge of the positron, D the dielectric constant, k Boltzmann's
constant, and T absolute temperature. The quantity $1/\alpha$ is the "thickness
of the equivalent double layer," according to Gouy (1910).

   When two solid plates are placed parallel and opposite to each other
so that the distance τ between them is not very great compared with $1/\alpha$,
then the ionic layers cannot fully develop. This causes repulsion between
the two solids (see, for instance, reference[22], p. 400). An approximate
expression for this repulsion is

$$f = 64c \frac{kT}{A^2} e^{-\alpha\tau} \qquad (1.18)$$

f is the repulsive force per unit area, c is the number of ions in 1 cm^3 far
from the interface, A is a pure number defined by the relation

$$A = \frac{\exp \dfrac{\varepsilon\Psi}{2kT} + 1}{\exp \dfrac{\varepsilon\Psi}{2kT} - 1} \qquad (1.19)$$

and the other symbols are explained above.

All soaps and many synthetic surfactants are ionic compounds. Their molecules consist of one small ion (usually Na^+ for anionic and Cl^- for cationic surfactants) and one large organic ion (e.g. oleate, $C_{17}H_{33}COO^-$, or trimethylhexadecyl ammonium, $C_{16}H_{33}N(CH_3)_3{}^+$). In the crust of a foam lamella, the organic ions usually form a "palisade" or a unimolecular mesomorphous phase, which is analogous to the solid plate referred to in the preceding paragraph. The charge on the wall is negative for soaps and positive for quaternary ammonium salts. The compensating charges (such as Na^+ or Cl^-) are spread in the liquid interlayer. When the thickness of the latter is less than, say, $20/\alpha$, the corresponding charges (also known as counter-ions) adjacent to the two opposite crusts repel each other more or less according to equation 1.18.

    §14.   Several experiments have been performed to check the existence of the effect predicted in §13, that is, of the mutual repulsion of electric double layers in foam lamellae.

    The principle of the earliest test[27] is indicated in Fig. 1.9. *A* and *B* are quartz bulbs submerged in a solution. Each has a small opening (e.g., 0.2 cm in diameter) located in the bottom of *A* and at the top of *B*, so that one is exactly above the other. When the bulbs through the tubes *a* and *b* are connected with a source of compressed air, bubbles appear at the two holes; they are indicated by dashes in the figure. Because the bubbles are convex to the liquid and are practically spherical, the pressure in them is by $2\gamma/r$ greater than that in the adjacent liquid; *r* is the radius of each

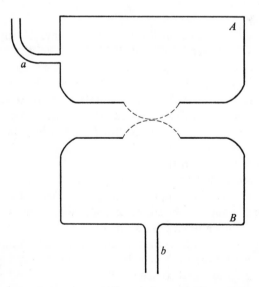

Fig. 1.9. Measurement of the mutual repulsion of electric double layers. Because of this repulsion, the two films (*dashes*) formed at the openings of bulbs *A* and *B* do not coalesce. Compressed air is injected through capillaries *a* and *b*.

bubble. This excess pressure is measured in the gas phase. In fact, so far the method is very similar to the maximum-bubble-pressure method of determining surface tension, as outlined, for instance, in reference[22], p. 30.

When now bulb A is lowered or bulb B is raised so that the bubbles press against each other, these do not coalesce if the liquid is well chosen; the boundary between them acquires the shape of a thin horizontal film, as indicated in Fig. 1.10. This lamella is observed with a microscope through the flat roof of bulb A, and its thickness δ can be calculated from the interference pattern recorded (see §126). As the lamellae are nearly plane, their ability to coexist with bubbles filled with compressed air cannot be attributed to capillary pressure. Another pressure (Π) must be active in the film to counteract that present in the bubbles.

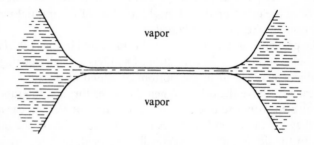

vapor

vapor

Fig. 1.10. Enlarged view of the boundary between the two bubbles of Fig. 1.9.

This Π usually reached measurable values (say, 200 dynes/cm^2) at film thicknesses δ between 100 and 1000 Å. When gas pressure and, consequently, Π increase, δ decreases. For instance, when one part of undecanoic acid $C_{10}H_{21}CO_2H$ was present for each million parts of water, Π was 660 and 1860 dynes/cm^2 at $\delta = 1500$ and 1050 Å. When the surfactant concentration increased (e.g., from 0.2 to 6 parts per million for undecanoic acid), the Π increased (for instance, by a factor of three) at a constant thickness δ. Extremely dilute sodium oleate solutions showed a similar dependence of δ on Π.

If it is assumed that the whole pressure Π is of electrostatic origin only and several plausible approximations are made, the electrostatic potential Ψ for the above undecanoic acid solutions comes out to be near 0.05 to 0.08 volt; thus it has a possible magnitude.

§15.   Fig. 1.11 indicates the principle of another arrangement[28,29]. T is a vertical glass tube of 0.4-cm internal diameter. A liquid drop D closes it near the bottom end. When small amounts of liquid are gradually withdrawn through the side capillary C, a horizontal microscopic plane film (e.g., of a radius equal to 0.01 cm) forms in the center of the drop. The thickness δ of the lamella does not vary for a time sufficient for its

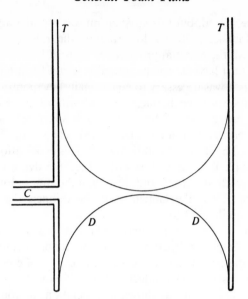

Fig. 1.11. Mutual repulsion of electric double layers. Drop *D* in tube *T* is made thinner by withdrawing liquid through capillary *C* until the two interfaces start repelling each other.

measurement (with a microscope placed above *T*) and may be considered to correspond to an equilibrium.

The liquid in the film is subject to the suction by the Plateau borders (§7). If the two air-liquid interfaces of the drop outside the lamella are more or less spherical, their radii of curvature are all equal and presumably equal to the radius r of the tube. Hence the capillary pressure $2\gamma/r$ was $2 \times 73/0.2 = 730$ g/cm·sec^2; the surface tension of the liquid was 73 g/sec^2. This pressure was pulling the liquid out of the lamella. Since δ remained constant, an equal pressure must have been present in the film; it was the Π of §14.

The major part of the experiments was performed on water containing 5 parts of saponin per million and variable amounts of potassium chloride. Saponin liberates very few ions in an aqueous solution; consequently a crust of pure saponin would give rise to a low potential Ψ. Presumably it adsorbs K^+ more than Cl^-, or Cl^- more than K^+, and electric double layers are formed in this manner. Since $\Pi(= 2\gamma/r)$ was almost constant in a given tube, the δ had to change when the salt concentration and Ψ varied. The δ decreased from 1400 to 500 Å when the concentration of potassium chloride increased from 0.0001 to 0.0020 M. The Ψ calculated from these data was approximately 0.09 volt.

In a third arrangement[30], the lamella was vertical. The frame carrying it was withdrawn at different speeds. When the speed decreased, the film thickness also decreased, but this attenuation was not unlimited: At some characteristic thickness, further lowering of the rate of frame ascent ceased to diminish the δ. An approximately equal thickness was achieved also when a vertical lamella was permitted to drain for a longer time. In both experiments it was necessary to guard against evaporation. When the rate of vaporization was negligible, the equilibrium δ of sodium dodecyl sulfate lamellae between about 100 and 1000 Å was a definite function of the counter-ion concentration c (which varied between 10^{-4}M and M), as would be expected by the electrostatic theory. This expectation is based on the concept of §13. Because α rises with c (i.e., the depth of the double layer increases with dilution) and repulsion becomes noticeable in a distance of about $20/\alpha$, the lamella must be thinner, the higher c, to manifest electrostatic repulsion.

A criticism of the above experiments was raised because sodium dodecyl sulfate is partly hydrolyzed to dodecanol; consequently, the composition of the crust presumably was different in different lamellae[31].

This hydrolysis may have affected also the experiments[32] in which the equilibrium thickness δ was shown to depend on the counter-ions. At 25°C, δ was 172, 112, and 54 Å, respectively, for 0.01, 0.10, and 1.0 N solutions of lithium dodecyl sulfate, and 175 and 105 Å for 0.01 and 0.1 N sodium dodecyl sulfate. The above concentrations are all greater than the critical micelle concentration of the salts. It is seen that the difference between the lithium and the sodium salt was small; and this was true also at 45° and for other salts. At 45° in 0.02 N solutions, δ was 93, 89, 83, 87, and 96 Å for the lithium, sodium, potassium, rubidium, and cesium salts of dodecyl sulfate, and the surfactant concentration (0.02 to 0.50 N) had only a very moderate effect on δ, contrary to the behavior at 25°. Addition of urea or sucrose greatly raised the δ values. For instance, δ of 0.01 N sodium dodecyl sulfate plus 0.25 or 0.50 M urea was 308 or 372 Å, instead of 175 Å in the absence of organic nonelectrolytes. The δ of films of p-octylphenyl poly(ethylene-oxide) $C_8H_{17} \cdot C_6H_4(OCH_2CH_2)_{11}OH$ at 25° was about 90 Å, independently of the surfactant concentration and the electrolyte added [NaCl, $MgCl_2$, $La(NO_3)_3$]. None of these results contradicts the electrostatic theory of stability, but none can be used to prove it.

In a fourth method[33], the repulsion is calculated from the vapor pressure p. When lamellae of aqueous solutions of the trimethyldecyl ammonium salt of decyl sulfuric acid are permitted to drain until a constant thickness δ is attained, this thickness varies when the relative humidity p/p_s of the gas phase is altered; p_s is the pressure of the saturated water vapor at the test temperature. The repulsion Π is supposed to be equal to $(RT/v) \ln(p/p_s)$; R is the gas constant, T absolute temperature, and v the partial molar volume of water. In this manner a relation between Π and δ is obtained. In a solution of 0.5 mmole of the above surfactant and 0.05 mole of sodium bromide per liter, Π was 4×10^6 dynes/cm² at

$\delta = 110$ Å, 20×10^6 at $\delta = 80$ Å, and varied between 8 and 12×10^8 dynes/cm^2 when δ was near 46 Å. When the vapor phase was almost saturated, the equilibrium thickness was about 150 and 50 Å when the sodium bromide concentration was, respectively, 0.05 and 2 M; thus the film was thinner when the ion concentration was greater—in agreement with the theory.

§16.   Surface viscosity, surface rigidity, the Marangoni effect, and the mutual repulsion of overlapping double layers are most commonly referred to in the discussion of the relative stability of foam films. However, there may be additional causes of persistence.

Electric charges distributed in a medium of a high dielectric constant (such as water whose $D = 80$) are repelled by a medium of low D (such as air whose $D = 1.0$). Consequently, the concentration of an aqueous electrolyte solution is smaller very near the air-liquid interface than in the bulk of the liquid[34]. There are two air-liquid interfaces in a lamella, so that the dissolved electrolyte must accumulate in the middle portion of it[35]. This accumulation is resisted by osmotic forces. As relative accumulation is greater the thinner the lamella, osmotic forces try to prevent thinning. In a crude approximation intended to be valid for films whose thickness is commensurate with $1/\alpha$ (see §13), the pressure which opposes thinning is $kT\alpha^3/6\pi$ and thus ought to be independent of δ. In thicker lamellae this pressure is approximately proportional to $1/\delta^3$, that is, rapidly decreases when δ increases. It acts also when the potential Ψ of §13 is zero and no electric double layer exists in the film.

In some publications, e.g.[27,33] the possibility was considered that the mutual approach of the two film surfaces may be opposed by the hydration of the ions immobilized in the two crusts. If the radius of the hydration zone of an ion is h cm, some water molecules must be removed from this zone whenever the thickness of the central liquid layer is reduced to below $2h$. This removal is believed to require some work. The idea that hydration layers prevent contact of two colloidal particles is many decennia old. Unfortunately, it has never been clearly formulated and expressed in a mathematical form convenient for comparing theory and experiment.

§17.   It has many times been stated that the equilibrium thickness of a lamella is determined by the balance between the repulsion of electric double layers (§§13 to 15) and the van der Waals attraction in the film; mathematical expressions for this theory can be found in the literature[36,37]. It seems to the present author that, when this approach is used, the classical explanation of the instability of foam films (based on the decrease in surface energy, §8) cannot be employed. The molecular forces to which nowadays the name of van der Waals is pegged are in no way different from those treated by Laplace. Laplace proves that the experimental property of surface tension is a direct and necessary consequence of their existence. Hence, whoever says that surface energy is the main cause of bursting means that molecular attractions are the real cause; and he should not count these attractions twice: once in surface tension and then directly.

Rayleigh[38] pointed out that planets are nearly spherical, although their shape certainly is not determined by surface tension. They are spheres because the mutual attraction of matter (i.e., gravitation) tends to give a compact shape to every ponderous body. When a lamella is transformed into a drop, the average distance between all molecules decreases. The energy gain achieved in this process is what is perceived by the observer as a decrease in surface energy.

In principle, the surface tension of extremely thin films should be smaller than that of the bulk liquid. According to Laplace, surface tension exists because a material point situated near a liquid-vapor interface is attracted by the (dense) liquid more than by the (rare) vapor; thus work is needed to move a molecule from a central location in the liquid to its surface. But this effect is weakened when the thickness of the lamella is less than the diameter Δ of the "sphere of molecular action" (in which sphere forces emanating from the central molecule cannot be neglected). In these films, each molecule belongs to both surface layers, and there is no bulk liquid whose attraction is great. It appears then that surface tension of lamellae whose $\delta < \Delta$ should be small and, consequently, the tendency to collapse would be attenuated. This effect has never been observed because no film thinner than Δ could be prepared (see §184).

§18.   The cause of bursting is clear, but its mechanism has not been elucidated. It is explained in §34 by what means capillary pressure makes long cylinders succumb to the slightest disturbance. An analogous effect apparently does not occur in plane parallel films; every disturbance in these systems, so far considered, requires work to start and to propagate. If the surface is indented (as the first step in making a hole) (Fig. 1.12), not only the Marangoni effect (§11) becomes operative but the capillary pressure in the liquid near summits A and C and also under the plane surface will push the liquid toward point B, where this pressure is negative.

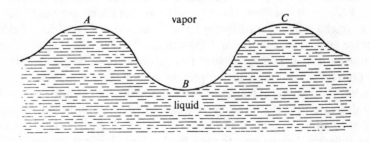

Fig. 1.12. Rupture of a plane lamella is resisted by the capillary pressure at A and C which pushes liquid toward B and fills the indentation.

Thus a flat surface has a tendency to heal also in the absence of the Marangoni effect.

Plateau[3] blamed external agents for the observed rupture of foam lamellae. Air blast is a mechanical agent of this kind. A speck of dust which adsorbs the surfactant and thus originates violent vibrations in the film would be an agent of the physicochemical type. For the experimental support of Plateau's views see §47.

Recently it was attempted to clothe the process of rupturing a lamella by vibration in mathematics[39-42]. Unfortunately, too many assumptions are needed to derive the final equations. It is concluded that the lifetime of a film is proportional to δ^5 and that the coefficient of proportionality is independent of the surfactant concentration. Future experiments will determine the area of applicability of these theories.

If lifetime is really proportional to δ^5 no limiting value of δ exists below which the lamella suddenly becomes unstable. Several other investigators[43-44], believed in the existence of a critical thickness. This belief obviously is not warranted for lamellae capable of thinning to black films (§19), which often persist for a long time. Many other films burst long before they even show interference colors; thus if a critical thickness is not a myth, it certainly is extremely variable from one to another surfactant. Perhaps the experiments described in §121 supply an explanation for the poor stability of thin lamellae.

§19. *Black films.*   The brilliant colors of soap films and bubbles are perhaps their most conspicuous property. However, some films contain black areas or are black all over. This alone would not arouse any interest: Every lamella, which is so thin that the optical paths of the beams reflected from the front and from the back interface are practically equal, would appear black (see Chapter 8). What is striking is that black patches sometimes appear in relatively thick lamellae, so that the thickness around the patch may be, say, 100 times as great as the patch thickness; and the transition between thin and thick areas, as a rule, is so sharp that the width of the transition zone cannot be discerned with a microscope.

Sharp steps cannot exist in the surface of a true liquid. Consequently, lamellae exhibiting black areas are not liquid in the usual sense of the word. Presumably they are mesomorphous[45] or, according to the old terminology, belong to liquid crystals[46]. Analogously to crystals, mesomorphous phases are anisotropic; and, analogously to liquids, they have no shear strength to speak of. In the smectic division of the mesomorphous state (*smegma* is "soap" in Greek), the molecules (usually elongated) are arranged in equidistant rows. Considerable force is needed to press two rows together or to bend them, but one row slides above the next almost without friction. In a pendant drop of a smectic phase the surface is covered with sharp steps, which presumably are analogous or physically identical with those surrounding a black patch. It is regrettable that the two branches of science—that of foam lamellae and that of mesomorphous phases—still advance independently of each other although their kinship was emphasized long ago[47].

The boundary between a black area and a thick part of the lamella often demonstrates a striking stability. In some vertical films, black spots appear first not at the top (the usual event) but somewhere in the middle. Then they may rise in the lamella to the top without mixing with the rest of the (identical!) liquid[48]. When two electrodes pierce a soap lamella containing black patches, and a weak direct current is sent from one electrode to the other, the liquid flows from the anode to the cathode, so that the lamella becomes thicker near the latter and thinner near the former, but the black areas in the path of the current maintain their thickness unchanged[49]. A horizontal lamella was produced in a low cylinder, and the cylinder was rapidly spun about its vertical axis. The centrifugal force drove the liquid toward the Plateau borders so that the central part of the film became black. When the centrifuging stopped, a part of the liquid started moving back toward the axis; this motion could be followed by color changes. However, no liquid penetrated the black area[50].

§20.   The peculiarity of the boundaries between a black film and a thick lamella or bulk liquid sometimes gives rise to small effects for which no satisfactory explanation is available yet. If a vertical film of 0.05% sodium dodecyl sulfate $+ x$M sodium chloride is permitted to drain in contact with a horizontal surface of the identical solution, and the weight of the (film + frame) is continuously recorded, it is seen that this weight suddenly decreases at the moment when the lower edge of the black field touches the Plateau border (§7), that is also the bulk liquid. This effect is noticeable whenever x is equal to or greater than 0.3, and becomes more pronounced the greater x, up to $x = 0.6$. No sudden weight decrease was observed when an octylphenyl poly(ethylene oxide) was the surfactant. It is suggested by Prins[51] that the effect is related to the contact angle between the black film, the bulk liquid, and the vapor phase. As indicated in §19, the surface of mesomorphous drops may have steps, the "contact angle" between which may be quite high (90°?).

An attempt[52] to determine the angle at which a black film joins the bulk liquid by using the Plateau border (at the bottom of the film) as an optical prism was not successful, because the profile of the liquid in the border zone is curved and the direction of this profile, exactly at the three-phase line (in which, supposedly, vapor, black film, and bulk liquid meet) cannot be ascertained.

Also a third arrangement[53,54] does not appear convincing. A bubble was formed at the tip of a capillary, so that a part of it lifted the liquid surface. In Fig. 1.13, *1* is the capillary, *2* indicates the top of the bubble (above which is the vapor phase), and *3* is the liquid. The capillary with the bubble was shifted a little, up or down, until the liquid surface appeared plane right to the three-phase line (if any). This is a legitimate method of determining contact angles in vapor-liquid-solid systems but requires testing in the absence of any solid. Unfortunately the thickness of the bubble roof *2* was not determined and it is not known whether this roof was a black film. To calculate the contact angle θ it was necessary to assume implicitly that the bubble had a spherical surface (of constant curvature) from the top *2* to below the water line; only if this unlikely hypothesis is accepted, is the ratio $d/2h$ equal to $\sin\theta/(1-\cos\theta)$; d is the diameter of the calotte, and h its height. The physicochemical conclusions from the observed value of $d/2h$ require, moreover, that the "roof" film should be in equilibrium, that is, no drainage should take place.

   §21.   Black films are very thin, and an exact determination of their thickness, either by optical (§125) or by electric (§130) means, is difficult and often unreliable. The uncertainty of the absolute values does not, however, change the fact that this thickness greatly varies from solution to solution and that a given solution can afford black films of more than one thickness.

   In the 19th century oleate soaps were the preferred foaming agents for film studies. A black film of a sodium oleate solution in aqueous glycerol had the thickness δ of 160 to 170 Å[55]. The δ of the most common black films of 2% aqueous sodium oleate was[56] 120 Å but a film of 60 Å also could be obtained[57]. An extensive series of δ determinations for solutions of 1 part sodium oleate plus x parts of water gave the following results[58].

x	30	40	60	80
δ	220	220	280	290 Å

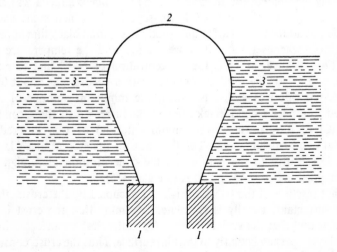

Fig. 1.13. Contact angle between a foam film (*2*) and bulk liquid (*3*). Air is introduced through capillary *1*.

When 1 g of sodium oleate was dissolved in 40 cm^3 of water and 27 cm^3 of glycerol, the δ was 250 Å.

Reinold and Rücker[58] made a remarkable observation which led to the theory (§13) of the effect of the electrostatic repulsion on the stability of foam. The δ of black films was smaller when a nonfoaming salt was added to a soap solution. Thus, when 100 g of aqueous solution, in addition to 2.5 g of sodium oleate, contained x g of potassium nitrate, the δ varied as shown below:

x	0	0.5	1.0	3.0	g
δ	220	145	135	124	Å

The explanation is given in §15: An increase in the salt content causes contraction of the diffuse double layer; consequently the thickness at which the two opposite layers start to overlap is smaller.

Similar results have more recently been obtained with synthetic surfactants. If the refractive index of the film is assumed to equal that of water, the "equivalent water thickness" δ_w is obtained. This δ_w was 110 Å for a solution of 0.5 g sodium dodecyl sulfate $+0.1$ mole sodium chloride per liter, 46 Å when the amount of NaCl was doubled, and 42 to 40 Å when this was doubled again[51]. It is possible, however, that the thicker and the thinner lamellae belonged to two different series.

The relative stability of the "first" and the "second" black film (of 120 and 60 Å in Johonnot's experiments) depends on the composition of the solution and the temperature[59]. An aqueous solution of 0.05% sodium dodecyl sulfate prefers to burst than to drain to a second black, but it affords second black films on evaporation. As long as evaporation was practically eliminated, the thickness of the first black lamella was about 130 Å. Inorganic salts lowered it to, say, 70 Å. When some sodium chloride was added to the surfactant solution, a second black film appeared on draining whenever the salt concentration and the temperature were right. For instance, at 15°C the concentration of NaCl had to exceed 0.12 M, and at 40° the minimum concentration needed was about 0.4 M. If a second salt was added to NaCl, the minimum amount of NaCl required to render second black films viable increased (if the second salt was LiCl or a quaternary ammonium chloride) or was lowered (by potassium and calcium salts). The thickness of the second black lamella was little sensitive to the nature and the concentration of foreign salts and remained between 40 and 50 Å (see also §15).

That fraction of the lamella thickness occupied by the crusts, that is by the surfactant, usually is calculated assuming that the crust is uni-molecular and that the surfactant ions (or molecules) are oriented in this "monolayer" perpendicularly to the film surface. Thus the crust, consisting of sodium dodecyl sulfate or rather dodecyl sulfate ions is generally believed[31,59] to be about 9 Å thick. The area occupied by one cation of hexadecyltrimethyl ammonium bromide is[60] about 44×10^{-16} cm^2, and the area taken up by a dodecyl sulfate anion is near 39×10^{-16} cm^2.

§22.   A whole series of black films was observed by Perrin[61] when a small horizontal lamella (e.g., 0.1 cm in diameter) was formed on a microscope stage and viewed with a vertical illuminator so that only reflected light reached the eye of the observer. To make the optical effects more striking, Perrin usually added organic dyes to his oleate solutions, but an ordinary rosin soap or a 5 % solution of sodium oleate[62] are just as suitable.

Apparently, rapid evaporation of the solvent (i.e., water) occurs in the thin lamellae heated by the concentrated light of the illuminator, and the remaining solute (i.e., a soap) forms stratified films which are analogous to mesomorphous (smectic) systems mentioned in §19 and to the fresh "built-up" multilayers reviewed, for instance, in reference[22], p. 232. In common with these systems Perrin's lamellae can have only discrete values of thickness.

This was demonstrated by two experimental procedures. One of these requires a microscope provided with a vertical illuminator and a source of monochromatic light. In such a light the lamella appears bright when the difference Δ between the optical paths of the beams reflected from the back and the front interface is $0.5\lambda_0$, $1.5\lambda_0$, etc.; and dark when Δ is 0, or λ_0, $2\lambda_0$, etc.; λ_0 is the wavelength of the radiation used, measured in a vacuum (not in the liquid). This means (see §123) that (when the incident beam is perpendicular to the film) the lamella is at its brightest when its thickness δ is equal to $\lambda_0/4n$, or $3\lambda_0/4n$, etc., n being the refractive index of the liquid. The minima of brightness are observed when δ is $\lambda_0/2n$ or λ_0/n, etc. The difference between the two δ, corresponding to two nearest maxima and minima, is $\lambda_0/4n$.

Perrin looked, for instance, at a film in the blue light of $\lambda_0/n = 0.354\,\mu$. When the film gradually thinned, he counted 16 different degrees of brightness between a minimum and a maximum, and again 16 steps between this maximum and the next minimum. Assuming that no step was overlooked and all steps were identical, each of these was

$$\frac{\lambda_0}{17 \times 4n} = \frac{0.354}{68} = 0.0052\mu,$$

or 52 Å. Thus the stratified lamella consisted of leaflets, each 52 Å thick, piled one on top of the other. The length of a C_{18} soap molecule in a crystal is near 23 Å, so that these leaflets appear bimolecular; this is seen in crystals as well.

In the second method the reflection coefficient, that is the ratio of the intensity I of the reflected light to that (I_0) of the incident light, was measured. This was achieved either by comparing the reflected beam with that which passed through two nicols, making a small angle with each other[61,62] or by duplicating the color of the reflected light with that of a beam retarded in a quartz wedge[63]. When the thickness of many lamellae was determined in this manner, it was found that all these thicknesses (for a given solution) were multiples of a unit thickness. This was, for instance, 41 Å for 5% potassium oleate solutions in water or aqueous glycerol.

Some mechanical properties of Perrin's films were as unexpected as those of the black spots referred to in §19. Human breath produced microscopic droplets on the lamellae. The weight of the droplets (whose diameters often were of the order of 100,000Å) did not cause any visible sagging of the film. The droplets readily slid along the lamella without causing any damage as long as the film thickness was constant, but they never crossed the boundary between two areas of different thicknesses.

References

1. Benning, C. J. *Plastic Foams*, two volumes. New York: John Wiley. (1969).
2. Valentin, F. H. H. *Absorption in Gas-Liquid Dispersions*. London: Spon Ltd. (1967).
3. Plateau, J. *Mém. Acad. Roy. Sci. Belg.* 33:5th and 6th series. (1861). Also in J. Plateau. *Statique experimentale et théorique des liquids soumis aux seules forces moléculaires*. Two volumes. Paris: Gauthier-Villars. (1873).
4. Boys, C. V. *Soap Bubbles*. New York: Crowell. (1962), p. 237.
5. Dickey, R. W. *Quart. Appl. Math.* 24:97. (1966).
6. Gibbs, J. W. *Thermodynamics*. New York: Dover. (1961), p. 300.
7. Rodewald, H. J. *Melliand Textilber.* 38:197. (1957).
8. Plateau, J. *Mém. Acad. Roy. Sci. Belg.* 23. (1849). Also in J. Plateau. *Statique experimentale et théorique des liquids soumis aux seules forces moléculaires*. Two volumes. Paris: Gauthier-Villars. (1873).
9. Dupré, A. *Théorie mécanique de la chaleur*. Paris: Gauthier-Villars. (1869), p. 350.
10. Culick, F. E. C. *J. Appl. Phys.* 31:1128. (1960).
11. Frankel, S., and K. J. Mysels. *J. Phys. Chem.* 73:3028. (1969).
12. Bikerman, J. J. *Foams*. New York: Reinhold. (1953), p. 169.
13. Vijayendran, B. R., G. Frens, and K. J. Mysels. Am. Chem. Soc., *Abstr.* 161st Meeting, Colloid Division 1 (1971).
14. de Vries, A. J. *Foam Stability*. Delft: Rubber-Stichting. (1957), p. 43.
15. Ranz, W. E. *J. Appl. Phys.* 30:1950. (1959).
16. McEntee, W. R., and K. J. Mysels. *J. Phys. Chem.* 73:3018. (1969). See also K. J. Mysels and J. A. Stikeleather. *J. Colloid Interface Sci.* 35:159. (1971).
17. Plateau, J. *Mém. Acad. Roy. Sci. Belg.* 37:49. (1869). See also J. Plateau. *Statique experimentale et théorique des liquids soumis aux seules forces moléculaires*. Two volumes. Paris: Gauthier-Villars. (1873).
18. Joly, M. *Recent Progress in Surface Science*. Vol. 2. Danielli etc., eds. New York: Academic Press. (1964), p. 1. Also *Abhandl. Deut. Akad. Wiss. Berlin, Kl. Chem., Geol., Biol.* 1966, 6:683. (1967).

19. Marangoni, C. *Nuovo Cimento* [2] 5–6· 239. (1871); [3] 3: 97, 193. (1878).
20. Rayleigh, J. W. S. *Proc. Roy. Soc. (London)* 48:363. (1890).
21. Ewers, W. E., and K. L. Sutherland. *Australian J. Sci. Res.* A5:697. (1953).
22. Bikerman, J. J. *Physical Surfaces*. New York: Academic Press. (1970), p. 72.
23. Rayleigh, J. W. S. *Proc. Roy. Soc. (London)* 47:281. (1890).
24. Mysels, K. J., M. C. Cox, and J. D. Skewis. *J. Phys. Chem.* 65:1107. (1961).
25. Prins, A., C. Arcuri, and M. van den Tempel. *J. Colloid Interface Sci.* 24:84. (1967).
26. Kretzschmar, G., and K. Lunkenheimer. *Ber. Bunsenges. Phys. Chem.* 74:1064. (1970).
27. Deryagin, B. V., and A. S. Titievskaya, *Kolloidn. Zh.* 15:416. (1953); *Discussions Faraday Soc.* 18:27 (1954).
28. Scheludko, A. *Kolloid-Z.* 155:39. (1957).
29. Scheludko, A., and D. Exerova. *Kolloid-Z.* 165:148. (1959).
30. Lyklema, J., and K. J. Mysels. *J. Am. Chem. Soc.* 87:2539. (1965).
31. Jones, N. M., and D. A. Reed. *J. Colloid Interface Sci.* 30:577. (1969).
32. Bruil, H. G. "Specific Ionic Effects in Free Liquid Films." Thesis. Wageningen. (1970). See also B. T. Ingram, *Faraday Trans. I*:2230. (1972).
33. Clunie, J. S., J. F. Goodman, and P. C. Symons. *Nature* 216:1203. (1967).
34. Wagner, C. *Physik. Z.* 25:474. (1924).
35. Smilga, V. P., and V. N. Gorelkin. *Issled. v oblasti poverkhnostnykh sil.* Deryagin, ed. Moscow: Nauka. (1967), p. 174.
36. Scheludko, A. *Dokl. Akad. Nauk SSSR.* 123:1074. (1958).
37. Felderhof, B. U. *J. Chem. Phys.* 49:44. (1968).
38. Rayleigh, J. W. S. *Phil. Mag.* [5] 30:285. (1890).
39. Scheludko, A. *Advan. Colloid Interface Sci.* 1:391. (1967).
40. Scheludko, A., and E. Manev. *Trans. Faraday Soc.* 64:1123. (1968).
41. Vrij, A. *Discussions Faraday Soc.* 42:23. (1966).
42. Ivanov, I. B., et al. *Trans. Faraday Soc.* 66:1262. (1970).
43. Rehbinder, P., and A. Trapeznikov. *Acta physicochim URSS* 9:257. (1938).
44. Brady, A., and S. Ross. *J. Am. Chem. Soc.* 66:1348. (1944).
45. Friedel, G. *Ann. phys.* [9] 18:273. (1922).
46. Gray, G. W. *Molecular Structure and Properties of Liquid Crystals.* London: Academic Press. (1962).
47. Reference 12, p. 156.
48. Lawrence, A. S. C. *Soap Films.* London: Bell and Sons. (1929).
49. Reinold, A. W., and A. W. Rücker. *Phil Mag.* [5] 19:94. (1885).
50. Hagenbach, A. *Arch. Sci. Phys. Nat.* 35:329. (1913).
51. Prins, A. *J. Colloid Interface Sci.* 29:177. (1969).
52. Princen, H. M. *J. Phys. Chem.* 72:3342. (1968).
53. Uuisman, F., and K. J. Mysels. *J. Phys. Chem.* 73:489. (1969).
54. Scheludko, A., B. Radoev, and T. Kolarov. *Trans. Faraday Soc.* 64:2213. (1968).
55. Drude, P. *Ann. Physik* [3] 43:158. (1891).
56. Johonnot, E. S. *Phil. Mag.* [5] 47:501. (1899).
57. Johonnot, E. S. *Phil. Mag.* [6] 11:746. (1906).
58. Reinold, A. W. and A. W. Rücker. *Trans. Roy. Soc. (London)* 184:505. (1893).

59. Jones, M. N., K. J. Mysels, and P. C. Scholten. *Trans. Faraday Soc.* 62:1336. (1966).
60. Jones, M. N., and G. Ibbotson. *Trans. Faraday Soc.* 66:2394. (1970).
61. Perrin, J. *Ann. Phys.* [9] 10:180. (1918).
62. Wells, P. V. *Ann. Phys.* [9] 16:69. (1921).
63. Marcelin, R., see A. Marcelin. *J. Chim. Phys.* 28:605. (1931).

CHAPTER 2

FORMATION AND STRUCTURE

§23.   Foams, being colloidal systems, can be prepared either by dispersion or by condensation, which is also known as agglomeration. In the dispersion methods, the future disperse (or discontinuous) phase is initially available as a large volume of gas, and this is then comminuted and mixed with the dispersion medium (often designated as the continuous phase), which, in a foam, is a liquid. In the condensation methods, the future dispersed material originally is present as a solute, that is, as molecules dissolved in the liquid. When these molecules combine to larger aggregates (i.e., bubbles), foams may be obtained.

Both foam types are common. Dispersion foams are observed, for instance, when a surfactant solution is shaken or whipped, or when air is injected into it; they are encountered in laundering, in making "mechanical" fire-fighting foams, and so on. Foams on beer and soft drinks, on boiling liquids, and "chemical" fire-fighting foams are instances of those produced by condensation. Apparently, both types are used in flotation.

A dispersion foam can be built-up by making soap bubbles at the bowl of a pipe or by rapidly moving a metal ring spanned by a soap film (see §5); many such bubbles can then be combined to a cluster simply by mutual contact. It seems that the mechanism of formation of a closed, practically spherical bubble from a plane film across the orifice of a pipe or a metal ring has not been scientifically studied since Plateau's time. Apparently, it is still unknown at what width of the pipe bowl or the ring the above formation becomes impossible and what is the minimum speed or acceleration needed to throw the bubble off. High-speed cinematographic reproduction of the detachment process would be instructive.

Dispersion Methods

§24.   A popular subdivision of the dispersion method involves injection of gas into a suitable liquid through one or many orifices. Voluminous literature exists on this procedure. First, very slow bubble formation at a single orifice will be examined here. The problem is analogous to that of drop formation (also slow and at one opening only) but, unfortunately, the knowledge of the latter phenomenon has not been fully utilized for understanding the former.

The elementary theory of the formation and detachment of bubbles is obviously incorrect, but it not only continues to be employed in calculations but also affords values which frequently agree with the experimental data. Fig. 2.1 shows a bubble (*1*) still attached to solid *3* in liquid *2*. Buoyancy tends to lift the bubble but surface tension forces act in the opposite direction. When the bubble grows, the buoyancy increases more rapidly than does the surface force, and it finally overtakes the latter; at this moment detachment takes place or starts. The buoyancy force is $vg\rho$, if v is the bubble volume, $\rho = \rho_2 - \rho_1$, and g is the acceleration due to gravity; ρ_1 and ρ_2 are, respectively, the densities of gas and liquid. The force resisting the ascent is said to be $2\pi r\gamma$; r is the radius of the orifice, and γ is the surface tension of the liquid. Thus, detachment should start as soon as

$$vg\rho = 2\pi r\gamma \tag{2.1}$$

or, more exactly, $vg\rho > 2\pi r\gamma$. An additional and unnecessary approximation is introduced by many authors, and the bubble is treated as a sphere of radius R. Then $(4\pi/3)R^3g = 2\pi r\gamma$ when separation becomes possible; that is,

$$R^3 = \frac{3r\gamma}{2g\rho}. \tag{2.2}$$

A description in a better agreement with the science of capillarity is given in this paragraph. As long as the bubble is being formed very slowly, the gas pressure p in it may be considered uniform in the whole volume. Also the law of capillary pressure (§3) will be approximately obeyed. Let

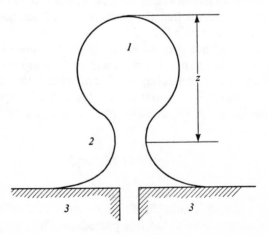

Fig. 2.1. Gas bubble (*1*) in a liquid (*2*) on a solid surface (*3*). The vertical distance between the top of the bubble and the nearly cylindrical stem is z cm.

the hydrostatic pressure in the liquid just above the top of the bubble be p_0. Then the curvature of the gas-liquid boundary at this top is determined by the equation

$$\frac{p - p_0}{\gamma} = \frac{1}{R_1} + \frac{1}{R_2},$$

in which both radii of curvature (R_1 and R_2) are treated as positive. At a level z below the summit, the hydrostatic pressure in the liquid is $p_0 + g\rho z$ and the two radii of curvature (R_3 and R_4) are given by the relation $(p - p_0 - g\rho z)/\gamma = (1/R_3) + (1/R_4)$; i.e.,

$$\gamma \left(\frac{1}{R_1} + \frac{1}{R_2} - \frac{1}{R_3} - \frac{1}{R_4} \right) = g\rho z. \tag{2.3}$$

As Fig. 2.1 illustrates, the "stem" of the bubble at level z is almost cylindrical; this means that the radius of curvature (say, R_4) swinging in the plane of the drawing is very large. Consequently, R_3 (which is approximately the radius of the cross-section of the cylinder) must be small. Thus the cylinder making up the stem must be narrow. As demonstrated in §34, a fluid cylinder becomes unstable whenever its length starts to exceed its circumference. Hence, the above narrow cylinder "necks down," and separation between the main body of the bubble and the gas adjacent to the solid takes place. Hence, buoyancy causes rupture of the bubble not by overcoming surface tension forces but by stretching the bubble stem until these forces give the fatal blow. The bubble is not spherical, and would have no tendency to separate if it were spherical. The radius of the orifice is not equal to the radius of the stem, and in many instances has only an indirect effect on the detachment process.

Fig. 2.2 is a simplified reproduction of eight photographs[1] made during the growth and detachment of an air bubble released in aqueous glycerol of viscosity 7 g/cm·sec; the volume of each bubble was about 2.5 cm^3, and the time interval between two succeeding bubbles was 15 to 20 sec. It is obvious that the bubble shape is very far from being spherical. It is seen also that the stem does not break as a hollow wire or filament does; consequently, its resistance is not likely to be equal to $2\pi r\gamma$, even if r now means the radius of the stem. When drops fall, usually a microscopic drop (originating in the central portion of the stem) follows the main drop; presumably, an analogous effect occurs also in bubble detachment, although it cannot be detected in the original photographs. On the other hand, it is clearly seen that a nascent bubble remains on the solid when the major part of the sessile bubble begins its ascent.

§25.   The ratio $vg\rho/2\pi r\gamma$ should be unity according to the (inexact) Equation 2.1. This prediction was fulfilled[2], for instance, for air bubbles released from vertical glass capillaries in water as long as the internal

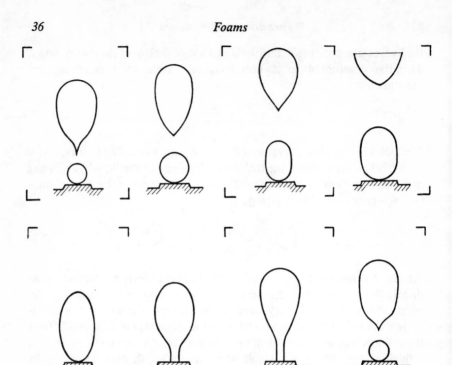

Fig. 2.2. Eight consecutive shapes of bubbles rising from a pore in a solid surface (free drawing after a photograph in reference[1]).

radius r of the orifice varied between 0.0007 and 0.0112 cm. A reasonable confirmation of Equation 2.1 was observed also when a much wider capillary (of inside diameter equal to 0.115 cm) was employed[3]. As long as not more than one (air) bubble was released (in water) in 20 sec, its volume v was 0.0216 cm^3; that is, the above ratio was near 0.82.

When Maier[2] used a waxed capillary, the v was, for instance, 20 times as great as when the capillary was hydrophilic. This undoubtedly was the effect of wetting. Air flowing out of the capillary tube displaced water from the external wall of the capillary; a similar displacement for a pore in a solid plate is indicated in Fig. 2.1. Hence the stem of the bubble was thicker, and the critical ratio of the length to radius of the stem was achieved when the bubble was larger than was the case with the hydrophilic walls. As would be expected, the material of the capillary was irrelevant as long as the degree of wetting was constant[4].

In the following tables (2.1 to 2.3) some additional data on the ratio $vg\rho/2\pi r\gamma$ are collected. The experiments on which Table 2.1 is based have been performed by Guyer[5]; the first column shows the orifice diameter, and the following give the above ratio for three liquids, namely water ($\rho = 1.00$ g/cm^3, $\gamma = 72$ g/sec^2), carbon tetrachloride ($\rho = 1.58$ g/cm^3,

$\gamma = 26$ g/sec^2), and diethyl ether ($\rho = 0.71$ g/cm^3, $\gamma = 16$ g/sec^2). The time between the release of two successive bubbles was 1 sec.

Table 2.2 contains some of the data published by van Krevelen[6]; the ρ of the oil was 0.877 and its γ was 35. The bubbling frequency was from one bubble in 0.7 sec to one in 3 sec. Table 2.3 refers to measurements[7] on air bubbles in water; again a vertical glass capillary was employed. The frequency was not specified, but apparently was about 1 bubble per second. It is not known why the ratio $vg\rho/2\pi r\gamma$ in these experiments was so much greater than in those of Tables 2.1 and 2.2. The v of bubbles rising from horizontal capillaries was smaller (for instance, only half as large) than that of Table 2.3.

If (the unjustified) equation 2.1 (§24) is correct, the ratio $vg\rho/2\pi r\gamma$ ought to be independent of temperature. This conclusion was confirmed[8] for air bubbles released in water from a glass capillary ($2r = 0.293$ cm)

Table 2.1. Volume of Air Bubbles in Different Liquids

	$vg\rho/2\pi r\gamma$		
$2r$(cm)	Water	CCl$_4$	Ether
0.0045	0.98	0.61	0.68
0.008	0.96	0.76	0.66
0.032	0.60	0.76	0.69
0.100	0.78	0.68	0.72
0.264	0.59	0.74	0.61

Table 2.2. Volume of Air Bubbles in Different Liquids

	$vg\rho/2\pi r\gamma$	
$2r$(cm)	Water	Transformer oil
0.15	0.90	0.88
0.95	1.05	—

Table 2.3. Volume of Air Bubbles in Different Capillaries

$2r$(cm)	0.022	0.095	0.132	0.179	0.211	0.296	0.450	0.519
v(cm^3)	0.0038	0.032	0.045	0.054	0.064	0.100	0.140	0.228
$vg\rho/2\pi r\gamma$	0.74	1.44	1.46	1.29	1.30	1.45	1.33	1.88

when the temperature ranged between 13°C and 60°C; the above ratio was 0.95 to 1.04. It remained near 1.0 also when the density of the liquid was raised to 1.486 g/cm^3 by using a concentrated zinc chloride solution instead of water. In agreement with the theory of §24, the ratio was not altered by increasing the viscosity of the liquid (experimentally, by substituting aqueous glycerol for water)[7].

§26. *Small bubbles.*   In many instances, production of small bubbles is desired. For fundamental studies, minute bubbles have the advantage of being nearly spherical. Molecular forces which result in capillary pressure and surface tension tend to reduce the area of the gas-liquid interface to its minimum; that is, they favor formation of spheres. Gravitation distorts the ideal shape of an immobile bubble and when the bubble is in motion, this shape is distorted even more by the resistance of the liquid. These two effects are less pronounced the smaller the bubble. The importance of the second (hydrodynamic) distortion depends on the Reynolds number:

$$(Re) = \frac{2Ru\rho_2}{\eta}. \tag{2.4}$$

Here $2R$ is the equivalent diameter of the bubble, u the linear rate of its advance, and ρ_2 and η are the density and the viscosity of the liquid. (Re) is a pure number. The smaller it is, the less the deviation of the bubble from the spherical shape. When high precision is not required, $(Re) < 0.01$ may be arbitrarily set as the necessary condition for the above deviation being negligible (see also §31).

From the theory of §24, minute bubbles will be obtained when their stems are very narrow, and this is likely to happen when the orifice has a very small diameter. This principle was used[9] to obtain bubbles as small as 16 μ in diameter; i.e., having volumes near 2×10^{-9} cm^3; the nearly horizontal capillary at whose tip the bubbles formed, was a hollow glass fiber with an internal radius of about 4 μ (= 0.0004 cm).

Many minute bubbles could be produced[10] in a liquid capable of frothing when this liquid was spread over a plate containing a fine orifice (e.g., 0.008 cm in diameter) and the air under the plate was kept at a relatively high pressure (e.g., 2 bars or 2 atmospheres). In this arrangement, however, the diameter of the pore certainly was not the only reason for the smallness of the bubbles. A rapid stream of bubbles causes turbulence in the liquid, and the bubbles are torn off not by buoyancy alone but mainly by the shearing force.

This force was deliberately used by Maier[2] to achieve small bubbles. He forced air through fine pores (e.g., $2r = 0.04$ cm) in a drum which could spin between stationary vanes. As long as the drum was also stationary, bubbles 0.35 to 0.40 cm in diameter rose from it. When it was making about 30 revolutions per second, the bubble diameter was only 0.04 cm; that is, its volume was about 3×10^{-5} cm^3.

Bragg[11] worked with a capillary having an orifice of $2r = 0.0038$ cm. According to Equation 2.2, this would result in bubbles of, perhaps, 0.05 cm in diameter. However, when the capillary was placed almost horizontally and the orifice kept under the surface of a surfactant solution moving with a velocity of 180 cm/sec, the diameter was near 0.014 cm. Thus while shearing lowered the bubble diameter by a factor of 10 in Maier's experiments, the factor here was presumably about 4. The air used was under a pressure of 190 cm of water. To achieve the desired speed of the liquid, the low cylinder containing it was rapidly rotated; in this manner over 1000 bubbles could be made in 1 sec. If the rotation continues for a longer time, the centrifugal force tends to throw the liquid out of the vessel; thus the present author found the method too cumbersome for his needs.

Smith's arrangement[12] was similar. Air was injected into a flowing soap solution through a 0.0085-cm orifice. At this value of $2r$ and assuming the surface tension of the liquid to be one-third that of water, the bubble diameter in stagnant liquid would be, say, 0.05 cm. A similar value (0.058 cm) was obtained when the liquid flow was slow (the driving head of water was only 9 cm), but bubbles of 0.024 cm formed in the solution driven by a pressure of 100 cm of water.

Neppiras[13] employed glass capillaries with inside diameters between 0.0025 and 0.0050 cm. According to Equation 2.2, the finer of these capillaries would give rise, in water, to bubbles about 0.1 cm across. The bubbles were simply hit several times every second. This caused detachment of unripe bubbles, and the smallest radius measured was 0.003 cm.

The growth of bubbles can be hindered by placing a curved solid directly above and very near the orifice[14]. The bubbles slide along the solid and escape long before their theoretical time. The volumes achieved in this manner ranged between 10^{-5} and 10^{-3} cm^3, and from 100 to 2000 bubbles were obtained in 1 sec. It is seen that so far use of extremely fine capillaries was most successful in lowering the bubble volume[9] but the volume of foam produced in this manner in unit time was very small indeed.

§27. Rate of bubbling. When the bubble formation is not very slow, the theory of §24 ceases to be valid even semi-quantitatively. One of the reasons for this breakdown of the theory is that it completely disregards the motion of the liquid. It is clear, however, that, every time a bubble becomes detached, liquid rushes in to occupy the initial space, and this flow may upset the next bubble, as the shearing force of §26 does. To have a clearer notion of what *slow* means in this context, again the Reynolds number is needed. This number is a function of a linear velocity, a density, a viscosity, and a "characteristic" length. In the process of bubble formation there are several velocities (of air, of growing bubble, of liquid rushing in, and so on), at least two densities (of the air in, and the liquid around, the bubble), two viscosities (of air and liquid), and several lengths which may be claimed as characteristic (e.g., the diameter of the orifice, the width of the bubble, the bubble's height, and the length of the stem). Thus, some arbitrariness is unavoidable when (Re) is calculated for bubbling. Several investigators, including Valentin[15] consider $2R$ of Equation 2.4 to be the diameter of the orifice (that is, $2r$); ρ and η the density and the viscosity of the gas; and $u(= u_1)$, the linear velocity of the gas in the orifice.

It is as easy to measure V as u_1; V is the volume rate (cm^3/sec) of gas flow. It can be determined with a commercial flowmeter (rheometer), or the gas transferred with the bubbles to the upper surface of the liquid layer can be collected in a vessel, as indicated in Fig. 2.3, and its volume G can be determined after a suitable time t; $V = G/t$. A small correction may be advisable, since the gas in the vessel G is at the atmospheric pressure, while the pressure in the bubbles is the sum of atmospheric pressure, hydrostatic pressure, and capillary pressure. For the corrected V, the u in Equation 2.4 is $V/\pi r^2$, so that

$$(Re) = \frac{2V\rho}{\pi r\eta}. \tag{2.5}$$

As long as this (Re) is small (say, less than 1), the theory of §24 is an acceptable approximation.

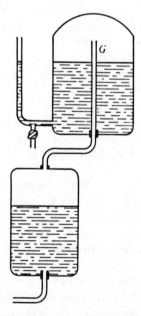

Fig. 2.3. Determination of gas volume bubbling through a liquid in a given time. G is a calibrated vessel in which gas is collected.

§28.   The volume rate V is the product of the volume v of a single bubble and the number N of the bubbles released per second; the dimension of N thus is sec^{-1}. When V increases, v can increase at a constant N, or N can increase at a constant v, or both quantities can change simultaneously. In some studies it was possible to observe three or four different regimes succeeding each other when V was increasing. In the first region valid for (Re) below, say, 100, N increases with V more rapidly than v does. This behavior is illustrated in Table 2.4; air was injected into water through an orifice 0.022 cm wide[1]. It is seen that, while V and u_1 increased in the ratio 250:1, N increased in the ratio of about 120:1 and v only doubled its magnitude. The minimum of v seen in Table 2.4 near $V = 0.9 \times 10^{-4}$ cm³/sec probably is not caused by an experimental error. Analogous observations have been recorded by Maier[2] and Remy[16] but a quantitative account for the minimum could not be found in the literature.

Fig. 2.4 presents some experimental data by Davidson[1] for higher gas flow rates and an orifice of 0.0334 cm. Air was injected into two aqueous glycerol solutions; the viscosity valid for the lower curve of v versus V was 5.1 g/cm·sec, and for the upper curve $\eta = 10.4$ g/cm·sec. When V increased from 0.3 to 2.5 cm³/sec (that is, the linear velocity increased from 342 to 2850 cm/sec), the v for the lower curve rose from 0.05 to 0.22 cm³ and N from 6 to 12 sec^{-1}; thus the relative increase of v was steeper than that of N, contrary to the data of Table 2.4. From Equation 2.5 the (Re) at the lowest point of the curve was about 76 (for air, not for the glycerol solution!) The curves in Fig. 2.4 have been drawn by the present author from one to the next experimental point without any attempt to smooth them out.

Bubble volumes at even higher values of V are shown in Figs. 2.5 and 2.6, redrawn from Ramakrishnan[17]. Note that the coordinates represent the logarithms of v and V rather than v and V themselves. Curves *1*, *2*, and *3* (Fig. 2.5) refer to water; the orifice diameters were,

Table 2.4. Bubble Frequency and Bubble Volume at Different Speeds

V(cm³/sec)	u_1(cm/sec)	N(sec⁻¹)	v(cm³)
0.4×10^{-4}	0.105	0.017	24×10^{-4}
0.9	0.24	0.083	11
1.75	0.46	0.15	12
7.38	1.94	0.42	18
36.1	9.5	1.0	36
60.0	15.8	1.5	40
101.0	26.6	2.08	49

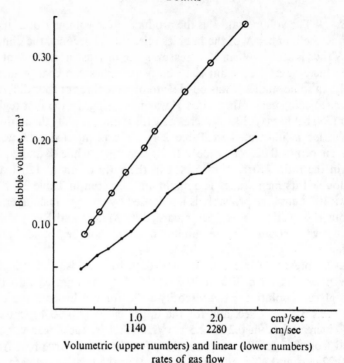

Volumetric (upper numbers) and linear (lower numbers)
rates of gas flow

Fig. 2.4. Dependence of bubble volume (ordinate, cm³) on volume rate (abscissa, upper numbers, cm³/sec) and linear rate (abscissa, lower numbers, cm/sec) of air flow. Aqueous glycerol of viscosity 5.1 g/cm·sec for the lower, and 10.4 g/cm·sec for the upper curve (data of reference[1]).

respectively, 0.14, 0.37, and 0.59 cm. Curves *4* and *5* (Fig. 2.6) summarize data obtained with aqueous glycerol whose viscosity was 5.52 g/cm·sec; the orifice diameters $2r$ were 0.37 and 0.70 cm. Because the values of $2r$ were so large, the linear velocity of air was similar to those of Fig. 2.4; for the capillaries of 0.14, 0.37, 0.59, and 0.70 cm, u_1 was, respectively, 650, 93, 37, and 20 cm/sec when V was 10 cm³/sec; naturally it was 10 times as high at $V = 100$ cm³/sec. It can be seen from the graphs that, when V increased 50-fold from 2 to 100 cm³/sec, the v in water at $2r=0.37$ cm increased in the ratio of 18:1, while N increased only from 17 to 45 sec^{-1} (i.e., 2.6:1). Thus, there is a qualitative agreement among Figs. 2.4, 2.5, and 2.6; all belong to the second regime, in which v increases with V more steeply than N does. This regime sometimes is, rather optimistically, referred to as the "constant-frequency bubbling."

When the volume rate V increases beyond those of Fig. 2.6, the bubble volume depends above all on the turbulence of the liquid and less on the orifice diameter and the physical properties of the liquid; turbulence may break bubbles formed initially and also cause coalescence of two or more

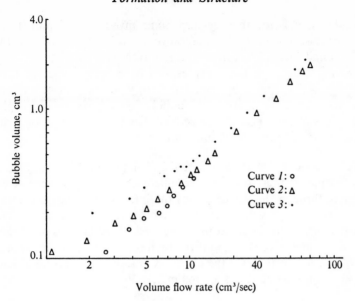

Fig. 2.5. Dependence of bubble volume (ordinate, cm³) on volume rate (abscissa, cm³/sec) of air flow. Water. Orifice diameters: 0.14, 0.37, and 0.59 cm for curves *1*, *2*, and *3* (data of reference[17]).

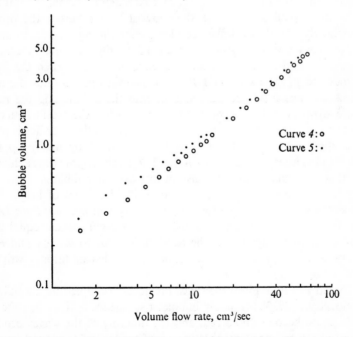

Fig. 2.6. Dependence of bubble volume (ordinate, cm³) on volume rate (abscissa, cm³/sec) of air flow. Aqueous glycerol, η 5.52 g/cm·sec. Orifice diameters: 0.37 and 0.70 cm for curves *4* and *5* (data of reference[17]).

bubbles[19]. This is the third regime. Some investigators count it as the fourth because they consider the range of volume velocities between Table 2.4 and Fig. 2.4 to correspond to a transition zone[20].

§29.   Although many expressions have been advocated for the function $v = f(V)$, none appears satisfactory to the present author; in many treatments, for instance, the bubbles are approximated as spheres at all stages of their growth. Three obstacles to formulation of a quantitative theory of rapid bubbling are mentioned below.

(1) Before a bubble starts growing, the air-liquid interface may be considered a plane; thus, the pressure difference across it is zero. The air streaming through the orifice causes the interface to bulge out into the liquid; hence the pressure in the bubble becomes greater than the hydrostatic pressure in the liquid at the same level. If the latter is p_0, if the gas pressure in the orifice is p, and if the radii of curvature at the given level are R_1 and R_2, then (when the air influx is slow) the pressure difference driving air into the bubble is approximately

$$p - p_0 - \gamma \left(\frac{1}{R_1} + \frac{1}{R_2} \right) ;$$

see §24. When the bubble becomes (approximately) a hemisphere, the two R's become equal and assume their lowest values during the process; consequently the above difference also passes through a minimum. The classical method of measuring surface tension γ by the maximum bubble-pressure method is based on this sequence of events. When the bubble continues to grow, its R_1 and R_2 also increase and so does the above difference. Thus the force pumping air into the bubble and the rate of bubble growth oscillate between a minimum (when the radii of curvature have their minimum values) and a maximum (when the interface is plane). These oscillations can be rendered less important by making p much greater than either p_0 or any $\gamma[(1/R_1)+(1/R_2)]$; the regime thus achieved is sometimes designated as the constant-rate-of-flow bubbling.

(2) When a bubble of volume v enters a liquid column of height h and cross-section A, the center of gravity of the system is raised by $v/2A$ cm. Thus the work of forming a bubble contains a component equal to 0.5 $vhg\rho_2$ g·cm²/sec²; again g is the acceleration due to gravity and ρ_2 the density of the liquid. Some consequences of this additional work are discussed by Potter[21].

(3) The flow of liquid into the space previously occupied by the bubble is mentioned in §27. When many bubbles rise one soon after the other, the whole liquid becomes involved, and the intensity of the waves produced obviously must depend on the shape and volume of the vessel in which bubbling takes place. These waves may shear the bubble off prematurely, as indicated in §26.

Qualitatively, the increase in bubble volume v with the volume flow rate V of air, as long as V is small, is easy to comprehend, and was understood long ago for falling drops. Bubble detachment occurs because of the necking of the stem described in §24. This process cannot be seriously influenced by the rate of air influx, unless this is excessive. Let the time needed for the collapse of the stem be t_0 sec; in this time, Vt_0 cm^3 of air enters the bubble; hence the bubble volume is greater the greater V.

A very peculiar correlation[6] was derived by dimensional analysis. It was claimed that v in the "constant frequency" region depended only on V and the acceleration g (due to gravity), not on any density. Thus $v = f(V, g)$. This equation can be dimensionally correct only if $v = kV^{6/5}/g^{3/5}$, k being a numerical constant.

The bubble volume v is treated as a function of V in §§28 and 29. It can be considered also as depending on the pressure p driving the gas into the liquid. In a 0.4% sodium oleate solution the equivalent bubble diameter increased from 1.85 to 4.7 mm when the driving pressure was raised from 21 to 270 mbar, and in a 0.2% peptone solution the increase was from 1.9 to 3.8 mm when p advanced from 19 to 160 mbar[22]; 1 mm Hg is equal to 1.33 mbar.

§30. Of the properties of the liquid, surface tension γ and density ρ_2 are important for the slow formation of bubbles (see §24). When formation is not slow, also the viscosity η of the liquid would be expected to influence the bubble volume v. At high volume velocities V, this v seems to be practically independent of γ. Thus, $v = f(V, \eta, g\rho_2)$. Since the dimensions of these four quantities are, respectively, cm^3, cm^3/sec, g/cm·sec, and g/cm^2 sec^2, the above function is possible only as long as

$$v = k \left(\frac{V\eta}{g\rho_2} \right)^{\frac{3}{4}};$$ (2.6)

k is a numerical constant. A similar equation has been derived earlier[1].

Experimental indications of the effect of η on v are visible in Figs. 2.4 through 2.6. When η was roughly twice as great, the v (at $V = 2$ cm^3/sec) was about 1.8 times as great. At $V = 10$ cm^3/sec and the orifice diameter of 0.37 cm, the bubbles had a volume of 0.35 cm^3 when η was near 0.01, and 0.89 cm^3 when η was 5.5 g/cm·sec. The agreement between the two series of experiments is not encouraging.

According to Guyer[5], an increase in bubble frequency N from 0.2 to 10 per second had no effect on v in carbon disulfide ($\eta = 0.0035$); in olive oil ($\eta = 0.84$) v very slightly increased with N; and in castor oil ($\eta = 9.5$ g/cm·sec) v increased about fourfold in this range of frequencies.

The effect of surface tension on v is particularly interesting for a student of foam because so many solutions capable of foaming have low values of γ. When bubble formation is slow, v increases with γ, as expected. This can be concluded, for instance, from the inspection of Table 2.1, §25.

From diethyl ether to water, γ increases in the ratio 4.5:1, and v is greater in the ratios ranging between 1.4:1 and 1.7:1. The increase of v with γ sometimes can be observed[23] at a volume rate as high as 2 cm³/sec, corresponding to a linear rate of 41 cm/sec; the v in water ($\gamma = 72$) was 1.4 times as great as in an aqueous solution of 2-propanol having $\gamma = 37$ g/sec². As stated above, the effect of γ on v appears to vanish at high rates V.

When the surface tension of water is lowered by a typical surfactant, an additional effect seems to be present. In an aqueous solution of commercial sodium dodecyl benzenesulfonate the v was not only smaller than in pure water, but also increased with V at volume rates so small that hydrodynamic effects surely were not responsible. Thus[3] v was independent of N in water as long as N was below 0.05 sec^{-1} (that is, one bubble in 20 sec), but in a surfactant solution (0.1 g/l.) the v was, respectively, 0.014 and 0.012 cm³ at N of 0.05 and 0.0055 sec^{-1}. In another series of tests[20], V had no effect on v in water up to V of 0.017 or 0.067 cm³/sec (depending on the orifice), but v increased with V at all velocities tested when the liquid was a 0.08% solution of a commercial surfactant. For additional data see reference[8].

The effect of formation time (i.e., $1/N$) on v in surfactant solutions in all probability is related to the dynamic surface tension defined in §11. The force retarding the escape of a bubble or causing the collapse of the bubble stem (see §24) is proportional to γ, whatever the frequency of bubbling. Thus this force changes in the same direction as γ does. The diffusion of large surfactant molecules or ions to the air-liquid interface is slow. During the gradual expansion of this interface, the supply of the surfactant to it usually is not fast enough to maintain the γ near its equilibrium (i.e., low) value. In other words, when $1/N$ and V are small, then γ also is small (because nearer to the equilibrium) and bubbles detach when their volume and buoyancy are small; and when $1/N$ and V increase, so do both γ and v. A method of determining the dynamic surface tension of solutions, based on this mechanism, is described, for instance, in a recent monograph[24].

§31.   The rate V of air injection affects not only the volume but also the shape of the bubbles. Minute bubbles are almost spherical immediately after detachment and remain almost spherical during their ascent through the liquid. When V increases, also the volume v of the rising bubbles is bigger, and big bubbles usually become flattened during their ascent so that their horizontal dimensions are greater than the vertical (see also §26). At a given v, the degree of deviation from the spherical shape is more pronounced the larger the dimensionless ratio $u\eta/\gamma$; u is the linear velocity of ascent (cm/sec), η the viscosity of the liquid (g/cm·sec), and γ its surface tension (g/sec²). As the u of big bubbles is greater than that of small ones (see Equation 1.3), their deformation is more drastic. When the bubble volume in water is, say, 1 cm^3, the bubbles often have the shapes of inverted dishes or spherical caps. Many acquire a toroidal shape similar to the smoke rings emitted by a skilled tobacco smoker. None of these geometries survives when the bubble leaves the liquid and becomes an element of the foam.

In a liquid incapable of foaming, the value of u restricts the volume rate V at which still discrete bubbles result rather than a continuous jet of air streaming through the liquid. At the moment a bubble starts its ascent, the next bubble starts growing. If the linear rate of upward movement of the summit of the growing bubble is greater than the speed of rise of the bottom of the first, the two may collide and an air jet be obtained[6]. In water, it is claimed[8,18], not more than about 50 bubbles can be produced in 1 sec, whatever the rate of air injection. In a solution of a foaming agent the two bubbles would remain separated by a foam lamella.

When the bubble is definitely not spherical, the rate of its ascent cannot be predicted by theory, as in §2; for an empirical equation see, for instance, Efremov[60]. When a stream of bubbles ascends in a vessel filled with liquid, additional complications arise, some of which are indicated here.

Let the orifice be situated in the center of the vessel bottom so that the bubbles rise along the vertical axis. Then the mixture of liquid and gas near the axis has a lower specific gravity than the liquid layer nearer to the walls. Consequently the liquid is maintained in continuous circulation: It rises along the axis and flows down along the periphery. Hence, the rate of ascent of a bubble relatively to the walls is greater than relatively to the liquid in which the bubble is embedded. For a hydrodynamic treatment of this process see Rietema[25].

When the rate of air injection is constant and the liquid neither foams nor evaporates, then in the steady state the vessel contains, in addition to the initial volume v_2 of the liquid, a volume (v_1) of gas. Consequently, the liquid layer is in the ratio $(v_1 + v_2)/v_2$ deeper than before bubbling, and each bubble spends more time—again in the ratio $(v_1 + v_2)/v_2$—in the liquid than would a single bubble of identical size and shape. Equation 1.3, §2, refers to the steady-state velocity of a single bubble; but it is seen on a closer examination that no constant velocity is ever acquired. At the bottom of the vessel the air in the bubble is compressed by the liquid above it. Hydrostatic pressure decreases as the bubble rises; the diameter of the latter increases, and so does the velocity u. These problems have been discussed by Melikyan[26].

If any asymmetry is present, bubbles whose diameters exceed about 0.1 cm often rise in zigzag lines or along spirals. A review of this and related effects will be found in Valentin's book[15].

§32. *Multiple orifices.* The number of bubbles and the volume of foam which can be produced in a reasonable time with a single capillary often are too small for the application intended. In these instances sieve plates, diffuser stones, porous tubes, and similar devices, often denoted as spargers, are employed. In a sense, all these instruments are inefficient. When a sparger has n pores, each $2r$ across, it does not give rise simultaneously to n bubbles whose volumes correspond to the orifice diameter equal to $2r$. It is practically impossible to make all pores so exactly alike that the air stream is equally distributed among them; in every instance only a fraction of the pores is active. Moreover, when two or more active pores are near each other, their bubbles tend to coalesce before detachment. Thus the number of bubbles is smaller, and each bubble is bigger that would have been obtained with n independently fed capillaries.

Otherwise, there is a marked similarity between the performances of single and multiple dispersion devices. When the gas injection rate V (for all pores together) increases, three or four regimes analogous to those enumerated in §28 can be discerned[27]. It is stated above that at very high V values the bubble volume becomes independent of the physical properties of the liquid; this seems to be true also for sieve plates[28]. Other similarities between the effects of single pores and porous disks are pointed out by Bowonder[29]. On the other hand, different spargers (spinnerettes and porous plates) gave rise to foams of different properties[30].

If the sparger is small in comparison with the width of the vessel in which the bubbles rise, these form a column which gradually expands (in the horizontal direction) on approaching the upper air-liquid interface. It is said[27] that the diameter Δ of this column, near its bottom, depends only on the volume rate V cm^3/sec and the acceleration (due to gravity) g cm/sec^2; dimensional analysis shows, analogously to §29, that the equation must be $\Delta = kV^{\frac{2}{5}}g^{\frac{1}{5}}$; k is again a pure number.

In some aerators, air is injected through wide pipes, and the large bubbles (or the air jet) obtained are (is) broken down to smaller bubbles by agitation (see §34). The bubble diameter $2R$ in this process is claimed[31] to depend only on surface tension γ, liquid density ρ_2, and the power input (of the impeller) per unit volume of the liquid; this quantity is symbolized by P/v_0. From dimensional analysis,

$$2R = \frac{k\gamma^{\frac{3}{5}}}{(P/v_0)^{\frac{2}{5}}\rho_2{}^{\frac{1}{5}}}, \qquad (2.7)$$

k being a constant.

In all examples described in this chapter so far, air was the material of the gas phase, and in almost all examples compressed air supplied the energy necessary to mix the two phases. Recently so-called aerosol foams have become popular; they are used for cosmetic purposes, as antiseptics, as contraceptives, and so on. Instead of compressed air, gases such as propane (whose vapor pressure is 9 bars at 22°) or 2-methylpropane serve as propellants. The aerosol cans contain these propellants and a liquid (e.g., a soap solution + the physiologically active ingredients). When the valve is opened, the vapor expands and gives rise to foam, as in the ordinary bubbling. For additional data on "aerosol foams" see §187.

   §33.   Of the many methods for dispersing a gas in a liquid, the injection of the gas through a single orifice (§24) is probably the easiest to control. The injection through many mutually connected orifices would occupy the next place, and various agitation methods give the least-controllable results. Some examples of agitation are:

1. A cylinder partly filled with the liquid is shaken up and down
2. The liquid is whipped or beaten, as in a blender
3. A perforated plate rapidly and frequently enters the liquid and is withdrawn from it
4. The liquid is poured from one into another vessel.

   A more detailed description of some agitation methods will be found in Chapter 3, as they are utilized for measuring the foaming capacity of a liquid. Their mathematical theory is, understandably, in a rudimentary state; the paper by Gilberg[32] can serve as a starting point for study.

   One of the earliest papers on agitation foams describes a strange process[33]. Mercury was covered with a thin water layer, and a jet of water was directed against it. In this manner some air was incorporated in the mercury. Air bubbles encased in mercury films rose through the water and floated on its surface. Some of them were as large as 1.5 cm across and contained 0.5 g of mercury so that their specific gravity presumably was near 0.4.

   The degree of dispersion achieved in many agitation devices is not considered satisfactory, and the gas-liquid mixture obtained is subjected to a secondary treatment, which may simply mean stirring (as mentioned in §32) or forcing the gas emulsion through a fine gauze[34], the latter process is analogous to homogenizing emulsions.

   §34.   Stirrers and homogenizers reduce the average volume of gas bubbles because of the very general effect which renders the stem of a bubble unstable (§24) and governs a host of other phenomena, ranging from emulsification to the formation of ocean spray. The effect was described and explained by Plateau[35]. The simplified proof given below is from reference[24].

   Approximate a bubble produced by agitation as a cylinder $4L$ cm long and of radius R. Because of random disturbances in the liquid, it is deformed. Its shape now is approximated by assuming the generatrix to be a sine curve, $y = k \sin (\pi x/2L)$. The directions of the two coordinates (x and y) are shown in Fig. 2.7, which represents only a part of the deformed cylinder. The diameter of the deformed cylinder at $x = 0$ is still

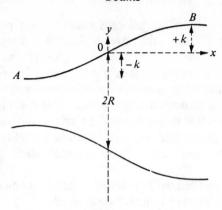

Fig. 2.7. Scission of elongated bubbles. Gas is between the two curved lines drawn as sine curves $y = k \sin (\pi x/2L)$. The pressure in the gas underneath point A is less than that underneath point B, (so that depression at A deepens) whenever the length $4L$ of the deformed cylinder exceeds its initial circumference $2\pi R$.

$2R$ (no local deformation), and the derivative $dy/dx = 0$ at both points A and B. From the general equation for the radii of curvature, the radius swinging in the plane of the drawing at point A is $-4L^2/\pi^2k$, and at point B it is $+4L^2/\pi^2k$. The ordinates of A and B are, respectively, $-k$ and $+k$. Consequently, the radius of curvature swinging in the cross-section of the cylinder (i.e., in a plane perpendicular to the drawing) is $R-k$ at A and $R+k$ at B. The depression at A tends to become deeper (and to result in splitting the cylinder) when the capillary pressure at A exceeds that at B, however small the disturbance k. Thus, necking-down occurs whenever

$$\frac{1}{R-k} - \frac{\pi^2 k}{4L^2} > \frac{1}{R+k} + \frac{\pi^2 k}{4L^2} \; ; \qquad (2.8)$$

that is when $4L^2 - \pi^2R^2 + \pi^2k^2 > 0$. When k is small in comparison with R, this condition is equivalent to $4L^2 > \pi^2R^2$. Hence $2L$ must be greater than πR, or $4L$ must exceed $2\pi R$. When the length of a cylinder exceeds its circumference, the cylinder ceases to be stable.

A bubble becomes elongated in the direction of flow whenever the surrounding liquid is sheared, as in stirring or whipping. When the elongated bubble happens to drift into a more or less quiescent zone, then the capillary pressure acts unchecked and the bubble may separate into two or more nearly spherical bubbles. When a large bubble is squeezed through a gauze, as mentioned in §33, a similar sequence of events unfolds.

Several foam-generating apparatuses employing dispersion methods have been patented. These patents are not reviewed here. An example of equipment suitable for fire-fighting foams is described in reference[36].

Condensation Methods

§35.   The gas in a corked bottle of champagne or beer is under enhanced pressure. When the bottle is uncorked, bubbles appear in the liquid. The liquid was saturated with compressed gas; when the pressure dropped, the solution became supersaturated, and part of the excess gas gathered together to form bubbles. This effect causes also the *caisson disease* and the *altitude sickness*, about which some literature is listed by Weatherford[37].

In the above examples, bubble formation is facilitated by the motion of the liquid; for instance, the surge of beer toward the bottle orifice. The process of creating bubbles in a liquid by violent motion of the latter is known as *cavitation*. It is not treated in this book because cavitation apparently is not important for foam making.

If the liquid is not disturbed and does not contain suspended particles (dust, gas bubbles, and so on), it usually can remain supersaturated for a protracted time. This sluggishness generally is attributed to surface effects. Let F be the decrease in free energy of the system when 1 g of the dissolved gas returns to the gas phase. If R is the bubble radius and ρ_1 the gas density in it, the decrease per bubble is $(\frac{4}{3})\pi R^3 \rho_1 F$. On the other hand, the surface area of the bubble is $4\pi R^2$ and its surface energy is $4\pi R^2 \gamma$, γ being again the surface tension of the liquid. Bubble formation is energetically favorable only when $(\frac{4}{3})\pi R^3 \rho_1 F$ is greater than $4\pi R^2 \gamma$—that is as long as $R\rho_1 F > 3\gamma$. This inequality, and much more elaborate treatments dispersed in the literature[38], show that extremely minute bubbles for which $R \approx 0$ cannot form, as long as γ has its normal value. However, it follows from Laplace's theory of capillarity that γ should decrease when R decreases. Crudely speaking, if the bubble is so small that liquid molecules can attract each other *across* the bubble, the difference between the two attractions—toward the main liquid and toward the bubble—is small and so is the surface tension, which is caused by the above difference; a quantitative account of this phenomenon can be found in reference[24]. Thus the improbability of satisfying the inequality $R\rho_1 F > 3\gamma$ is not as great as was long believed. Still, the necessity of overcoming surface forces when forming a bubble should always be kept in mind.

Another way of indicating this difficulty starts from the concept of capillary pressure. Pressure in the bubble is by $2\gamma/R$ greater than that (p_0) in the liquid around it. Let p be that pressure in the liquid, below which supersaturation takes place; this means that the dissolved gas tends to join the gas phase whenever p_0 is less than p. Thus, bubbles can grow when $p > p_0 + (2\gamma/R)$. When $R \approx 0$ and γ has its normal value, this inequality is impossible to satisfy. If γ decreases with R, again the obstacle is not as formidable as it appeared to earlier investigators.

§36.   The formation of gas nuclei in a dust-free liquid usually is designated as homogeneous nucleation. In agreement with the theory of §35, it is not a common occurrence, and it is difficult to prove that the observed bubbles formed really without any contact with foreign particulate matter. An attempt to achieve homogeneous nucleation starts with achieving supersaturation. Examples for the three main methods of obtaining supersaturated solutions of gases in liquids are given below.

Water, ethanol, and 2-propanol were (separately) heated in closed vessels until the vapor pressure of the liquid was, for instance, 0.3 bar higher than the atmospheric pressure outside. Then the vessel was opened and simultaneously the central part of the liquid observed. Bubbles appearing in it, far from the wall, presumably have been homogeneously nucleated[39].

Bubble formation can be induced by altering the dissolving ability of the liquid. Thus, Krause[40] mixed, among others, 1 cm³ of water saturated with nitrogen with 10 cm³ of 19 N sodium hydroxide solution. Sodium hydroxide (and many other electrolytes) "salts out" nitrogen (and many other gases), that is, depresses its solubility in water. In this manner bubbles as small as 100 Å across could be produced. Minute bubbles were observed[41] also when caustic soda was added to water saturated with helium.

Gas can be liberated in a chemical reaction, for instance, by adding a few drops of hydrogen peroxide to a concentrated solution of sodium hydroxide or by mixing sodium carbonate solutions with an acid. A complicated example of this type has recently been patented[42]. Hydrogen peroxide and then silver oxide deposited on alumina, for instance, are introduced into an aqueous soap solution. Not only is oxygen evolved, giving rise to a foam, but the foam is warm because the chemical reaction is exothermal. In this system, however, the nucleation may well have been heterogeneous at the surface of silver or alumina.

Heterogeneous nucleation is much more common than homogeneous nucleation. It is easier for a bubble to start growing at a solid wall or particle than in a dust-free liquid far from any wall. Presumably, this is not due to any fundamental influence of the solid but is simply the consequence of the fact that solids contain occluded or adhering gas which acts as a pre-existing nucleus. According to the fascinating historical review by Gernez[43], this view was advocated by Deluc in 1772, Schönbein in 1837, and several other scientists.

More recent observations confirm this concept. When water is supersaturated with air but still is practically bubble-free, copious bubble formation can be initiated by plunging almost any solid into the liquid; but when the solid (e.g., fibrin or rubber) had previously been kept in de-gassed water, its "catalytic" activity ceased[44]. It has often been noticed that rough solids promoted the evolution of bubbles more efficiently than did very smooth specimens. This difference presumably also is not of a fundamental nature. It is more difficult to displace the adhering air from a rough solid than from a plane. If the experimenter takes the trouble to remove practically all air from a rough solid, this ceases to precipitate gas evolution in water supersaturated with air or heated to the boiling temperature[44,45].

If the ability of solids to facilitate bubble formation were their inherent property, then a solid immersed in a supersaturated solution of a gas would have been covered with bubbles all over. In reality, there is a finite number of "active sites" (on which nucleation occurs) on any solid surface. When a liquid is heated in a vessel, boiling (that is, evolution of vapor bubbles) invariably starts from one or few points; this is a common laboratory experience. Active sites can be prepared deliberately. For instance, when a metal plate was scratched with diamond, placed under water, and heated, bubbles appeared only along the scratch, and about 10 min of boiling was necessary to deactivate the latter[46]. The number of active sites is greater when the heating temperature is higher. For instance, Jakob[47] counted five spots from which steam bubbles rose when 7500 kg·cal/m²hr was supplied to a horizontal polished metal surface submerged in water; and when the energy input was four times as great, 24 spots gassed.

§37. The two preceding examples refer to nucleate boiling. This phenomenon has been investigated mainly from the viewpoint of heat transfer, which is outside the scope of this monograph. However, the problems of bubble frequency N and bubble volume v, which are important in the dispersion processes (§28) must be considered in reviewing boiling as well.

If the amount of energy Q (ergs or joules per sec) is transferred to the boiling liquid in unit time from the heat source, the amount Q/λ (g or kg per sec) of the vapor must form and an equal amount of liquid disappear; λ is the latent heat of evaporation expressed in joules/g, or any other equivalent unit. The magnitude of Q/λ determines the total volume V cm³/sec of the bubbles produced in 1 sec. Contrary to §28, here $V = N_s N v$, N_s being the number of active sites, and the question to answer is, what happens to N_s, N, and v when V increases or decreases with the heat input rate Q.

The increase of N_s with Q is mentioned in §36; apparently it has not been studied systematically. As long as V is very small, N may be supposed to be independent of it. For this regime, the review by Cole[48] recommends the equation

$$N = \frac{(g\rho)^{\frac{3}{4}}}{\rho_2^{\frac{1}{4}} \gamma^{\frac{1}{4}}}. \tag{2.9}$$

As before, ρ is the difference between the densities of liquid and vapor, and ρ_2 is the density of the liquid. In the same regime of boiling, the volume of the detached bubble is said to be

$$v = \beta\left(\frac{\gamma}{g\rho}\right)^{\frac{1}{2}}.$$

(2.10)

β is a numerical constant whose value depends on the contact angle θ mentioned, for instance, in §11. As there is no pore or orifice, the v is not a function of r (see Equation 2.1). If both Equations 2.9 and 2.10 are correct, then V can increase on an increase in Q only by altering θ or by having a greater number (N_s) of active sites in the vessel.

Experimentally, the relation

$$Nv^{\frac{3}{2}} = K$$

(2.11)

sometimes is found[49]; K is a constant having the dimension cm²/sec. According to the old observation by Jakob[47], the volume of steam bubbles rising from a hot chromium-plated surface increased in the ratio $1:8:27:64:125$ when the bubble frequency decreased from 73 to 48, 35, 29, and 25 bubbles/sec. This frequency apparently was N alone, not NN_s. Hence the product Nv^3 increased with v.

§38.   The coefficient β in Equation 2.10 is a function of the contact angle θ. It is easy to show that the greatest volume of a bubble, which still can cling to a horizontal solid surface in spite of buoyancy, must be a function of θ. As long as the bubble shape is determined by surface and gravitational forces only, the variety of these shapes is restricted.

Fig. 2.8, redrawn from Wark[50], illustrates six of the permissible bubble profiles. For purely mathematical reasons, the summits of all bubbles are placed at the origin of the coordinates; i.e., at $x = 0$ and $z = 0$; z is the vertical and x is the horizontal distance. The numbers from 0.1 to 0.7 at the profiles mean the ratio $g\rho R_1^2/\gamma$ (a pure number). R_1 is a radius of curvature at the summit; as the bubble is supposed to be symmetrical about the vertical axis, the other radius of curvature at this point also is equal to R_1. Qualitatively, if the radii near the summit are short, the whole bubble is small, as would be expected.

The contact angle (in the liquid!) is the external angle which the profile makes with the horizontal. For instance, at the depth of 0.8 below the summit, the line marked 0.7 shows $\theta \approx 146°$, whereas for the same curve θ at $z = 0.4$ is near 100°. The θ along this curve is nowhere smaller than about 95°. The smallest θ made by the curve 0.6 is approximately 90°, and the smaller the ratio $g\rho R_1^2/\gamma$, the smaller is the least θ. This relation is easy to understand. When θ is large, air readily displaces liquid from the solid surface; consequently, the bubble has an extensive base, and the biggest bubble which can still adhere to the solid also is large. When θ is small the liquid tends to displace the gas so that the base and the volume

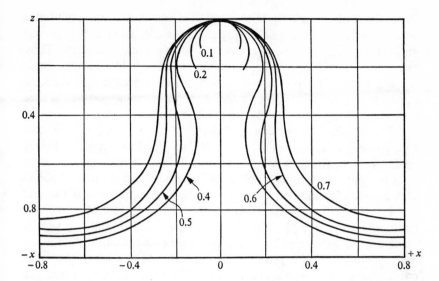

Fig. 2.8. Examples of possible profiles of sessile bubbles. If the liquid is water, the units shown along the coordinates are centimeters (after reference[50]).

of the maximum bubble are small. Both by calculation and experiment[47,51], the maximum volume is

$$v_m \approx k_1 \theta^3 \qquad (2.12)$$

between $\theta = 0°$ and $\theta = 100°$; k_1 is a constant for a given liquid. The properties of the liquid are, in principle, eliminated when this equation is altered to

$$\frac{v_m}{a^3} = k_2 \theta^3; \qquad (2.13)$$

a is Laplace's capillary constant equal to $(2\gamma/g\rho)^{0.5}$, and k_2 is a numerical constant which depends, of course, on whether θ is measured in degrees or in radians.

§39.　　Fig. 2.8 and Equation 2.13 refer to bubbles in equilibrium and are based on the assumption that contact angles exist and have a definite value. Presumably, none of these conditions is generally valid for vapor bubbles in a boiling liquid. First, as is seen in Fig. 2.2, §24, the bubble breaks in two before the upper part starts rising; a sizable part of the original (still-adhering) bubble remains on the solid after the detachment. Thus, θ of the equilibrium drop does not directly determine the volume of the loose bubble. Moreover, as indicated above, every motion in the liquid is likely to lower this volume.

A third difficulty is related to the ambiguity of contact angle values. When a liquid invades a solid surface, that is, displaces gas from the latter, the contact angle θ_A (in the liquid!) is greater than the equilibrium contact angle θ; θ_A is the advancing contact angle. The receding contact angle, θ_R, is smaller than θ; it is observed when gas displaces liquid. Thus, as long as the bubble grows, θ_R is relevant. When a bubble escapes, liquid tends to occupy its position, and for this process the angle θ_A is valid. As θ_A may exceed θ_R by 30° or even more, the value of contact angle to be used in calculating bubble volumes is difficult to choose.

The equilibrium contact angle between a vapor (or gas), a liquid, and a solid strongly depends on the nature of the solid. This is true also for the volume v of the rising bubble. When the solid was oiled[47], the θ of water on it was much greater than on "clean" solid and also the steam bubbles were bigger. Sometimes v decreased when the surface roughness of the solid increased. This may have been caused by a greater number of active sites on a rough metal (see §36). Greater roughness usually results in a smaller contact angle if the equilibrium contact angle on a smooth surface is less than 90°; this is another possible reason for the above effect of roughness on v. Hysteresis of wetting (that is, the difference between θ_A and θ_R), which usually increases with the degree of roughness, is another possible cause. Jakob's observations were confirmed in the experimental boiler of Leaf[52]; bubbles leaving a hot copper surface were big when the metal was smooth, and small when it was deformed by multiple punctures.

A relation between contact angle and some properties of the solid would be expected as long as there is a true contact between the solid and the gas. However, it was suspected long ago that a thin film of liquid often is present over the whole (or at least the major part of the) visible interface between the gas on one hand and a solid or a liquid metal on the other, so that the gas and the metal phases have only a small contact area or none at all. Recently this concept became popular also for steam bubbles on a heater surface. It agrees with the observations on the heat transfer from the solid to the bubble more closely than does the older notion of an extensive gas-solid interface[53].

Jawurek[54] determined the thickness of some of these films by means of interferometry. The flat bottom of the vessel was made of an electrically conducting glass, that is, a glass covered with a stannic oxide coating. Electric current passed along this coating and heated the liquid in the vessel. The liquid was methanol or ethanol. When the greatest horizontal dimension of the vapor bubble was, say, 1.2 cm, a "dry patch," along which the thickness of the liquid film (if any) was below about 800 Å, was visible in the center of the bubble base. From the center toward the periphery this thickness increased to, for instance, 4 μ. During the growth of a bubble, the diameter of its "dry patch" rose from zero to, for instance, 0.2 cm and then dropped again to zero as the detachment took place and liquid invaded the glass surface.

A study of this process by Kotake[55] led to the conclusion that such films were possible from the standpoint of hydrodynamics. Their thickness is said[53] to be approximately $0.8(\eta t/\rho_2)^{0.5}$; η is viscosity and ρ_2 the density of the liquid, and t is the duration of bubble growth (i.e., $t = 1/N$). The form of the function can be readily derived by dimensional analysis, but the numerical coefficient can be valid for only one distance from the bubble axis. Its value was estimated from experimental data on boiling toluene and 2-propanol.

At present, it would be rash to conclude that true gas-solid contact in boiling is a rare occurrence or does not occur at all. Some common observations cannot at present be reconciled with this view. Thus, excessive bubbling in boiling liquids, which causes "carry-over" and "priming" (see §145) may in favorable circumstances be suppressed by additions which, as far as is known, act only by making the solid more hydrophobic, that is, raising the contact angle (in the liquid!) as did the oil in Jakob's experiments. A greater θ resulted in bigger bubbles, which had a higher mortality than dwarfs (see §47).

§40. *Bubbles at interfaces.*   The volume of the bubble when it reaches the upper surface of the liquid (in which bubble formation took place) is always different and sometimes quite different from that in the moment of detachment. The small difference has been mentioned already in §31: the bubbles near the top of the liquid are under a lesser hydrostatic pressure than before their rise and thus are a little bigger. The big difference is caused by coalescence. This is a frequent event in liquids incapable of foaming; when a foaming agent is present, merging of two bubbles is much less probable because the film between them is not different from those reviewed in Chapter 1, and there shown to be relatively long-lived.

When the bubble reaches the upper surface of the liquid, and the liquid has no foaming tendency, the bubble bursts at once; that is, the film separating it from the bulk gas phase immediately ruptures. When the liquid contains a foaming agent, the above film has a significant persistence, and the bubble lifts a "dome," as illustrated in Fig. 2.9.

The shape of the bubble under the dome has been repeatedly discussed in scientific literature, first by Plateau himself[56]. There are several difficulties. Equation 2.3, §24, determines the shape of any vapor-liquid

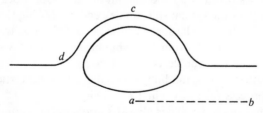

Fig. 2.9. A bubble under a liquid surface. The hydrostatic pressure at a must (in equilibrium) be equal to that at b; this condition partly determines the curvature of the "dome" at c.

interface in the gravitational field, provided that the system is in equilibrium and that no other force is operating. In reality, buoyancy is always present and hydrodynamic forces may be affecting the bubble shape, as stated in §31.

As already stated, when the bubble is minute, it is nearly spherical, and its deformation, at a first approximation, may be disregarded. In the liquid immediately underneath the bubble (point a in Fig. 2.9) the hydrostatic pressure in equilibrium must be identical to that under the plane surface at the identical level (point b in Fig. 2.9); that is, it must be approximately equal to $2Rg\rho_2$, if R is the bubble radius, g acceleration due to gravity, and ρ_2 the density of the liquid. But the hydrostatic pressure at a, neglecting the weight of the film above the bubble, is only $2Rg\rho_1$; ρ_1 is the gas density in the bubble. The difference must be compensated by the capillary pressure of the dome $2\gamma/R_1$; R_1 is the value of the two identical radii of curvature of the dome roof at its highest point—that is, point c in Fig. 2.9. Thus, using ρ for $\rho_2-\rho_1$, the approximate equation

$$R_1 \approx \frac{\gamma}{g\rho R} = \frac{a^2}{2R} \tag{2.14}$$

is obtained. The square (a^2) of the Laplace capillary constant for water at room temperature is about 0.14 cm^2. Thus, if $R = 0.05$ cm, then R_1 from Equation 2.14 is 1.4 cm. The dome is rather flat. The curvature of the dome between the summit and the plane liquid surface changes with height, according to Equation 2.3, §24. In fact it is identical with that of liquid under which a (well-wetted) marble of the shape of the bubble is suspended.

Equation 2.14, of course, is only an approximation. It implies that the upper end of the gas phase in the bubble is on the level of the plane surface far from the bubble, and it neglects the weight of the dome. If the thickness of the latter at its highest point is δ, the neglected term $g\rho_2\delta$ is equal to that considered ($2Rg\rho_1$) when $\delta = 2R\rho_1/\rho_2$. The thickness δ must be much smaller for Equation 2.14 to be admissible.

The bigger the bubble, the more its shape deviates from the spherical and the more the diameter Δ of its base increases in comparison with its height h. This is so because, if the summit of the bubble does not rise, the center of gravity of the system still can descend by shifting the liquid from the level of the bubble to the space below the bubble; this transfer can be achieved by pushing the bottom of the bubble up, that is, by shortening h. As the volume of the bubble is almost constant, a shorter h means a bigger Δ. In other words, the bubble is flattened in the vertical direction because this process raises *its* center of gravity.

The summit of the bubble is almost stationary (in a liquid capable of foaming) because the capillary pressure of the dome acts against buoyancy. An equation analogous to Equation 2.14 can be derived also for large bubbles. Let h_1 be the depth of the bottom of the bubble below the plane liquid surface. The roof and the ceiling of the dome above a large bubble are almost parallel to each other. Thus, the radii of curvature of the two are almost identical; let R_1 be their value for the top of the dome. The two radii of curvature at the lowest point of the bubble are equal since the bubble is symmetrical about its vertical axis; they may be denoted by R_2. Thus, to make the pressure in the liquid at level h_1 uniform everywhere,

$$\frac{4\gamma}{R_1} - \frac{2\gamma}{R_2} = g\rho h_1 \tag{2.15}$$

must be approximately valid. In big bubbles because of buoyancy R_2 generally is much longer than R_1 so that, as a crude approximation, $R_1 h_1 \approx 4\gamma/g\rho$. A large bubble, before it bursts or becomes an element of foam, is more like a hemisphere with a horizontal bottom than a sphere.

The rules derived for an equilibrium cannot be exact because, as stated above, the bubbles as a rule never stop rising and the dome never stops thinning. The liquid present in the dome flows down, both because of gravitation and because of suction by Plateau's borders near points d in Fig. 2.9. This flow is discussed in Chapter 6.

Foam Structure

§41. If a bubble tarrying in the surface of a liquid happens to be near the (wetted) wall of the vessel in which the liquid is contained, it is attracted to and remains clinging to the wall. If the domes of the two bubbles happen to be near each other, they also are mutually attracted, and if there are many bubbles, the whole liquid may become covered with a sheet of bubbles, one bubble thick, which is sometimes given the inexact designation "two-dimensional foam."

The above attraction exists because of capillary pressure. The liquid surface between two bubbles is concave to the gas phase, as illustrated in Fig. 2.10. Hence the pressure at point a is less than at points b, b, situated at the same level in the liquid. The greater pressure from right and left

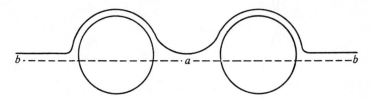

Fig. 2.10. Capillary attraction between two bubbles floating in a liquid surface. Because pressure at points b,b is greater than that at a, the bubbles are pushed together.

pushes the floating bubbles together. A more mathematical review of capillary attraction can be found, for instance, in reference[24].

When two bubbles continue to come closer, finally only a thin liquid film separates the two gas volumes. The shape of this septum, again because of capillary pressure, depends on the relative dimensions of the two neighbors. This is shown in Fig. 2.11. For the sake of simplicity, gravitation is disregarded so that the bubbles are spherical except for the lamella between them. Let R_1 be the radius of the bigger, and R_2 that of the smaller bubble. Then the pressure on the septum from the left is $p+(2\gamma/R_1)$ and from the right, $p+(2\gamma/R_2)$; p is the hydrostatic pressure in the liquid. The septum must bend to compensate for this pressure difference. It becomes concave to the right with the radius of curvature R' such that $(2\gamma/R_1)+(4\gamma/R') = (2\gamma/R_2)$; that is,

$$R' = \frac{2R_1R_2}{R_1-R_2}. \qquad (2.16)$$

The coefficient is 4 in the term $4\gamma/R'$ because the septum has two parallel surfaces and identical surface tensions γ act in each. If the bubbles are not submerged, as shown in Fig. 2.11, but are in the gas phase, then each is coated with a film (having two surfaces) so that mechanical equilibrium is achieved when $(4\gamma/R_1)+(4\gamma/R') = (4\gamma/R_2)$ and

$$R' = \frac{R_1R_2}{R_1-R_2} \qquad (2.17)$$

Also this reasoning is found in Plateau's work[56].

It follows from Equations 2.16 and 2.17 that the septum is plane only as long as the bubbles are equal. When R_2 is much smaller than R_1, then, from Equation 2.17, $R' \approx R_2$, so that very small bubbles in a foam suffer almost no deformation when they come in contact with a much bigger one.

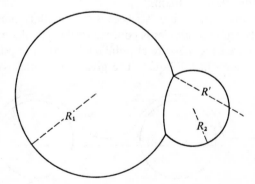

Fig. 2.11. Two bubbles (of radii R_1 and R_2) in contact. The pressure difference bends the film between them so that its radii of curvature (R') satisfy Equation 2.17 or 2.18.

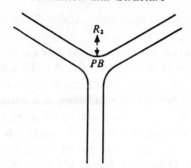

Fig. 2.12. Contact of three lamellae. To achieve mechanical equilibrium, the three angles between them must each be equal to 120°.

§42. When a second layer of bubbles forms under the first and is followed by subsequent layers, the bubbles shift to assume positions dictated by surface tension γ and capillary pressure. The main feature of the structure thus achieved is that the three lamellae radiating from a Plateau border (*PB*) make equal angles (of 120° each) with each other[57], as indicated in Fig. 2.12. The necessity of this arrangement can be demonstrated in two ways which employ, respectively, γ and capillary pressure.

(1) Six vapor-liquid interfaces are seen in Fig. 2.12. Each tries to contract with the force γ per centimeter; thus each lamella pulls at the Plateau border with the force (per unit length) of 2γ. Three equal forces acting in one plane can balance each other only if the three angles between them are equal. This gives 120° (or $2\pi/3$ radians) for each.

(2) Of the two radii of curvature of the interface between a Plateau border and the adjacent gas (in the bubble), one (R_1 swinging in a plane perpendicular to the plane of Fig. 2.12) is approximately equal to or bigger than the bubble radius, while the other radius R_2 is indicated by dashes in Fig. 2.12 and is very short. Consequently, the capillary pressure in the border is determined by the ratio γ/R_2 almost independently of the value of R_1. This means that R_2 is almost independent of the bubble dimensions and thus is almost identical on all three sides of the border. This is possible only if the angles between the three lamellae are equal. —Of course, both proofs presuppose equilibrium.

The forces exerted by different lamellae on a Plateau border can balance each other also when there are four lamellae making angles of 90° (or $\pi/2$ radians) with each other, and so on. However, when more than three lamellae join in a border, the equilibrium is not stable. Suppose that the four bubbles occupy the north, east, south, and west positions. If, from whatever cause, the pressure in the east bubble exceeds that in the north bubble, the septum will shift toward north-northeast, and the system of four bubbles will be transformed into two systems of three bubbles each. Consequently, in a reasonably stable foam three and only three lamellae meet in a liquid vein.

Four such veins (or Plateau borders) may come together at one cross-point, which may be visualized as a droplet. This arrangement is stable only as long as the veins form, with each other, angles of about 109°, identical with those between the valencies of a carbon atom or between the lines connecting the center of gravity of a regular tetrahedron with the four summits. The angles between the veins must be nearly identical because the surface tension forces acting along each liquid column are identical (ideally!). Only four liquid veins radiate from one droplet, because a greater number would result in an unstable system analogous to that in which four lamellae meet in one Plateau border.

According to Desch[58], who studied foams of gelatin, soap, and rosin, regular pentagonal dodecahedron is the crystal shape nearest to that of an ideal foam bubble. Such a crystal has 12 identical faces, each of which is a regular pentagon. When many identical crystals of this shape are brought together and compacted to the maximum density, their faces meet in one line and form angles of about 116°, and not more than four edges meet in one point. The similarity of an agglomeration of dodeca-hedrons to foams is, however, incomplete because these crystals (contrary to, for instance, cubes) cannot completely fill the space without any voids or pores; the relative void volume is stated to be 0.03. This drawback of the model is not particularly relevant because at any rate no foam consists of exactly identical bubbles and not all films between the bubbles are plane. If a cm is the length of each pentagon edge, the internal volume is[59] approximately 7.66 a^3 cm^3, and the internal surface is 20.64 a^2 cm^2; in other words, they are practically equal to volume and area of a cube whose edge is 1.885 a to 1.971 a. Let the thickness of the lamella be δ cm and include one-half of it in the volume of the bubble. The ratio of the surface area to this (larger) volume is the specific surface area A cm^2/cm^3 of the foam. It is $A = 2.692/\ a(1+0.363\,\delta/a)^3$; when δ is much smaller than a, then $A \approx 2.692\ /a$. It is equal to the area/volume ratio of a cube whose edge is 2.23 a cm.

From purely geometrical considerations, the equation $4C+2S = 3E+2n$ must be valid in every foam treated as an agglomeration of polyhedrons[61]; n is the number of bubbles, C the number of corners, S the number of sides, and E the number of edges (that is, Plateau's borders).

References

1. Davidson, J. F., and B. O. G. Schüler. *Trans. Inst. Chem. Eng.* 38:144 (1960).
2. Maier, C. G. *U.S. Bur. Mines Bull.* 260 (1927).
3. Okun, D., and J. K. Baars. *Proc. Intern. Congr. Surface Active Subst.* 4th Congress, Bruxelles (1964) 2:57 (1967).

4. Kovalski, W. von, and A. Wacker. *Ber.* 63:1698 (1930).
5. Guyer, A., and E. Peterhans. *Helv. Chim. Acta* 26:1099 (1943).
6. Krevelen, D. W. van, and P. J. Hoftijzer. *Chem. Eng. Progr.* 46:29 (1950).
7. Datta, R. L., D. H. Napier, and D. M. Newitt, *Trans. Inst. Chem. Eng.* 28:14 (1950).
8. Coppock, P. D., and G. T. Meiklejohn. *Trans. Inst. Chem. Eng.* 29:75 (1951).
9. Bikerman, J. J. *J. Appl. Chem. (London)* 18:266 (1968).
10. Achorn, G. B., and J. L. Schwab. *Science* 107:377 (1948).
11. Bragg, L., and J. F. Nye. *Proc. Roy. Soc. (London)* 190A:474 (1947).
12. Smith, C. S. *J.Appl. Phys.* 20:631 (1949).
13. Neppiras, E. A. *J. Acoust. Soc. Am.* 46:587 (1969).
14. Pattle, R. E. *Trans. Inst. Chem. Eng.* 28:32 (1950).
15. Valentin, F. H. H. *Absorption in Gas-Liquid Dispersions.* London: Spon Ltd. (1967).
16. Remy, H., and W. Seemann. *Kolloid-Z.* 72:3 (1935).
17. Ramakrishnan, S., R. Kumar, and N. R. Kuloor. *Chem. Eng. Sci.* 24:731 (1969).
18. Eversole, W. G., G. H. Wagner, and E. Stackhouse. *Ind. Eng. Chem.* 33: 1459 (1941).
19. Kupferberg, A., and G. J. Jameson. *Trans. Inst. Chem. Eng.* 47:T241 (1969).
20. Kippenhan, C., and D. Tegeler. *A.I.Ch.E.J.* 16:314 (1970).
21. Potter, O. E. *Chem. Eng. Sci.* 24:1733 (1969).
22. Ostwald, W., and A. Siehr. *Kolloid-Z.* 76:33 (1936).
23. Satyanarayan, A., R. Kumar, and N. R. Kuloor. *Chem. Eng. Sci.* 24: 749 (1969).
24. Bikerman, J. J. *Physical Surfaces.* New York: Academic Press. (1970), p. 70.
25. Rietema, K., and S. P. P. Ottengraf. *Trans. Inst. Chem. Eng.* 48:T54 (1970).
26. Melikyan, R. A. *Zh. Prikl. Khim.* 29: 1792 (1956); 42:2733 (1969).
27. Baxter, R. T., and A. E. Wraith. *Chem. Eng. Sci.* 25:1244 (1970).
28. Calderbank, P. H. *Chem. Eng. (London)* No. 212:CE209 (1967).
29. Bowonder, B., and R. Kumar. *Chem. Eng. Sci.* 25:25 (1970).
30. Hofer, M. S., and E: Rubin. *Ind. Eng. Chem., Fundamentals.* 8:483 (1969).
31. Lee, J. C., and D. L. Meyrick. *Trans. Inst. Chem. Eng.* 48:T37 (1970).
32. Gilberg, D., and R. A. Anderson. *J. Appl. Phys.* 19:127 (1948).
33. Melsens, *Compt. Rend.* 20:1658 (1845).
34. Fry, J. F., and R. J. French. *J. Appl. Chem. (London)* 1:425, 429 (1951).
35. Plateau, J. *Mém. Acad. Roy. Sci. Belg.* 37 (1869), 11th ser.
36. Peterson, H. B., R. R. Neill, and E. J. Jablonski. *Ind. Eng. Chem.* 48:2031 (1956).
37. Weatherford, W. D. *J. Colloid Interface Sci.* 34:197 (1970).
38. Hirth, J. P., G. M. Pound, and G. R. St. Pierre. *Met. Trans.* 1:939 (1970).
39. Florschuetz, L. W., C. L. Henry, and A. Rashid Khan. *Intern. J. Heat Mass Transfer* 12:1465 (1969).
40. Krause, A., and K. Kapitanczyk. *Kolloid-Z.* 80:273 (1937).
41. Kapitanczyk, K. *Chem. Abstr.* 44:7119 (1950).
42. Start, J. F., L. Seglin, and B. R. Franko-Filipasik. U.S. 3 423 330 (1969).
43. Gernez, D. *Ann. Chim. Phys.* [5] 4:335 (1875).

44. Dean, R. B. *J. Appl. Phys.* 15:446 (1944).
45. Larson, R. F. *Ind. Eng. Chem.* 37:1004 (1945).
46. Sernas, V., and F. C. Hooper. *Intern. J. Heat Mass Transfer* 12:1627 (1969).
47. Jakob, M. *Mech. Eng.* 58:643 (1936).
48. Cole, R. *A.I.Ch.E.J.* 13:779 (1967).
49. Bewilogua, L., R. Knöner, and H. Vinzelberg. *Cryogenics* 10:69 (1970).
50. Wark, I. W. *J. Phys. Chem.* 37:623 (1933).
51. Gunther, F. C. *Trans. Am. Soc. Mech. Eng.* 73:115 (1951).
52. Leaf, W. B., et al. *Am. Ry. Eng. Assoc.* 45. *Bull.* 441:58 (1943).
53. Cooper, M. G., and A. J. P. Lloyd. *Intern. J. Heat Mass Transfer* 12:895 (1969).
54. Jawurek, H. H. *Intern. J. Heat Mass Transfer* 12:843 (1969).
55. Kotake, S. *Intern. J. Heat Mass Transfer* 13:1595 (1970).
56. Plateau, J. *Mém. Acad. Roy. Sci. Belg.* 33 (1861), 6th ser.
57. Plateau, J. *Mém. Acad. Roy. Sci. Belg.* 33 (1861), 5th ser.
58. Desch, C. H. *Rec. Trav. Chim.* 42:882 (1923).
59. Manegold, E. *Schaum.* Heidelberg: Strassenbau, Chemie und Technik, (1953), p. 98.
60. Efremov, G. I., and I. A. Vakhrushev. *Izv. Vysshikh Uchebn. Zavedenii neft i gas* 6:79 (1968).
61. Schwarz, H. W. *Rec. Trav. Chim.* 84:771 (1965).

CHAPTER 3

MEASUREMENT OF FOAMINESS

§43. After a discussion of the formation and structure of foams, a chapter on their properties would be expected, but the decay of liquid foams is so much more striking than the classical traits (such as density, compressibility, and so forth) that the lifetime or persistence is dealt with separately and first. As usual, a review of the measuring methods precedes that of the results obtained; unfortunately, the first is unusually long as compared with the latter. This is so because many students of foam use arbitrary methods and arbitrary instruments to obtain data on foam persistence which cannot be securely compared with other data obtained by an equally arbitrary method. An ideal measurement of foaminess would result in a number independent of the apparatus and the procedure employed and being as characteristic for the solution tested (at the given temperature, etc.) as, say, viscosity or surface tension. Often it is not even recognized that such an ideal exists and ought to be aspired to. Each of the established properties has its immutable dimension (e.g., "mass/ square of time" for surface tension), but the dimension of foaminess is still disputed or, worse yet, entirely disregarded.

This chapter could have been, but will not be, very useful. It can inform many authors of future methods for measuring foaminess that similar methods have been described before; these authors may, perhaps, realize that substitution of a 1-liter flask for a 0.5-liter flask or an analogous alteration of a minor detail is an insufficient advance (if any at all) to justify an additional publication.

At least three processes take place during the whole life of a typical liquid foam. The films rearrange their positions so that the numbers and the linear dimensions of the bubbles change in time. Then the liquid present in the lamellae and the Plateau borders flows down because of gravitation and from places in which capillary pressure P_c is greater to those of a smaller P_c. Finally, the lamellae burst. Usually there is a fourth process which influences the other three; this is evaporation of the solvent. However, evaporation can be greatly reduced by a suitable design of the experiment, so that it is not as fundamental as the other three phenomena.

The persistence of a foam is determined by the frequency of collapse of the lamellae. When all lamellae have collapsed, that is, all bubbles have burst, no foam remains. The other processes mentioned in the preceding paragraph alter the structure, the density, and all other properties of the colloidal system, but do not directly cause its disappearance.

Films and Bubbles

§44. Single films. The lifetime of single lamellae described in Chapter 1 is the simplest indication of the stability of a foam. To be a property of foam only, it has to be independent of the material of the frame. Plateau[1], who was the first to measure the persistence of films, found that iron wire treated for a minute in dilute nitric acid gave rise to lamellae that lasted longer than those stretched across a frame of an untreated wire (see also §5). Presumably, the cleaning technique was not as advanced in Plateau's time as it is now; at any rate, Foulk[2] had no difficulty in using rings of platinum, chromel, or glass. The authors quoted in §15 and §§19 through 21 employed glass or metal frames and recorded no effect of the material.

The organic liquids tested by Foulk[2] had a very low foaming ability; presumably because of this, the results were sensitive to minute changes in the operating conditions that would be of secondary importance for soap lamellae. Usually a horizontal platinum ring (0.05 cm wire diameter, 1.3 cm ring diameter) was immersed in the liquid and then pulled up by a rubber spring; this was repeated 100 times for each liquid, but only in x instances out of 100 was a film seen across the withdrawn ring. This x was a measure of the foaming capacity of the liquid. It decreased when the ring was less exactly horizontal and the withdrawal was too rapid or too slow.

For an aqueous solution of a commercial detergent, persistence was about three times as great when the ring was kept horizontal after the withdrawal than when immediately after the withdrawal it was turned into the vertical position[3].

A more elaborate equipment was employed for unstable films of aqueous solutions of inorganic salts (such as sodium chloride or sodium nitrate). The film was formed (apparently) between two concentric nickel rings kept at different electric potentials. The weak current in the film had the frequency of 50 cycles/sec, and its oscillograph trace was photographed. The number of oscillations until the bursting (which stopped the current) could readily be counted[4].

The film persistence method can be used also for boiling liquids[5]. An aqueous solution was heated to boiling, with a reflux. A small, apparently horizontal ring was lowered through the condenser into the liquid and rapidly lifted again to remain a few millimeters above the vapor-liquid interface. The lamella in it could be seen from outside the flask, and its lifetime thus could be determined.

Plateau[6] obtained soap films also by rolling suitable solutions in a cylindrical tube and suddenly stopping the motion. Such lamellae are plane and perpendicular to the cylinder wall because the curvatures of the two Plateau borders (along the two sides of the lamella) must be equal. If the tube is conical rather than cylindrical, the lamellae still are perpendicular to the walls but are not plane; they are concave toward the apex of the cone and tend to glide because of the difference in capillary pressure. Films in a cylinder are readily produced by shaking a stoppered flask containing a small amount of a surfactant solution or by blowing (through a fine capillary) air on the surface of the liquid[7].

The importance of vaporization is pointed out in §15. Fig. 3.1 shows an apparatus in which a vertical film can be kept in a closed vessel and a saturated atmosphere[8]. Two parallel platinum wires (*b*) are sealed in a glass tube (*a*); *c* is a wire attached to them. The solution is introduced through *m* until it covers wire *c*. Then a part of the liquid is transferred (by pressure or suction) into the companion tube *d*, while air escapes through tube *l* and rheometer *k*. A lamella remains between the three wires and the liquid covering the bottom of *a*. The persistence of lamellae

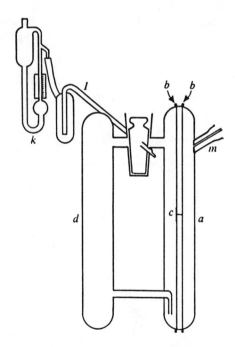

Fig. 3.1. Measurement of film persistence in a controlled atmosphere. The film remains between the two wires *b* below wire *c* when liquid is pushed from vessel *a* into vessel *d* by gas entering through *m*; *k* is a flow meter and *l*, a trap (from reference [8]).

remaining in contact with the mother liquid usually is longer than of those spanning a frame, but no systematic study of this difference could be found by the present author.

§45.   The area of the lamella remained constant during the tests reviewed in §44. The foaminess can be estimated from an area instead of from a time. Thus a vertical wire triangle was slowly withdrawn from a solution so that the film area gradually increased[9]. The area at which bursting took place served as a measure of stability. In a similar method[10], air was injected (at a slow constant rate) into a bubble, and the time until the bubble broke was noted; from this time the maximum volume and the maximum area of the viable bubble can be calculated. The more recent procedure[11] is practically identical with that patented in 1949. It is illustrated in Fig. 3.2. The end *a* of a tube is immersed in the test solution to a constant depth for a constant time, after which the tube is inserted into bottle *b* and the air in the latter is compressed by running water from buret *c* into the bottle. In these tests the rate of air injection had to exceed 20 cm^3/sec to obtain reproducible results. The air pressure caused the bubble *d* to grow. The maximum dimensions of the bubble immediately before bursting are recorded.

From the scientific point of view, the three methods of this section are less satisfactory than the direct determination of the persistence of

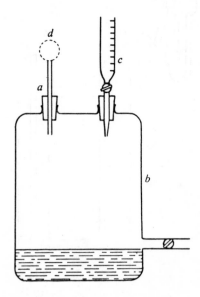

Fig. 3.2. Measurement of the greatest bubble volume before bursting. Capillary *a* contains a droplet of the liquid under test; when water flows from *c* into *b*, the droplet is transformed into a bubble (*d*), whose dimensions are measured (after references [10],[11]).

lamellae, as described in §44. When the film (plane or spherical) grows in area, its thickness decreases because of drainage; and in the second and the third methods the thickness may also decrease because the initial volume of the liquid was insufficient for a thicker film. Thus the test results must depend on the rate of withdrawal of the triangle or the rate of air injection and on the initial volume of the liquid used to make the bubble. Also the vibrations caused by the unavoidable motion may lower the lifetime of the lamella.

It is mentioned in §12 that a suggested procedure for measuring the Marangoni effect consists in comparing the weights of a vertical film during withdrawal and immersion. Matalon[12] believes that the foaming ability of a liquid is greater, the greater the difference between the two weights. This view recently found additional defenders[13]. The above difference depends on the rate of the vertical motion of the wire frame, in which the lamella is suspended, and thus cannot be an absolute measure of foaminess. At a given rate of motion the rate of drainage depends on the mechanical properties of the "crust." In addition, the surface tension of the film depends on the rate of achieving the equilibrium between the crust and the central layer of the lamella. Thus these observations are not as fundamental as Plateau's method of simply waiting for the film to burst.

§46. Single bubbles. Bubbles stopped at the liquid-vapor interface are discussed in §40. Their lifetime also was measured first by Plateau[6]; it seems to be as well suited for absolute determinations as the persistence of plane lamellae. The persistence of any film depends on its thickness, and from this point of view plane horizontal, plane vertical, and highly curved lamellae belong to three different classes. In the first, considerable areas of the film may have a constant thickness; in the second class, the thickness more or less regularly increases from top to bottom; and in the third, the thickness in some instances seems to be even less uniform. Because the film (i.e., the dome of §40) is nearly vertical far from the summit and is horizontal at the summit, the downward flow of the liquid (between the ceiling and the roof) tends to have a higher linear rate nearer the plane surface of the liquid around the bubble. As a result, the thickness may be greater at the pole than at a lower latitude of the (approximate) hemisphere. "Dimples" of this kind are known when a drop approaches a rigid surface (see, for instance, reference [14]), but are likely to exist in bubbles as well.

When bubbles are very small and the liquid has a poor foaming ability, the notion of bubble persistence becomes hazy. The velocity of ascent of a minute bubble is small; it may take the bubble several seconds to advance 1 μ. In an obvious method of determining low persistencies, the liquid surface is photographed n times every second.[15] If the same bubble is seen on m successive frames, its lifetime is counted as m/n sec. However, the usual depth of focus is such that a bubble whose front end

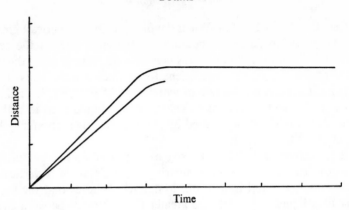

Fig. 3.3. Bubble rise in a liquid capable of foaming (upper curve) and in one incapable of it (lower curve). The distance from the orifice (shown along the ordinate) remains practically independent of time (plotted on the abscissa) when the bubble floats in the surface.

is still 1 μ below the plane surface is well visible on the photograph, and the question arises as to what part of the m/n sec was spent on the rise, determined only by gravitation and hydrodynamics, and what part was due to surface effects. In other words, when is the bubble below, and when is it in, the interface? For bubbles which need only milliseconds to traverse a 1-μ path and then remain almost stationary for many seconds, this difficulty obviously is negligible.

This reasoning may be illustrated by a graph. In the schematic Fig. 3.3, the ordinate is the distance of the bubble from the orifice (at which the bubble originated), and the abscissa is the time since the detachment. The lower curve is for a bubble in a liquid incapable of foaming, while the upper curve is for a solution of a foaming agent. Each curve has a kink when the bubble "hits" the surface, but the upper one has, in addition, a long practically horizontal branch so that the value of the persistence τ is very little affected by the duration of the kink. On the other hand, the persistence of the first bubble is impossible to define with precision because the beginning of the kink is indefinite, and only an arbitrary decision can be made as to what fraction of the "kink" belongs to "persistence."

A method which the present author considers to be misleading was used by Gleim[16]. Single bubbles were produced at a capillary covered with a solution kept in a closed flask. The air inlet tube was connected with a tilted mercury manometer. The pressure in the tube rapidly dropped every time a bubble broke off the capillary. The corresponding shift of the mercury meniscus interrupted an electric current, and each interruption was recorded by a counter. If there were n interruptions per minute, the average persistence τ of bubbles was said to be $60/n$ sec. Obviously, this time interval was the time needed by the bubble to grow at the orifice from zero to the maximum volume; it was different from, and presumably greater than, the true lifetime of bubbles almost immobilized in the interface.

Gleim[17] constructed also another apparatus which was more complicated than those generally used for measuring τ; it appears unobjectionable to the present author. The liquid was contained in a stainless steel cylinder provided with glass windows at the top and the bottom so that the liquid surface could be seen in transmitted light. Single bubbles were released from a capillary, and their τ was determined visually. The cylinder could be pressurized to 7 bars (or 7 atmospheres), and τ was lowered, for instance, from 2 to 0.9 sec by this increase in pressure. The liquid was 0.7 N sodium carbonate solution at 20°.

It is fortunate that, generally speaking, preparation of foam lamellae and bubbles and their direct observation are easy, because the τ of many specimens must be measured before a dependable average value can be calculated. As Plateau pointed out (see §18) rupture of a film is a chance phenomenon so that the standard deviation of τ values in an experimental series is large. Individual results usually are not mentioned in the literature, but the paper by Trapeznikov[18] is an exception. As an example, 10 values of τ for bubbles under a unimolecular layer of ethyl palmitate were, respectively, 10, 10, 17, 4, 11, 8, 5, 15, 9, and 12 sec. Errors of such magnitude in the measurement of "classical" properties (such as density or viscosity) are unthinkable. In a more recent paper[99], in which also an instrument for measuring τ is described, a histogram for 172 measurements (bubbles in distilled water) is shown; the values ranged between 0.05 and 0.7 sec.

§47. Sample dimensions.   Single films and bubbles are particularly suitable for a study of the dependence of persistence on sample dimensions. Again, in the beginning there was Plateau[19]. One of his soap solutions gave films which lasted up to 1 hr in an iron ring of 7-cm diameter and up to 12 hr in a 2-cm-wide ring; unfortunately no average values were reported. It is seen that these maximum values of τ are inversely proportional to the film area: The ratio $(7:2)^2$ is equal to 12.25. In another investigation[7], an identical relation was observed for the mean values of τ. The films were obtained and preserved in stoppered flasks, the necks of which had internal diameters of, respectively, 1.0 and 3.5 cm. Thus the ratio of the film areas again was roughly 12. The average τ for the smaller flask, for 0.43% solution of disodium α-tocopheryl phosphate monohydrate, 3.7% solution of sodium laurate, and 7-week-old beer was 420, 260, and 390 sec, while the lifetimes in the wider flask were 35, 23, and 35 sec. The ratios (14.0, 11.3, and 11.1) are remarkably near 12.

Dewar[20] measured the lifetime of only three films 40 cm across and only two films of 30 cm, so that the average values of τ cannot be expected to be reliable. Nevertheless their ratio (53 days to 96 days) was very near to $(30:40)^2$. The solution used consisted of water 50, glycerol 50, and ammonium oleate 5 parts. Qualitative indications of an increase of τ when film area decreases are common; see, for instance, references [3] and [8].

The results for bubbles are less clear-cut, perhaps because bubbles can be observed also in solutions that give no free lamellae at all. Table 4.1, §65, shows[15] that the τ of aqueous solutions of common inorganic salts is little affected by the bubble diameter when this varies 10-fold from 60 to 600 μ. The higher τ values observed for the smallest bubbles (16 to 60 μ) presumably can be accounted for by the limitations of microscope observations (see §46). The increase of τ at diameters exceeding 600 μ is likely to be caused by the longer duration of drainage in bigger bubbles.

On the other hand, the persistence of bubbles in the surfaces of liquids capable of foaming generally is greater the smaller the bubble. This was noticed first by Plateau but in a qualitative manner only. Later[22] the τ of bubbles on 0.005 M sodium oleate solution was recorded as 23 and 4.5 min for bubbles having diameters $2R$ of 0.1 and 0.4 to 0.5 cm respectively; in this instance, τ was approximately proportional to $1/R$. The τ of mercury bubbles of the Melsens type (see §33) is said[23] to be proportional to $1/R^2$. An approximate proportionality between τ and $1/R$ is visible in the graph given by Geffken[24] for air bubbles in glycerol; unfortunately, neither the details of the experimental technique nor the degree of purity of the glycerol employed are indicated in the original publication. For some bubbles (apparently 0.3 to 0.5 cm in diameter) in aqueous emulsions of mineral oil, τ seemed[25] to increase with R and to be a linear function of $1/R$. These experiments should be repeated.

The empirical rule

$$\tau = \frac{K}{A}, \tag{3.1}$$

A being the film area and K a constant, is in agreement with the probability theory of the lamella strength hinted at in §18 and developed in more detail in §133. If more extensive tests confirm its validity for many systems, it would be advisable to use $K = \tau A$ as a measure of foam stability, independent of sample dimensions. The dimension of K is, of course, cm²·sec. If bubbles on solutions of foaming agents really have a persistence inversely proportional to R, then the product τR cm·sec would be a logical measure of their stability.

§48. *The environment.*   The persistence τ of a plane film or a bubble greatly depends on the surroundings. Apparently the rate of evaporation is a most important parameter and determines, at least partly, the difference between the τ values in open and closed vessels.

Plateau[6] observed bubbles sitting in a wire frame; the liquid was a solution of commercial soap in aqueous glycerol. Because of the hygroscopic nature of the latter, the thickness of the bubble walls increased when the bubbles were kept in humid air and, presumably, decreased when they were maintained above calcium chloride. Although the common observation is that thin films have a shorter lifetime than thick lamellae, in this series the persistence in dry air was greater. The τ was particularly great for plane films in closed tubes. The τ of the above bubbles in air usually was about 6 hr, but a comparable plane lamella lasted for 10 days. Also the large and long-lasting (months!) soap lamellae of Dewar[26] were preserved in a closed bottle (and kept in a horizontal position).

Vertical films of aqueous ethanol on wire triangles ruptured at a greater area in an open than in a closed vessel[9]. The τ of single bubbles in water-ethanol mixtures had a maximum value (in 3% solution) when the vessel was closed, but no maximum occurred in open air; and the τ of 0.4 N aluminum chloride solution was twice as great in an open trough as in a closed space[27]. Also air bubbles in 0.00092 and 0.00184 M isoamyl alcohol solutions lasted in an open vessel longer (e.g., 5 to 10 times as long) than in a closed one[22]. Apparently in these systems vaporization enhanced stability. This effect may have a simple explanation. The foaming ability of many solutions has a maximum at a (more or less definite) concentration c_m. If the concentration of the initial solution was less than c_m, evaporation of the solvent raised it and, consequently, also the lifetime. However, the τ of aqueous soap solutions often shows the opposite behavior.

Striking results, which deserve a careful checking, have been obtained on nonaqueous systems[28]. In a mixture of 0.4 mole octane and 0.6 mole nitrobenzene, air bubbles of 0.3 cm in diameter lasted 3 sec when the gas phase was the saturated vapor of the mixture. When the vapor was not saturated or the gas phase contained, as far as possible, nitrobenzene only, τ sometimes exceeded an hour. Diethyl ether in the atmosphere drastically lowered τ.

It would not be correct to attribute the difference between the lifetimes in closed and open spaces to evaporation alone. In a closed space the bubble or the lamella is better protected from external foes, such as dust or oil particles, or wind gusts; this effect presumably was the most important in the instance of Dewar's Methuselahs. Plateau noticed that vibrations tended to shorten the persistence of his soap bubbles (see also §50).

§49. *Cleanness.*　　As reported in more detail in §81, unimolecular films on the surface of water may increase the persistence τ of bubbles by factors reaching, say, 100. The surface concentration of such films may be of the order of 10^{-10} moles/cm^2. Let the area of the vapor-liquid interface be 100 cm^2 and the liquid volume be 1000 cm^3. Then, at 10^{-10} moles/cm^2 the total amount of contamination is about 10^{-8} moles and its overall concentration is 10^{-8} M. Evidently extremely small amounts of foreign substances can have a very strong effect on foaminess. Thus the cleanness of the surface, of the vessel, and so on, is highly important in all measurements of film and bubble persistence.

Insufficient cleanness presumably was responsible for the high values of τ in water and aqueous solutions of inorganic salts as reported by Talmud[27]. According to Table 4.1, §65, the bubble persistence in salt solutions usually is of the order of 0.1 sec; the τ on pure water could not be defined, not only measured (see §46). Talmud observed τ of, for instance, 20 sec for 0.1 N KBr and 40 sec for N potassium sulfate. His bubbles were 0.15 cm in diameter, that is, a little bigger than the biggest bubbles of Table 4.1, but this difference is not likely to be responsible for the large discrepancy. Water films in a ring suspended above the surface of boiling water lasted in Gleim's experiments[17] as long as 55 sec; and the lifetime was further raised by sodium hydroxide, sodium carbonate, sodium sulfate, or sodium chloride. A repetition of these measurements would be highly welcome. The τ of bubbles in a pressurized vessel is mentioned in §46, and the numerical values given there refer to the arrangement in which the sodium carbonate solution was confined in a glass vessel and had no contact with steel. When the solution was kept in a steel container, the τ was greater (e.g., 5 sec) at the high pressure (of 7 bars). The authors[17] attributed the difference to the presence of iron and chromium compounds in the second series of samples, but no analysis was performed.

Two popular methods of surface cleaning may be mentioned here. In one, a double funnel is used; the liquid moves up through the stem of the inner funnel, runs over the brim, and flows to waste through the external funnel[29]. In the other, the liquid is kept in a trough filled almost to overflowing, and the surface is swept with a nonwettable bar[30]. In some instances, the liquid may be condensed from the vapor directly in the vessel used for bubbling.

When the liquid is capable of foaming, the composition of its surface layer is different from that of the bulk, and considerable time may be needed to achieve equilibrium. This presumably is the cause of the observation[22] that the τ of a bubble (under the surface of a soap solution) is different according to whether a previous bubble did, or did not, recently burst at the same spot; the difference vanished when bubbles were released not more often than once in about 2 min. Also cleaning the surface after each bubble was helpful. When bubbles of 0.002% trimethylcetyl ammonium bromide or 0.02% terpineol solution were observed in fresh surfaces, their τ was about 2 and about 12 sec, respectively, but when the solution was left undisturbed for approximately 20 min before the first bubble was released, the τ was, respectively, 19 sec and 22 sec; thus aged surfaces stopped the rising bubbles for a longer time[31].

§50. *Several bubbles.* A two-bubble instrument was devised by Foulk[2,32] and is schematically represented in Fig. 3.4. Air is supplied

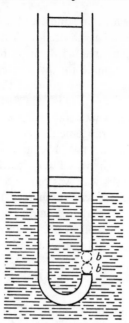

Fig. 3.4. Two-bubble instrument for estimating foam persistence. Bubbles *bb* form at two orifices, one above the other, and the frequency of their coalescence is an inverse measure of persistence (after reference [2]).

simultaneously to two vertical capillaries of identical cross-sections. The lower end of one of the capillaries is bent up so that the orifice is situated exactly under the orifice of the straight twin. The instrument is partly immersed in the test liquid so that both orifices are submerged. The bubbles rising from the bent tube meet the bubbles growing under the upper orifice and, when the liquid is incapable of foaming, immediately coalesce. In other liquids they break off and rise as independent bubbles. The percentage of bubbles which escape coalescence is taken as a measure of foaminess. As would be expected, it depends on the distance between the orifices[33], the precision of the alignment of the two bubbles, and other circumstances. It is clear that, if the centers of two identical bubbles are exactly one above the other, symmetry considerations would preclude any movement to the right or to the left, and the bubbles will never escape at all; thus the values obtained show first of all how asymmetric the arrangement is. Nevertheless, the method is useful for comparative estimates of different liquids and was employed also by other investigators[34]. An advantage of the method is that the behavior of the bubbles is observed far from the main vapor-liquid interface, which is so readily contaminated with dust, laboratory fumes, and so on.

In Schütz's method[35], many bubbles are produced, but only one is utilized. Bubbles are blown at a submerged capillary until the whole surface (A cm^2) of the liquid is covered with a one-bubble-thick layer. Then a wire ring enclosing an area 0.5 A cm^2 is fixed above the center of the surface, and the time required for the smaller area to become free of foam is noted. In other words, the lifetime of the most persistent bubble among an indefinite number of these is recorded, while all other bubbles are disregarded; thus this is a wasteful procedure. The effect of environment was striking in this arrangement, probably more so than if the average τ had been determined. In 0.01 % saponin solution the maximum τ was too long to measure if the vessel was closed or protected from draft by a 40-cm tall wall; when the wall was only 2.5 cm tall, the τ was only a few seconds. When the orifice was submerged more deeply so that the bubble had to traverse 76 rather than 6 cm to reach the surface, the τ was raised from 6.5 to 15.5 sec; this effect presumably was caused by the slow diffusion of saponin to the bubble surface.

Foams

§51. To produce a three-dimensional (i.e., more than one bubble thick) foam for stability determination, any of the methods described in Chapter 2 can be employed, and the systematics adopted there could be used in this chapter also. However, another classification seems to be more instructive here; it is based on whether the mixing of gas and liquid is, or is not, controlled.

It is pointed out in the literature again and again that the "foaming ability" is different from the "foam stability". Experimentally this means that a "standard" agitation may produce a more voluminous foam on liquid A than on liquid B, but the lifetime of the former may be shorter than that of the latter. To understand this effect it may be helpful to refer to the molecular mixing of a gas or a solid with a liquid. Suppose that a heap of crystals is agitated in a "standard" manner with a liquid and the concentration of the solution is then determined. This concentration will be a (crude) measure of the *rate* of dissolution and will greatly depend on the volume of liquid, the particle size and the perfection of the crystals, etc. If the experimenter waits long enough, saturation eventually will be reached and the concentration of the solution will be independent of any agitation or any fleeting property of the crystals in the heap; it will be equal to the *solubility* of the solid. It happens quite often that the rate of dissolution of solid A is greater than that of B, but the solubility of B is greater than that of A.

When a gas is introduced into a liquid through a capillary or a porous solid, the total volume of the gas injected can be readily controlled, and also bubble dimensions can be (more or less) predetermined. Hence the stability of the foam obtained may, ideally, be expected to be a property of the liquid and the gas almost as characteristic for them as solubility (see §53). Also in the condensation processes the total volume of gas in the bubbles is almost quantitatively determined by pressure during saturation and the smaller pressure causing liberation of bubbles. On the other hand, in all agitation methods, the first step is forcing the gas into the liquid, and the degree of mixing achieved depends on numerous external conditions (shape of paddle, rate of shaking, etc.), which just as little affect true foam stability as the volume of the crystal heap influences equilibrium solubility. Naturally the volume of gas stirred in depends on the properties of the liquid also but this dependence is related to the external conditions.

This is easy to comprehend when it is remembered that many hydrodynamic phenomena are definite functions not of the absolute velocity u but of the Reynolds number $(Re) = u\rho_2 l/\eta$, introduced in §26; ρ_2 is the density of the liquid, η its viscosity, and l is a characteristic length whose magnitude is determined by the process under consideration. Suppose that in a "standard" shaking method, l is the constant amplitude of vibrations and u, the constant average speed. Obviously, the (Re) will be much smaller for a liquid having a very small ratio ρ_2/η (such as a viscous oil) than for one whose ρ_2/η is large. Thus a method in which u and l are kept constant for different liquids is "standard" in quotation marks only, since (Re) values vary from one to another liquid. Glycerol must be shaken in a tank 8 m across to render the shaking comparable to that of water in a test tube 1 cm in diameter.

The real situation is even more complicated because bubble formation implies an increase in surface area; this process is hindered by the surface tension γ of the liquid. If, after a "standard" shaking, less air is incorporated in a mercury phase than in an aqueous system, the result has perhaps no connection with true foaminess; simply, it is more difficult to mix air with mercury than with water because the γ of the former is by far the bigger of the two and also because the density of the mercury is very high and gravitation pushes air out of the metal with a greater force than out of water. These difficulties do not arise when air bubbles form at the submerged tip of a capillary.

§52. *Pneumatic methods.* This group includes the methods in which a gas is injected into a liquid through separate capillaries[36,37], through a cloth filter[38], sintered glass spargers[39-41], diffuser stones[42], spinnerettes, Pasteur-Chamberland filters, and so on. The use of single capillaries has the advantage of giving rise to relatively uniform bubbles and the disadvantage of a relatively slow foam formation.

Fig. 3.5 is an early example of the experimental setup[39]. Water from aspirator *6* displaces air from the graduated cylinder *3*; the displaced air is dispersed in the porous plate *1* and rises as a foam in the graduated tube *2*; *4* and *5* are manometers; and *7*, *8*, *9*, and *10* are stopcocks. The apparatus can be improved in several respects. A Mariotte bottle may be substituted

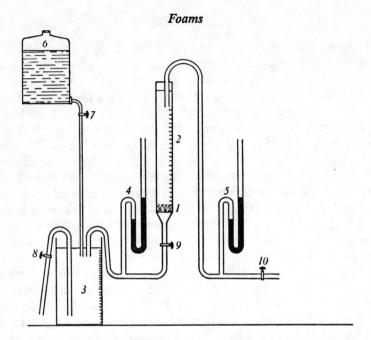

Fig. 3.5. Pneumatic determination of foam stability. Foam forms at porous plate *1* and rises in tube *2*; *3* is a graduated cylinder, *4* and *5* are manometers, *6* is an aspirator with water; *7, 8, 9,* and *10* are stopcocks (after reference [39]).

for the aspirator to maintain the pressure constant. A cylinder of compressed air (or another gas) may be used instead of both *3* and *6*; in this instance, the high-pressure gas first fills a large jar (if of glass, a protective screen is recommended) to facilitate the control of the gas pressure. The jar is provided with a pressure gauge and a flowmeter is inserted between the jar and the porous septum. The measuring tube (*2*) can be surrounded with a glass jacket through which a thermostated water flows, as was done, for instance, by Imhausen[43] and Moroi[44].

§53. *The dynamic methods.*   This is a subdivision of the pneumatic method: the foam height or the foam volume is measured while the gas continues to produce bubbles. For many systems (e.g., for very dilute solutions of many surfactants) the measurement of foam height presents no difficulty, as the top of the foam column in the graduated tube (see Fig. 3.5) does not vary along the circumference of the tube more than by one or a few millimeters; the error introduced by this uncertainty is not serious when the total height reaches, say, 50 cm or more. The bottom of the foam column often is not sufficiently definite, that is, there is no sharp boundary between the gas emulsion below and the true foam above; in these instances, the level of the gas-liquid interface before foaming is taken as the lower boundary of the foam. Also the error caused by this approximation is unimportant as long as the steady foam height h_0 is considerable. A method for increasing h_0 is indicated below (see Equation 3.2).

In other systems—for instance in some protein solutions—the upper boundary of the foam sometimes markedly deviates from the horizontal. It happens also that instead of a uniform foam, single films spanning the whole width of the measuring tube rise in the latter; then the experimenter has to record the heights at which every film bursts and to take the arithmetic average.

When gas injection starts and continues at a constant rate, the foam height h first increases also with a constant linear velocity, but then the velocity gradually decreases, and after a time a constant height h_0 is reached. This h_0 cm is, within a range, proportional to the linear gas velocity u cm/sec (identical with u_1 of §27). For instance[40] for 1 % aqueous solution of 1-butanol, the ratio h_0/u was 4.7, 6.5, 6.8, 6.8, and 6.4 sec when u was, respectively, 0.130, 0.247, 0.395, 0.498, and 0.669 cm/sec.

In Sasaki's experiments[45] on an aqueous solution of 1-hexanol, not the ratio h_0/u but $(h_0 - a)/u$ was observed to be constant; a cm was a correction term. Maybe it was caused by inexact reading of the lower boundary of the foam column. It is clear that the proposed relation cannot be valid for small velocities u, since it implies that $h_0 = a$ when $u = 0$, that is, no gas is injected at all. At greater velocities there is practically no difference between the two equations $h_0/u =$ constant and $(h_0 - a)/u$ = constant. Dyakonov[46] measured h_0/u as a function of the Reynolds number (Re) defined as $\rho_2 u d/\eta$; again ρ_2 is the density of the liquid, η its viscosity, and d is the diameter of the measuring tube. In dilute soap solutions h_0/u was constant at small u up to about $(Re) = 100$, corresponding to $u < 0.25$ cm/sec as d was 4 cm, $\eta = 0.01$ g/cm·sec, and ρ_2 = 1.00 g/cm^3. Between $(Re) = 100$ and $(Re) = 250$, the ratio h_0/u increased with u, but it decreased again at $(Re) > 250$, presumably because the air blast broke the soap lamellae. Pattle[47] worked with 0.1 % solution of 2-methylpropanol; h_0/u was about 3.2 sec independently of u at all u values below 1.3 cm/sec but increased with u between $u = 1.3$ and 2.0 cm/sec.

In a measuring tube of a constant cross-section A, a constant foam height h_0 implies a constant foam volume v_0 ($= Ah_0$), and linear velocity u cm/sec means volume rate $V/t = uA$ cm^3/sec. As long as h_0/u is invariable, so is the ratio $v_0 t/V$. The question then arises as to which of the two expressions is more nearly constant when the cross-section is not uniform. Experiments have shown[40] that foam volume is more fundamental than foam height. Thus, two straight tubes and a tube provided with a bulb-like enlargement were employed for identical liquids (urine and 1 % butanol); h_0/u was altered by the bulb but $v_0 t/V$ remained unaffected. The value of $v_0 t/V$ was not changed by the diameter of the measuring tube; this was shown by using tubes 2.7 and 3.7 cm wide. It appears that an identical conclusion can be derived from the older experiments[36] on air injection into raw sugar solutions; when tubes of 2.5 cm and 4.0 cm were compared, h_0 varied in the ratio 3.14:1 but the quantity $v_0 t/V$ decreased only in the ratio 1.23:1.

It is mentioned in §50 that bubble persistence is greater when more time is given to the gas-liquid interface to approach the equilibrium state. This effect can be observed also in the dynamic method[40]. Thus the $v_0 t/V$ of 1% 1-butanol solution was 4.0, 5.6, 6.7, and 7.1 sec when the height H of the liquid layer above the porous plate was, respectively, 0.5, 1.0, 1.5, and 2.0 cm. In Dyakonov's experiments, h_0/u was supposedly proportional to $H^{0.8}$, which is rather improbable; as a rule, h_0/u or $v_0 t/V$ is independent of H when H is tall enough. Qualitative statements on the increase of h_0 with H have been published by Peters[48].

When H is so deep that its magnitude ceases to influence $v_0 t/V$ and the gas velocity is not too small (otherwise evaporation is too upsetting) and not great enough for the blast to rupture the lamellae, then

$$\Sigma = \frac{v_0 t}{V} \tag{3.2}$$

is independent of the gas flow rate, of the shape and dimensions of the measuring tube, and of the amount of liquid present. Thus so far Σ is the nearest approximation to the ideal "unit of foaminess" which would be as characteristic for the liquid as its density, viscosity, and so on (see also Equation 3.1 in §47).

It is not known how closely Σ depends, if at all, on the bubble volume. It proved to be independent of the average pore size of the sintered glass filter used[40] but the bubble diameter frequently has no direct connection with the pore size, as pointed out in §32.

Σ usually is expressed in seconds. It is equal to the time an average bubble exists in the foam before bursting[40]. This can readily be shown as soon as two restrictions are introduced, namely that, when a bubble bursts, its gas content is given to the atmosphere (rather than to another bubble) and that the volume of the liquid in the foam may be neglected in comparison with the gas volume. Let n bubbles, each of volume v', be formed at the gas dispersion tube in one second; hence the volume of gas introduced in Σ sec is $v_0 = nv'\Sigma$ cm^3. This is the greatest possible foam volume at the constant rate V/t because every bubble is supposed to perish as soon as it is Σ sec old. This rate V/t obviously is equal to nv' so that $v_0 = V\Sigma/t$; this formula is identical with Equation 3.2.

For the aqueous emulsions of mineral oil mentioned in §47, Σ appeared to be about 13 times as great as the τ of a single bubble; additional comparisons of this kind would be highly welcome.

§54.   The Σ is the most characteristic quantity which can be obtained from dynamic foam measurements. Several scientists, for instance Ostwald[37] and Ruyssen[49] did not wait to reach the steady state, and determined the foam height h as a function of time t of gas bubbling when h still was smaller than h_0. As long as t is small in comparison with Σ, the rate of increase of foam volume dv/dt is practically equal to that of gas injection dV/dt, still neglecting the volume of the liquid in the foam; the equality of dv/dt and dV/dt is easy to observe on solutions of commercial surfactants whose Σ may exceed 1 hr. It is clear that, as long as dv/dt is almost equal to dV/dt, its value affords no information on foam stability. When t is comparable with Σ, the equation

$$\frac{dv}{dt} = \frac{dV}{dt} - \frac{v}{\Sigma} \qquad (3.3)$$

seems to be a reasonable approximation. In the steady state $dv/dt = 0$; thus Equation 3.3 agrees with Equation 3.2 for a constant value of dV/dt equal to V/t. According to Equation 3.3, the value of dv/dt depends on dV/dt, v, and Σ, and thus cannot be readily used for characterizing foaminess. The method was criticized by Lederer[50] many years ago but is revived again and again.

In some methods an arbitrary volume V cm^3 of air is injected during t sec into a liquid, and the volume v of the foam is determined at once. Evidently this v is equal to v_0 of Equation 3.2 if the time t chosen is long (i.e., greater than Σ sec), but is smaller than v_0 when t is small in comparison with Σ; thus no information of fundamental nature can be extracted also from these measurements. An example can be found in reference[51]: 400 cm^3 of air was introduced in 120 sec into solutions of sodium dodecyl benzene sulfonate, and the foam volume measured. The two methods recommended by the American Society for Testing and Materials[42] are very similar to the above example but apparently refer to the steady-state volume v_0. Thus 5 l. of air is forced in 5 min through 145 cm^3 of an aqueous solution (of an antifreeze compound) present in a container of 500 cm^3, and the foam volume is recorded. If Σ were not much smaller than 5 min, the 5 l. of air would have given more than 0.355 l. of foam (0.355 l. is the difference between the total volume of 500 cm^3 and the liquid volume 145 cm^3) and the foam would have overflown the vessel. Apparently the foam volume usually is considerably smaller than this; and if 5 l. of gas give rise to only, say, 0.1 l. of foam, it is very probable that this volume represents the steady state. If a volume rate equal to 5000 cm^3 in 300 sec, i.e., $V/t = 16.7$ cm^3/sec, results in a v_0 of 100 cm^3, then the Σ of the solution is 6 sec, that is very much smaller than the time (5 min) employed for the test. In both ASTM methods a gas diffuser stone serves for bubble formation.

A modification of the above procedures was recommended by Nash[52].

A constant volume of air was injected into 100 cm³ of test solution (through a metal tube provided with several holes) during an 80-sec period. The foam formed was collected in a plastic bag and weighed. It is clear that the result depended on V/t, on the time of drainage, and so on.

An expression related to the term v/Σ of Equation 3.3 was derived by Aleinikov[39]. He gave the name "average rate of dynamic coagulation" to the ratio $U_d = (V - v)/t$. If of the gas volume V introduced only volume v remained in the foam, evidently $V - v$ cm³ has escaped ("coagulated"); as this event took place during t sec, the average rate of coagulation was equal to U_d. This derivation neglects the volume of the liquid in the foam. In 0.004 M aqueous p-toluidine, when t was 4, 8, 15.2, and 26.3 sec and v was 50, 90, 150, and 210 cm³, U_d was 10.9, 5.5, 9.2, and 11.4 cm³/sec, respectively. The physical meaning of this spread of U_d values is not known. In this method, as in the preceding ones, it may be necessary to correct V or v for pressure differences. If, for instance, V is measured at the high pressure necessary for injection but the foam is at practically atmospheric pressure (because bubbles are not very small), then a given volume V will give rise to a foam volume markedly exceeding V, even when the liquid volume is subtracted from v.

§55. Some quantities derived from pneumatic foam measurements and supposedly characteristic for the liquid tested have been introduced into the literature; Nash's method mentioned in §54 could be, with equal justification, treated here. For instance, the volume of liquid v_1 transferred into foam by unit volume of gas was called the "foam number"[50]. Obviously this number depends on the rate V/t of gas injection; if V/t is small, liquid has more time to drain down to the bulk, and v_1 also is small. This objection applies equally to the ratio v_1/m; m is the amount of solute present in the bulk liquid[53,54]. In a similar procedure[49], the volume v_1 of liquid transferred into foam after an arbitrary time of nitrogen injection at an arbitrary rate is plotted against the concentration of the foaming agent in the initial liquid. Extrapolation of the curve to $v_1 = 0$ gives the concentration at which no foam is formed. This concentration was, for instance, 4, 40, and 10 mg/l. for a saponin, a sodium dodecyl sulfate, and trimethylcetyl ammonium bromide. Naturally, the values will be different for another duration and another rate of gas injection.

It has been suggested also[51] to use sodium dodecylbenzene sulfonate (which unfortunately usually is a mixture) as the standard and to express the "foaming tendency" and the "foam stability" as that concentration of the standard at which the foam volumes before and after a 4-min aging are equal to those of the test liquid.

To measure the liquid volume in the foam, i.e., v_1 during the bubbling, an apparatus illustrated in Fig. 3.6 (not to scale) was constructed[49]. From a reservoir kept at a constant pressure, nitrogen is led into jar *1* and,

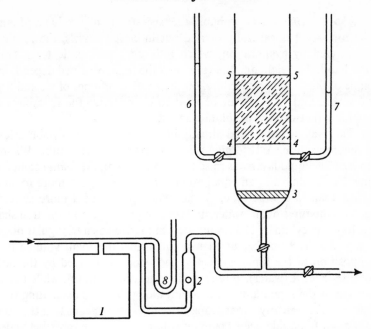

Fig. 3.6. Determination of the liquid volume in foam during bubbling. *1* is the nitrogen reservoir; *2* is the flow meter; *3* is the porous plate; *4, 4* and *5, 5* are the boundaries of the foam column; *6* is a buret; and *7* and *8* are manometers (after reference [49]).

through the flow-meter *2*, into the measuring tube; the gas is dispersed by porous plate *3*. Numbers *4,4* indicate the liquid level, and *5,5* is the upper boundary of the foam column. The test liquid fills buret *6*. When a part of the initial liquid is transferred into foam, the liquid level is maintained at *4,4* by adding the necessary volume of the solution from this buret; and the volume added is recorded. It is equal to the above v_1 and can be determined at any moment during the test. Tube *7* is a manometer; it too is filled with the test liquid. As the liquid level remains at *4,4*, any change in the manometer reading is caused by the pressure exerted by the foam column and can serve to determine this pressure. The pressure in *1* is measured with manometer *8*.

In surfactant solutions, as long as dV/dt was considerable (for instance, 0.8 cm^3/sec), both v and v_1 increased linearly with time so that the ratio v/v_1, i.e., the expansion factor of §1, was constant. When dV/dt was small (for instance, 0.067 cm^3/sec), the foam volume still rose almost linearly with t, but the increase of v_1 with t was almost negligible after about 15 minutes: the amount of liquid draining down from the foam (and evaporating) was nearly as great as that introduced into the foam with the bubbles, and the expansion factor gradually increased.

§56.   *Shaking* is probably the easiest way of producing foam by
agitation. As it is generally carried out in closed vessels, the results are
not affected by evaporation, which influences pneumatic tests. On the
other hand, the height and the volume of the foam obtained depend on the
details of the shaking procedure, on the relative volumes of air and liquid,
and so on, and thus cannot be used to characterize the foaminess of a
liquid in a (reasonably) absolute manner.

This was demonstrated already in one of the earliest publications[55]
on this procedure. A 10% glue solution was shaken in a tube. When the
depth of the liquid layer was raised from 5 to 25 cm, the foam height h rose
from 3 to 22 cm; and an equal volume of solution gave more foam in a
narrow than in a wide tube. Dumanskii[56] published a more systematic
study on mixtures of air, water, and 0.1% aqueous gelatin solution shaken
in a 100-cm³ cylinder. The results are summarized as a triangular diagram
in Fig. 3.7. The number at each point means the foam volume v. The
composition of the liquid system at each point is defined by the latter's
distance from A (water), B (air), and C (gelatin solution), while the composition of the liquid alone is constant along the lines emanating from B
(dashes). The v clearly varies along these lines; this means that the volume
ratio of air to liquid affects the foam volume, and it is impossible to decide
which ratio should be used for meaningful results. Hetzer[57] observed that
the foam volume (of a 0.14% soap solution) was raised by 25% when
50 cm³ of the liquid were shaken with 145 cm³ rather than 70 cm³ of air;

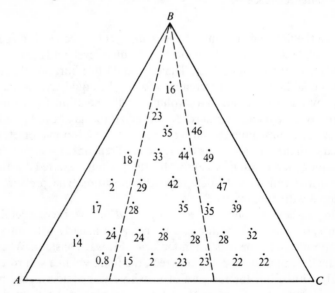

Fig. 3.7. Foam volume of mixtures of water (A), air (B), and 0.1%
gelatin solution (C). The numbers at the experimental points mean foam volumes
in arbitrary units (after reference [56]).

this difference, however, may have been affected by the carbon dioxide in the air.

Sasaki[58] was able to obtain either a water-in-oil emulsion covered with a foam layer or an oil-in-water emulsion without foam by changing the movement from swinging to up-and-down jerks. He worked with a stoppered test tube containing air, 2.33 cm^3 acetic acid, 3.79 cm^3 diethyl ether, and 3.88 cm^3 water. When the liquid contained equal volumes of water and benzene and only a small amount of acetic acid, so that two layers were present, the foam height depended on whether the shaking amplitude of the lower or the upper half of the test tube was greater.[101]

Numerous photographs of foam produced by shaking surfactant solutions can be inspected in Manegold's book[59]; naturally, the foam heights reproduced are specific for the shaking procedure selected and possess no general validity.

§57. The *pour test* is popular because it has been adopted by the American Society for Testing and Materials (ASTM)[60]. The procedure is similar to that by which beer foam is obtained: the liquid is rapidly poured from one into another vessel. However, the gas phase in beer foams originates mainly from the dissolved gas, while in applying the pour test to surfactant solutions the outside air has to be incorporated in the liquid; as indicated in §51, this is a process that is difficult to control and is related to properties quite different from the true foaminess.

In the ASTM test the lower vessel contains a prescribed volume v_i (of solution to be tested) before the pouring; the beer mug is not empty when more beer is run into it. Also the dimensions of the upper and the lower vessels are fixed. The volume v of the foam obtained in the receiver is at once recorded. Ross[61] found that v increased with height h_1 from which the liquid fell (that is, the height of the lower cylinder). For instance, when the distance h_1 was varied from 98 to 202 and 297 cm, v increased in the ratio 22:42:54; but there was almost no increase when the surfactant concentration was too low. When a larger volume of the liquid was poured, v also was greater. When, at constant values of the liquid volume and of h_1, the volume v_i was varied, the v first increased with v_i and then reached a constant value. The effect of the apparatus dimensions, the efflux time (which depends on the shape of the upper vessel), etc., on v was studied also by Gohlke[62].

Rosen[63] applied the pour test to solutions of sodium hexadecyl sulfate, several alkane sulfonates, etc., and noticed that the foam consisted of two distinct layers; the lower one contained small bubbles (0.01 to 0.18 cm across), while the bubbles in the upper (and thicker) layer had diameters between 0.05 and 0.4 cm. The total surface area A of the bubbles, calculated from their diameters, was remarkably insensitive to the nature of the foaming agent and to its concentration. The work required to create this surface is said to be γA g·cm^2/sec^2, γ being the equilibrium surface tension of the solution; this assumption neglects the difference between the surface tensions of fresh and aged liquids (see §11). The work performed by gravitation when mass M falls from height h_1 is gMh_1 g·cm²/sec²; g is the acceleration due to gravity. The ratio $\gamma A/gMh_1$ proved to be near 0.03. Thus about 97% of the energy available is wasted on churning the liquid and is transformed into heat. These valuable experiments deserve to be continued and expanded.

A circulating kind of pour test was described by Kelly[64] and by Oxé[65]. The liquid flows from the upper into the lower cylinder during, for instance, 4 sec; then it is pumped back, which takes again 4 sec; and then the new cycle starts. The two operations can also be performed simultaneously, so that the pump raising the liquid and the gravitation sending it down act uninterruptedly. The foam height can be read off at any time during the circulation; it usually reaches a steady-state value in less than 30 min.

Instead of pouring a liquid into a pool of identical solution, sand can be dropped into the pool; this method was recommended for beer[66].

§58.   *Stirring*, beating, and whipping tests seem to be even less meaningful than the pour tests of §57. The geometry of the instruments which can be used to incorporate gas into a liquid can be altered in an infinite number of ways, and it is impossible to decide which shape supplies the most fundamental or the least arbitrary results. For instance, one of the popular methods uses a perforated plate which is rapidly and frequently pushed up and down through the vapor-liquid interface. It is quite possible that, if the bottom side of the plate is provided with depressions (in which air would be trapped), the volume of air incorporated would be greater; thus the test results would depend also on the smoothness of the agitator parts.

An experimental evaluation of the perforated plate test was performed by Clark[67], who however did not alter the plate in a systematic manner. When a 5% protein solution was beaten, the foam volume v increased with the frequency of oscillations, for instance, in the ratio 2:1 when the frequency rose from 2 to 4 per second. When the amount of the liquid in the test vessel was doubled, v increased by 30% in one instance, but the relative increase depended also on the concentration of the solution. When oscillations continued for a longer time (for instance, 90 instead of 10 min), the final volume v was greater (for instance, 11 instead of 8 arbitrary units). In Götte's tests[68], the foam volume eventually reached a steady-state value; that is, as many bubbles burst or were broken by the plate as were formed in the given time interval. Thus the v of 0.05% solution of sodium palmitate in water at 60° was

200	300	450	520	580	580	590	cm³ after
10	20	30	40	60	80	100	oscillations.

A modification of this method aims at retarding the destruction of the foam by the vibrating plate[69]. The vessel used is 11.5 cm wide and about 40 cm tall, and is provided with a "well" (5.5 cm wide and about 12 cm deep) in its bottom, so that the instrument looks like an inverted bottle. The well is filled with the test liquid (200 cm³), and a perforated plate 4.8 cm in diameter is made to oscillate (up and down) in this well. The foam produced invades the wide part of the bottle. This contains a wire net which forms a cylinder, 5.0 cm in diameter, along the vertical axis, so that the foam has to penetrate the screen to spread over the whole cross-section; the screen presumably lowers the average size of the bubbles (see §33). The perforated disk contains 30 holes, 0.3 cm each. No fundamental data are likely to be generated in this method, but it was repeatedly used, for instance, by Machemer[70] and by Jensen[71].

Laboratory stirrers have been employed, for instance, by Weeks[72], Kaertkemeyer[73] and Sironval[74]. The foam volume generally increases with the frequency of revolutions of the stirrer (this frequency was sometimes as high as 57 per second), reaches the highest value in a few minutes, and is in some systems independent of the depth of immersion of the stirrer blades. The results were compared[74] with those obtained in the ASTM pour test, §57. Fig. 3.8 summarizes some data on solutions of several synthetic detergents (e.g., sodium dodecylbenzene sulfonate) in distilled water at 25°. The steady-state foam height (in centimeters) found after mechanical stirring is plotted along the abscissa, and the foam height

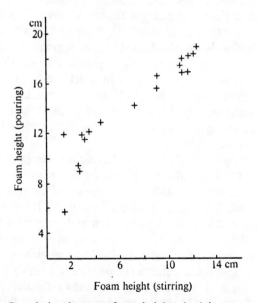

Fig. 3.8. Correlation between foam heights (cm) in a pour test and after stirring for several synthetic surfactants at 25° (data of reference [74]).

(again in centimeters) obtained in the ASTM test is indicated along the ordinate. It is seen that the two heights usually increase together, but no general quantitative correlation seems possible. The precision of the stirring test was several times better than that of the pour test[74]. The results of a similar comparison have been published by Sisley[75] but in his tests the foam volumes varied only in the ratio 1.3:1 when 22 commercial surfactants were employed.

A striking effect of the geometric conditions on foam height[73] deserves mention. The stirring was performed in a 1-l. beaker. When the stirrer was the only foreign body in the latter, the foam volume (of a commercial surfactant) was 330 cm^3 after 15 min of stirring; but when a mercury thermometer was present in the liquid and the foam, this volume was only 80 cm^3. Instead of a glass beaker of 1000 cm^3 capacity, Spangler[76] recommends a steel beaker of 1500 cm^3.

Frequently difficulties are encountered also when household beaters and blenders are employed. Bailey[77] reports that one "type of beater was found to be unsuitable for quantitative studies of the foaming power of egg white because the whipped white, by virtue of its stiffness, soon became immobilized around the beaters with the result that the blades rotated in a central cavity." A wire whipper eccentrically located on a rotating disk gave a copious foam. Ross[78] could not obtain any foam by beating aqueous ethylene glycol, although identical solutions formed tall and persistent foam columns in the pneumatic method.

A blender method was "standardized" by Sisley[75]. In an attempt to devise a test closely corresponding to the conditions of actual use, washing machines (dish-washing or laundering), more or less altered, sometimes are used[79]. Obviously no absolute values of foaminess can be expected from such observations. Even more "practical" are, that is a more restricted validity have, those methods in which the foam is produced by, for instance, stirring and then broken by cloth contaminated with a known amount of approximately known impurities; the amount of impurity needed for collapsing the foam is recorded[80]. Instead of dirty fabric, dirty dishes may be employed[81].

The lather applied to skin before shaving is made by moving a brush. Wilmsmann[82] fixed a polyester brush over a rotating sieve cylinder immersed in a surfactant solution and determined the rate of increase of the foam volume. The methods of estimating the foaminess of boiling liquids are reviewed in §145 in connection with priming and carry over. Here, an amusing semiquantitative procedure may be mentioned[83]. An edible oil is heated to, for instance, 200° and a slice of raw potato is pushed into it. The foam volume (in which presumably water vapor is the main ingredient of the gas phase) is visually estimated.

It is mentioned in §32 that in some machines air is injected into the liquid through a wide pipe, and the big bubbles are then comminuted by a stirrer. The foam thus obtained may overflow the vessel[84] and the volume of the overflow be used as a measure of foaminess[85]. When a suitable (hot) liquid is sprayed against a steel baffle, the foam can be caught in a jar and its volume measured[86].

§59. Foam collapse. A foam column, by whatever method produced, starts shrinking (that is, the foam height starts decreasing) as soon as the introduction of air has stopped. It might be argued that this shrinkage is an unreliable measure of foam stability. As long as bubbles coalesce only in the lower and the central parts of the column, its upper boundary may remain stationary, and the impression would be gained that the foam does not collapse although, in reality, many lamellae have ruptured. This objection is valid in theory but irrelevant for practically all real foams because in these, bursting starts in the top layer and gradually progresses downward. This observation in all probability is related to that referred to in §18, namely that thin lamellae are more likely to break than succulent ones. The top layer of a foam column usually is exposed to the atmosphere so that the evaporation in it is more rapid than in the underlying strata. It loses liquid also by drainage. Of course, drainage takes place in every lamella and every Plateau border in the bulk of the foam as well, but there the downward flow of liquid is almost compensated by the arrival of liquid from the higher strata, while no liquid comes from above to the top layer.

The conditions in the top layer can be rendered similar to those in deeper regions if the test liquid is continually sprayed over the foam column. This experiment has been performed by Manegold[59] on sodium alkylnaphthalene sulfonate solutions. As indicated in Fig. 3.9, foam is formed in a glass cup (*1*) provided with an outlet tube and a stopcock (*2*). Air is injected through capillary *3*. The lid of the cup is a flat dish (*4*) whose bottom is perforated and covered with a sheet of filter paper. The dish contains a small amount of the test solution. When the cup is filled with foam, air flow is discontinued and the faucet (*2*) opened. The liquid from the lid irrigates the foam which then (it is said) lasts indefinitely. Unfortunately no quantitative results are quoted in the book. If the sprinkling occurs only along the vertical axis of the foam column, the wet part does not, and the surrounding region does, burst, so that an initially cylindrical column becomes conical.

The great effect of irrigation on the rate of collapse would be expected to cause a marked variability of this rate. For instance, of two foam columns of equal volumes, the one of a smaller height would collapse more rapidly than the taller column. Observations confirming this conclusion have been made by Peters[48] and some earlier investigators (see §62). When about 500 cm^3 of a foam was permitted to collapse in vessels of different diameters, the decrease in volume was more rapid the wider the vessel[98].

Foams

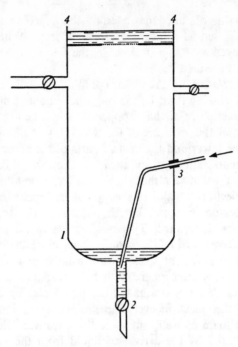

Fig. 3.9. Measurement of stability of foams kept humid. Liquid in *1* is transferred into foam by air flowing in through capillary *3*, and the foam is irrigated by the liquid seeping through the perforated bottom of trough *4*. *2* is a stopcock (after reference [59]).

The rate of lowering of the upper boundary of a foam column is smaller the lower the level observed; this also may be caused by different moisture contents. A foam column about 25 cm tall was produced by air injection[100]. The decrease in height from 20 to 19 cm (that is, 1 cm) took about 1 sec, but an equal decrease from 2.5 to 1.5 cm occurred in 2.5 sec. Unfortunately the liquids used were 0.5 M solutions of sodium chloride, sodium sulfate, and sodium nitrate, which are not likely to produce foam columns of 25 cm when properly purified.

§60. If the irrigation effect is disregarded and it is assumed that every bubble, whether near the top or near the bottom, at the wall or far from it, has an equal probability of rupture, then the number n of bubbles still intact after time t would be expressed by the exponential relation

$$n = n_0 e^{-mt}, \tag{3.4}$$

m (sec^{-1}) being a constant, and n_0, the number of bubbles at $t = 0$. This equation is not likely to be valid for any system. Manegold[59] tested the related equation $v = v_0 e^{-mt}$; v is the foam volume at time t, and v_0 the foam volume at $t = 0$. He found a poor agreement between it and the experimental results on 1 % saponin solutions. An equation containing, in addition to m, another arbitrary constant (v_m), namely $(v - v_m)/(v_0 - v_m)$ $= e^{-mt}$, seemed to be more successful. Schwarz[97] prefers another equation with two arbitrary constants, namely $v_0 - v_m = A(e^{Bt} - 1)$.

If the bubble is most likely to rupture when it is Σ sec old (see §53) and the probability of bursting decreases at greater and smaller times according to the normal distribution curve, then

$$ n = \frac{n_0}{\sigma\sqrt{2\pi}} \int_0^t \exp\left[\frac{-(t-\Sigma)^2}{2\sigma^2}\right] dt, \tag{3.5} $$

σ being the standard deviation. If this equation is followed, the curve of n versus t first slowly slops downward because there are only few short-lived bubbles; then n decreases almost linearly and rapidly and passes an inflection point; along this stretch many bubbles burst every second because their ages are not very different from Σ sec. Finally the decay of n is again slow when the few remaining oldsters pass away.

The experimental material on the gradual decay of foam volume or foam height is relatively extensive, but it is difficult to summarize and to compare it with the above theories. The curves published by Lubman[87] for aqueous solutions of isoamyl alcohol, α-naphthylamine, and m-cresol are not satisfactorily explained in the original. The rate of collapse in some systems decreased in time and reached a very small value when the foam column still was as tall as 1 cm. Did this behavior correspond to the third stretch of the theoretic curve $n = f(t)$ outlined in the preceding paragraph, or was it caused by impurities? Consider a solution containing two foaming agents, one weak and volatile, and the other nonvolatile and strong. Evaporation of the first agent proceeds concurrently with foam collapse, and the collapse becomes slow when almost only the strong foamer is left behind.

The question of impurities is particularly bothersome when observations are made on foams obtained from poorly characterized materials such as treated sulfate liquor, animal glue, humic acid, and commercial soap[88-91]. It is clear that, when different lamellae in a foam have compositions different and variable in time, also their average persistencies Σ or their probabilities of rupture will be different and variable; thus the experimental curve of foam volume v versus time t would be a resultant of several different functions $v = f(t)$ and difficult to decipher. The measurement errors in impure solutions naturally are great; for instance, the foam volumes of 10 % solutions of glue 1, 3, and 5 min after the end of shaking, ranged[90] from 6 to 84 cm³, from 5 to 70 cm³, and from 3 to 64 cm³. The collapse of 1 % soap solution[88] was better reproducible; the foam height was, respectively, 19.1, 18.8, 17.3, 12.1, 7.9, and 3.9 cm after an aging for 5, 10, 15, 120, 210, and 360 min.

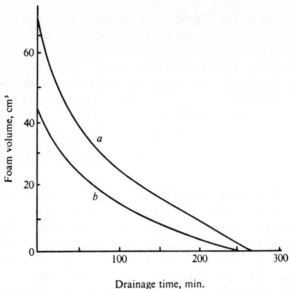

Fig. 3.10. Decay of pneumatic foam of 2% "Aerosol OT" in triethanolamine at 100°. *a* is volume of gas (cm³) and *b* is volume of liquid (cm³) in the foam
at different times (minutes) (from reference [92]).

Fig. 3.10 refers to a nonaqueous system[92]. Foam was produced by
injecting air into 2% solution of "Aerosol OT" (see §75) in triethanolamine $(HOCH_2CH_2)_3N$ at 100°; the expansion factor of the foam
was only 3. The volume of air in the foam is shown in the upper curve
(*a*) as a function of the time of aging, and the lower curve (*b*) refers to the
corresponding volume of liquid. The area under the upper curve divided
by the initial volume was termed "the average lifetime of gas in the foam"
and denoted by the sign L_g. For the foam of Fig. 3.10, this L_g was approximately 5300 sec. Similar collapse curves can be seen also in other publications, for instance, in Kaertkemeyer's paper[73].

In some systems the curves of *v* versus *t* exhibited steps; that is,
$-dv/dt$ was small for a protracted time, then great for a short period,
small again, and so on. Also this behavior may be indicative of impurities
in the initial solution. The foam column can act like a chromatographic
column: different components of the mixture are concentrated in different
strata. Each component has its own persistence, Σ sec, and the above steps
occur near the Σ of every major component. The 5% solution of magnesium oleate in benzyl chloride is an example observed by Manegold[59];
it is not known why the steps disappeared when a small amount of water
was added to this solution.

§61.   Some experimenters were (and are) satisfied with determining
only one or a few points on the curve of *v* versus *t*. Thus in the ASTM

Standard Method D 892-63 the foam volume is read immediately and 10
min after the cessation of air injection. In the ASTM Method D 1173-53
the aging time is 5 min, and in the ASTM Method D 1881-67 the time is
determined between the closure of the air valve and the appearance of the
first bare spot ("eye") on the liquid surface. The selection of these time
intervals is, of course, arbitrary. So is the 4-min waiting time preferred by
some other investigators.

The time until an "eye" appears in the foam-covered surface was
advocated as a measure of foaminess many years ago[39,87]; this time
interval is better reproducible than the time needed for all bubbles to
burst, because the bubbles clinging to the wall usually are more persistent
than those having no solid support; the cause of this phenomenon is not
well understood yet. Some scientists[93,94] waited until just two bubbles
remained. These methods obviously are very similar to that advocated by
Schütz (see §50). They are unsatisfactory because only the persistence of
the longest-lasting bubble among many is determined. An analogous
mistake would be to estimate the death rate of a population from the age
of the oldest (or the third oldest) inhabitant who died last year. Naturally,
this age is greater, the greater the sample examined. This means that a
foam containing more bubbles will appear more stable than an otherwise
identical foam of smaller volume. As reported in §54, some investigators
injected an arbitrary (constant) volume V of air into a solution; if then
the time t_e for an "eye" to appear or for only two bubbles to remain is
measured, this t_e will be found to increase with V. Thus t_e is not a suitable
quantity to characterize the foaminess of a liquid. Since the volume
achieved by a "standard" shaking depends on the viscosity η of the solution, time t_e will be a function of η also.

Nakagaki[95] measured the height h of the foam produced by shaking
at once after the motion stopped, and also the time (t_e) required for an
"eye" to appear and that (t_b) until the rupture of the most persistent
bubble. In this procedure the lifetime of two bubbles (rather than one
bubble) was determined—a modest advance. When the composition of the
solutions (mixtures of water, ethanol, and glycerol) was systematically
varied, usually h and t_e increased or decreased together, but t_b frequently
changed in the opposite direction.

§62.   An improvement on the methods of the preceding section
is the older suggestion[39] to divide the initial foam volume v_0 by the time
t_e until an "eye" appears. The ratio v_0/t_e cm^3/sec was denoted as the
"average rate of static coagulation" (see §54). This rate was measured[96]
for monodisperse foams made by injecting air through a single capillary
into a 0.37% solution of sodium 6-butylnaphthalene-1-sulfonate. Contrary
to t_e itself, the ratio v_0/t_e was independent of v_0 as long as the aging was
conducted in saturated vapor. The t_e and the above ratio were greatly
influenced by the environment, as was the case for the other methods of

measuring foaminess (see, for instance, §48). A 12-cm tall foam column gave t_e of 420 sec in air, 3700 sec in nearly saturated vapor, and 17,000 sec in the absence of an extensive gas phase. The last-mentioned system was formed by releasing air bubbles under an inverted beaker immersed in the test solution. The foam gradually displaced the liquid from the beaker and thus had no contact with the atmosphere. When foam columns of equal height but different widths (because they were obtained in beakers of different diameters) were compared, t_e was four times as great in a 1-cm-wide vessel as in one of 4 cm diameter or wider; this may have been a wall effect (see §61). The effect of bubble size on t_e was different depending on the ease of evaporation. The t_e of a monodisperse foam of aqueous 1% "Aerosol OT" in an open vessel rapidly decreased from 1000 to 20 sec when the bubble diameter $2R$ increased respectively, from 0.1 to 1.0 cm. In a closed vessel t_e was much greater and almost independent of $2R$; it reached 48 hr at $2R = 0.004$ cm and 22 hr at $2R = 1.0$ cm.

The "rate of static coagulation" v_0/t_e proved to be similar to the "rate of dynamic coagulation" of §54. The four p-toluidine foams mentioned there were permitted to collapse. Their values of v_0/t_e were 3.1, 4.8, 8.2, and 11.5 cm³/sec as compared with the U_d values of 10.9, 5.5, 9.2, and 11.4 cm³/sec. A qualitative agreement was observed[39] also for some additional solutions.

The L_g of §60 was compared with the Σ of §53 for three lubricating oils[92]. The former was, at 26°, 79, 104, and 27 min, and Σ at 56° was 7, 10, and 4 min. Thus the ratio of L_g to Σ varied only in the range between 7 and 11. Unfortunately, no more recent determinations of the ratio L_g/Σ could be found by the present author.

Perhaps the gradual decay of the total area of the gas-liquid interface in foam would be a better measure of the collapse process than the diminution of the foam volume or the foam height. Attempts in this direction have been made by Clark[67], Schwarz[97], and Rosen[63].

References

1. Plateau, J. *Mém. Acad. Roy. Sci. Belg.* 33 (1861), 5th series.
2. Foulk, C. W., and J. E. Barkley. *Ind. Eng. Chem.* 35:1013 (1943).
3. Kirschke, K. *Materialpruefung* 1:235 (1959).
4. Bashkirov, M. M., V. Ya. Khentov, and M. S. Ostrikov. *Kolloidn. Zh.* 32:822 (1970).
5. Gleim, V. G. *Zh. Prikl. Khim.* 26:1157 (1953).
6. Plateau, J. *Mém. Acad. Roy. Sci. Belg.* 36 (1867), 7th series.
7. Bikerman, J. J. *J. Phys. Chem.* 56:164 (1952).
8. Smirnova, A. M., and P. A. Rebinder. *Compt. Rend. Acad. Sci. URSS* 52:317 (1946).
9. Sameshima, J., M. Nakamuta, and R. Terada. *J. Chem. Soc. Japan* 63:691 (1942). *Chem. Abstr.* 44:10414 (1950).
10. Stokes, A. J. U.S. 2 473 553; *Chem. Abstr.* 43:6764 (1949).

11. Wilde, D. G. *Chem. Ind. (London)* 1966:764.
12. Matalon, R. *Compt. Rend.* 227:634 (1948). *Research (London) Suppl. Surface Chemistry* (1949), p. 195. *J. Soc. Cosmetic Chemists* 3:216 (1952).
13. Eydt, A. J., and H. L. Rosano. *J. Am. Oil Chem. Soc.* 45:607 (1968).
14. Platikanov, D. *J. Phys. Chem.* 68:3619 (1964).
15. Bikerman, J. J. *J. Appl. Chem. (London)* 18:266 (1968).
16. Gleim, V. G., I. K. Shelomov, and B. R. Shidlovskii. *Zh. Prikl. Khim.* 32:1046 (1959).
17. Gleim, V. G., B. R. Shidlovskii, and G. B. Ryabinina. *Zh. Prikl. Khim.* 37:209 (1964).
18. Trapeznikov, A. A. *Acta Physicochim. URSS* 13:265 (1940).
19. Plateau, J. *Mém. Acad. Roy. Sci. Belg.* 37 (1869), 9th series.
20. Dewar, J. *Proc. Roy. Inst. Gt. Brit.* 22:186 (1917). Quoted in reference 21.
21. Lawrence, A. S. C. *Soap Films.* London: Bell & Sons. (1929).
22. Rebinder, P., and E. Venstrem. *Kolloid-Z.* 53:145 (1930).
23. Rodewald, H. J. *Naturwissenschaften* 37:284 (1950).
24. Geffken, W. *Symposium sur l'affinage du verre.* Paris (1955), p. 251.
25. Gleim, V. G., V. M. Vilenskii, and S. G. Tereshchenko. *Kolloidn. Zh.* 32:849 (1970).
26. Dewar, J. *Proc. Roy. Inst. Gt. Brit.* 21:786 (1916).
27. Talmud, D., and S. Sukhovolskaya. *Zh. Fiz. Khim.* 2:31 (1931). *Z. physik. Chem.* 154A:277 (1931).
28. Teitelbaum, B. Ya., and E. E. Sidorova. *Kolloidn. Zh.* 14:372 (1952).
29. Röntgen, W. *Wied. Ann.* 46:152 (1892).
30. Hardy, W. *J. Chem. Soc.* 127:1207 (1925).
31. Wark, I. W. *Principles of Flotation.* (Austral. Inst. Mining Metall., Melbourne. (1938), p. 282.
32. Foulk, C. W., and J. M. Miller. *Ind. Eng. Chem.* 23:1283 (1931).
33. Hazlehurst, T. H., and H. A. Neville. *Ind. Eng. Chem.* 31:1084 (1941).
34. McBain, J. W., and J. V. Robinson. *Nat. Advisory Comm. Aeron., Tech. Note No. 1844* (1949).
35. Schütz, F. *Trans. Faraday Soc.* 38:85 (1942).
36. Šandera, K., and A. Mirčev. *Listy Cukrovar.* 51:83 (1932). *Chem. Abstr.* 27:2328 (1933).
37. Ostwald, Wo., and A. Siehr. *Kolloid-Z.* 76:33 (1936); 77:251 (1936); 79:11 (1937).
38. Taggart, A. F., and A. M. Gaudin. *Trans. Am. Inst. Min. Met. Eng.* 68:479 (1923).
39. Aleinikov, N. A. *Tsvetn. Metal.* 6:1546 (1931). *Kolloid-Beihefte* 36:82 (1932).
40. Bikerman, J. J. *Trans. Faraday Soc.* 34:634 (1938).
41. Clark, G. L., and S. Ross. *Ind. Eng. Chem.* 32:1594 (1940).
42. *ASTM.* Standard Methods D 892-63 (reapproved 1968) and D 1881-67.
43. Imhausen, A. *Kolloid-Z.* 85:234 (1938).
44. Moroi, Y., and R. Matuura. *Mem. Fac. Sci. Kyushu Univ.* C7:51 (1970).
45. Sasaki, T. *Bull. Chem. Soc. Japan* 13:517 (1938).
46. Dyakonov, G. K. *Zh. Tekhn. Fiz.* 12:302 (1942). *Chem. Abstr.* 37:5299 (1943).

47. Pattle, R. E. *J. Soc. Chem. Ind.* 69:369 (1950).
48. Peters, D. *Angew. Chem.* 64:586 (1952).
49. Lauwers, A., and R. Ruyssen. *Proc. Intern. Congr. Surface Activ.*, *3rd Congr.* 3C:146 (1960/62); R. Ruyssen. *Boll. Chim. Farm.* 101:105 (1962).
50. Lederer, E. L. *Z. Angew. Chem.* 47:119 (1934).
51. Carpenter, W. L., and I. Gellman. *Tappi* 50, No. 5:83A (1967).
52. Nash, T. *J. Appl. Chem.* (*London*) 8:440 (1958).
53. Henry, W. C., and A. D. Barbour. *Ind. Eng. Chem.* 25:1054 (1933).
54. Perri, J. M., and F. Hazel. *Ind. Eng. Chem.* 38:549 (1946). *J, Phys. Colloid Chem.* 51:661 (1947).
55. Trotman, S. R., and J. E. Hackford. *J. Soc. Chem. Ind.* 25:104 (1906).
56. Dumanskii, A. *Kollodchem. Beihefte* 31:418 (1930).
57. Hetzer, *Chem. Ztg.* 57:715 (1937).
58. Sasaki, T. *Bull. Chem. Soc. Japan* 14:63 (1939).
59. Manegold, E. *Schaum.* Heidelberg: Strassenbau, Chemie und Technik. (1953).
60. *ASTM.* Standard Method D 1173-53 (reapproved 1970).
61. Ross, J., and G. D. Miles. *Oil Soap* 18:99 (1941).
62. Gohlke, F. J. *Parfuem. Kosmetik* 45, No. 3:59 (1964). *Chem. Abstr.* 61:840 (1964).
63. Rosen, M. J., and J. Solash. *J. Am. Oil Chem. Soc.* 46:399 (1969).
64. Kelly, W. R., and P. F. Borza. *J. Am. Oil Chem. Soc.* 43:364 (1966).
65. Oxé, J. *Melliand Textilber.* 49:586 (1968).
66. Schuster, K., and W. Mischke. *Wochschr. Brauerei* 54:177 (1937). *Chem. Abstr.* 32:1396 (1938).
67. Clark, N. O. "A Study of Mechanically Produced Foam for Combatting Petrol Fires." London: Dept. Sci. Ind. Res. (1947).
68. Götte, E. *Melliand Textilber.* 29:65, 105 (1948).
69. Schlachter, A., and H. Dierkes, *Fette Seifen* 53:207 (1951).
70. Machemer, H., and K. Heinz. *Congr. Mondial Détergence Prod. Tensio-Actifs.* 1er Congres, Paris 1954. 1:278.
71. Jensen, A. V. *Dansk Tidsskr. Farm.* 31:109 (1957). *Chem. Abstr.* 51:18482 (1957).
72. Weeks, L. E., J. C. Harris, and E. L. Brown. *J. Am. Oil Chem. Soc.* 31:254 (1954).
73. Kaertkemeyer, L. K. *Proc. Intern. Congr. Surface Activity*, 2nd Congress London. 1:231 (1957).
74. Sironval, L., and H. Ramakers. *Ind. Chim. Belge. Suppl. Compt. Rend.* 31e Congr. Intern. Chim. Ind., Liège 2:543 (1958).
75. Sisley, J. P., and M. Loury. *Rev. Franc. Corps Gras* 1:390 (1954). *Soap Chem. Specialties* 31, No. 4:44 (1955).
76. Spangler, W. G. *J. Am. Oil Chem. Soc.* 41:300 (1964).
77. Bailey, M. I. *Ind. Eng. Chem.* 27:973 (1935).
78. Ross, S. *J. Phys. Chem.* 50:391 (1946).
79. Schwartz, A. M., and C. A. Rader. *J. Am. Oil Chem. Soc.* 42:800 (1965).
80. Hartwig, G. M. *J. Am. Oil Chem. Soc.* 45:661 (1968).
81. Edwards, G. R., and H. Stüpel. *Proc. Intern. Congr. Surface Active Subst.*, *4th, Brussels* 1964 (London 1967), 3:49.

82. Wilmsmann, H. W. *Fette Seif.*, *Anstrichmittel* 66:955 (1964).
83. Kajimoto, G., and K. Mukai. *Eiyo To Shokuryo* 16:425 (1964). *Chem. Abstr.* 61: No. 12551 (1964).
84. Walling, C., E. E. Ruff, and J. L. Thornton. *J. Phys. Chem.* 56:989 (1952).
85. Mortenson, M. *Kgl. Norske Videnskab. Selskab. Forh.* 7:11 (1935). *Chem. Abstr.* 29:4709 (1935).
86. Fineman, M. N., H. L. Greenwald, and C. G. Gebelein. *Soap Chem. Specialties* 31, No. 8, 43; No. 9, 50 (1955).
87. Lubman, N. M. *Tsvetnye Metally* 6:854 (1931).
88. Kionka, H. *Z. Deut. Oel- Fett-Ind.* 43:467 (1923).
89. Ostwald, Wo., and A. Steiner. *Kolloid-Z.* 36:342 (1925).
90. Sauer, E., and W. Aldinger. *Kolloid-Z.* 88:329 (1939).
91. Smirnov, V. A., and A. N. Bondarenko. *Zh. Prikl. Khim.* 20:97 (1947).
92. Brady, A. P., and S. Ross. *J. Am. Chem. Soc.* 66:1348 (1944).
93. Robinson, J. V., and W. W. Woods. *J. Phys. Colloid Chem.* 52:763 (1948).
94. Teitelbaum, B. Ya. *Kolloidn. Zh.* 12:375 (1950).
95. Nakagaki, M. *Bull. Chem. Soc. Japan* 21:30 (1948).
96. Savitskaya, E. M., and P. A. Rebinder. *Dokl. Akad. Nauk SSSR* 76:427 (1951). *Kolloidn. Zh.* 13:200 (1951).
97. Schwarz, H. W. *Fette Seif. Anstrichmittel* 66:380 (1964).
98. Mokrushin, S. G., and L. G. Zhidkova. *Kolloidn. Zh.* 21:336 (1959).
99. Vilenskii, V. M., and G. V. Goncharov. *Kolloidn. Zh.* 32:612 (1970).
100. Gleim, V. G., V. Ya. Khentov, and M. M. Bashkirov. *Kolloidn. Zh.* 30:503 (1968).
101. Sasaki, T., and S. Okazaki. *Kolloid-Z.* 159:11 (1958).

CHAPTER 4

RESULTS OF FOAMINESS MEASUREMENTS

§63. It was stated in the precursor of this book[1]: "If a single value independent of the apparatus and method used could be established for the foaminess of a liquid (at specified temperature and other external parameters), then this chapter would have consisted of tables such as are given in more fortunate books for surface tension, viscosity, and analogous physical quantities. Compilation of such tables for foaminess is impossible in the present state of our knowledge, and the values presented below may claim validity only for the particular instrument and the particular procedure employed."

Unfortunately these statements are as valid in 1972 as they were in 1953. Since no other way out seems possible, the numerous quantitative data of various investigators are reproduced with but little criticism and an only brief indication of the procedure employed. They are arranged according to the chemical composition of the solution tested; inorganic compounds in water come first, and highly complicated mixtures are the last to be reviewed. The gas composition and the temperature effect on foaming ability are discussed, respectively, in §83 and §84.

Poorly Foaming Liquids

§64. Pure liquids. Several times in the previous chapters, and also in the preceding section, foaming solutions have been referred to as a term equivalent to that of foaming liquids. This equivalence exists because it is generally agreed that pure liquids do not foam. This belief was established first by Plateau[2] for water and glycerol; the method of §46 (i.e., persistence of single bubbles) was used. However, no reliable method for drying liquids and purifying liquid surfaces was available to him; presumably because of this drawback he observed measurable lifetimes for air bubbles under surfaces of glacial acetic acid, ethanol, chloroform, ether, and so on.

More thorough tests were performed by Foulk[3], whose method is indicated in §44. Ordinary distilled water gave films, which existed long enough to be seen, in a horizontal platinum ring 8 times out of 100 tries; after another distillation this percentage was reduced to 4, and after distillation from potassium permanganate and potassium hydrogen sulfate, to zero. A sample of purified water kept in air for 12 hours gave 58 films in 100 tries. Evidently, whenever a water sample manifests some foaming ability, the sample surely is contaminated. In organic liquids, water seemed to be the most common contaminant causing foaminess; elimination of water resulted in elimination of foaming ability. Benzene and *n*-heptane lost almost all capacity for forming films after distillation over sodium, methanol after distillation over magnesium, and acetone when distilled over anhydrous potassium carbonate. Nitrobenzene and diethyl ether formed no films also without a special treatment, but were capable of foaming when 0.1 % or 0.02 % of water was added to the initial liquid.

The persistence of single bubbles (0.3 cm in diameter) was zero under the surface of *n*-octane, cyclohexane, and nitrobenzene[4]. Also during distillation no pure liquid was observed to give any foam[5].

According to Plateau[2], saturated solutions are as incapable of foaming as pure liquids. He tested aqueous solutions of potassium nitrate, calcium chloride, sodium carbonate, and some other salts. His rule seems to apply also to nonaqueous solutions. As neither pure liquids nor saturated solutions foam, a maximum of foaminess should generally occur at some medium concentration. Numerous confirmations of this prediction will be found in the following.

§65. Inorganic compounds. Older data on the foaming ability of aqueous solutions of simple inorganic salts are inconsistent. The persistence τ of single bubbles, 0.15 cm in diameter, under the surface of 0.05 to 1.6 N solutions of potassium thiocyanate, potassium hydroxide, potassium bromide, potassium chloride, potassium ferrocyanide, potassium sulfate, and aluminum chloride was said[6] to be several seconds (up to 72 sec) at room temperature. Even greater lifetimes have been claimed[7] for films in horizontal rings at the boiling temperature of the liquid; thus 0.5 N solutions of sodium hydroxide, sodium carbonate, sodium sulfate, and sodium chloride gave τ of 157, 100, 78, and 70 sec, but these large values certainly were caused by impurities, since pure water in the same arrangement afforded a τ of 55 sec. Much shorter persistences have been observed in the same laboratory[8] for single bubbles (e.g., 1.27 sec for 0.5 N NaCl), but even these are exaggerated because the time included that needed for the growth of the bubble (see §46). It is mentioned in §46 that higher pressure lowered τ from 2 to 0.9 sec, but see §49 for the effect of contamination.

Ravinskii[9] pulled liquid films from aqueous solutions or suspensions and determined the length at which the film broke. This length increased with the concentration of sodium sulfate from 10 mm in 0.1 M to 19 mm in 0.5 M solutions. When the liquid, in addition to sodium sulfate, contained suspended calcium carbonate and magnesium carbonate, maximum film length was greater and increased when the amount of suspended materials and the pH of the liquid increased.

In all probability, organic impurities were responsible for the high stabilities recorded in the preceding paragraphs. When better care was taken to keep the liquid surfaces reasonably clean, much lower bubble persistences τ were observed. Table 4.1 lists[10] some τ values (in seconds) obtained at 22° at different bubble diameters ($2R$ μ) and different salt concentrations (weight percent in water).

Table 4.1. Lifetime (Seconds) of Bubbles in Aqueous Salt Solutions

	Sodium chloride			Sodium bromide		Magnesium chloride	
$2R\,\mu$	3–3.5%	10–12%	20%	$2R\,\mu$	10%	$2R\,\mu$	10%
16–60	—	0.19	—	30–60	0.10	25–60	0.21
60–130	0.08	0.14	0.10	60–130	0.10	60–120	0.16
130–190	0.08	0.09	—	130–200	0.07	120–190	0.11
190–300	0.08	0.10	—	300–600	0.22	300–600	0.16
300–600	0.11	0.22	0.42	600–800	0.28	800–1400	1.25
900–1400	—	0.91	—	—	—	—	—

Similar persistences of the order of 0.1 sec have been obtained also for other concentrations of NaCl, NaBr, and MgCl$_2$, for some solutions of KCl, and for two mixtures of NaCl and MgCl$_2$. It should be remembered, as explained in §46, that very short persistence times (below, say, 0.05 sec) have no definite physical meaning. Thus the τ of small bubbles in inorganic salt solutions usually is near the low boundary of meaningful values.

The few data on the stability of three-dimensional foams (as distinct from single bubbles or films) do not appear trustworthy, as it is impossible to judge the degree of cleanness of the surfaces employed. Aqueous solutions of several electrolytes have been shaken in test tubes with an equal volume of air. When the concentration was 1 mole/l., the foam height was over 2 cm for sodium sulfate, about 2 cm for sodium chloride, a little smaller for potassium chloride and sodium nitrate, and 1 cm or less for ammonium nitrate and sodium hydroxide. Usually it was little affected by the concentration[11]. In the pneumatic method (see §52) the foam volume of sodium chloride solutions had[12] the greatest value when the concentration was as low as 0.01%.

Several series of data are available for the height and the rate of collapse of foams formed on boiling solutions. The main results are reproduced here, but it is not known how greatly they were falsified by

impurities. Not even the order of magnitude is certain. Solutions containing 2.57 g/l. of sodium hydroxide, sodium carbonate, and sodium
sulfate, or their mixtures, were subjected to boiling at such a heat input
that the foam rose at the rate of 5.7 cm/sec. The height h of foam in the
more or less steady state is shown[13] in the triangular diagram, Fig. 4.1.
It is seen, for instance, that h of 0.257% sodium hydroxide solution (the
right-hand corner) was less than 10 and greater than 5 cm (because there
is no curve for 5 cm near the right-hand corner); additional measurements
showed that this h was near 8 cm. Analogously the h of 0.257% Na_2CO_3
and 0.257% Na_2SO_4 was, respectively, 10 and 2.5 cm. The highest h was
observed near the mixture of NaOH, 35 parts; Na_2CO_3, 35 parts; and
Na_2SO_4, 30 parts; it was above 41 cm. The time of collapse of the foam
on boiling 0.2% solutions was 6 sec for sodium hydroxide, 5 sec for
potassium hydroxide, 3 sec for potassium chloride, and less than 2 sec
for potassium nitrate[14]. Were caustic alkalis better foaming agents than
the two neutral salts because they saponified the oily impurities in water?

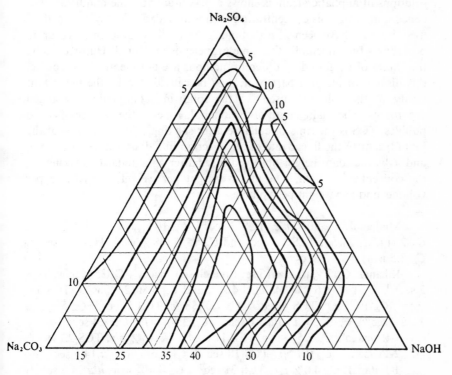

Fig. 4.1.   Foam height (cm) above boiling solutions (2.57 g/l.) of sodium
sulfate, sodium carbonate, sodium hydroxide, and their mixtures. The composition of each mixture is given by its distances from the three apexes, and the
curves connect compositions having identical foam heights, indicated by the
numbers outside the triangle (data of reference [13]).

Foam height h was observed to increase with the salt concentration; was this an example of "salting-out" the contaminants which acted as frothers? When the solute was a mixture of 6 parts of sodium sulfate, 1 part of sodium hydroxide, and 1 part of sodium carbonate, the h on the boiling liquid increased with the total concentration within the range 1.7 to 5.0% but was nearly independent of concentration above this range[15]. In boiling sodium chloride solutions h increased[16] with concentration up to 4%, was almost independent of it between 4 and 8%, and decreased again when the concentration was raised to 15%. The time of collapse of foam[14] on boiling 1% solutions of potassium chloride and potassium nitrate was, respectively, 4 and 3 sec; that is, longer than at the 0.2% concentration (see above).

It is clear from Fig. 4.1 that the h of mixtures has no simple relation to the h values of the constituent solutions. This can be seen also from the results found by Durov[17] and summarized in a book[18].

§66. *Alcohols.* The foaminess, however determined, of aqueous solutions of aliphatic alcohols shows a maximum at some middle alcohol concentration. If this concentration (C_m) were independent of the method used, it would have been clear that in all, externally so different, measuring procedures fundamentally the same property is measured. Unfortunately, the values of C_m found by different methods are not identical. Of course, this disagreement, or a part of it, may be caused also by the inconstant purity of the solutions used. In the following compilation, arranged according to the chain length of the alcohol, τ means the lifetime of single bubbles, TCS is the time of collapse after shaking, h/u is the ratio of the foam height to the linear rate of air injection in a dynamic method (§53), and h_5 is the foam height observed 5 sec after a "standard" shaking. In the concentration data, w/w means weight per weight, v/v volume per volume, and w/v weight per volume.

Methanol: C_m 90% v/v, TCS 3 sec[19]. C_m 10% w/w, TCS 1.5 sec[20]. C_m 2 M (3.2% w/v), TCS 1.5 sec, and C_m 15 M (48% w/v), TCS 1.5 sec[11]. C_m 3.5 w/w, h/u 4.7 sec[21].

Ethanol: C_m 3% w/w, τ 15 sec (in a closed space)[6]. C_m 35% w/w, TCS 2 sec[20]. C_m 1 M (4.6% w/v), TCS 4 sec, and C_m 6 M (28% w/v), TCS 1.5 sec[11]. C_m 10% v/v, TCS 4 sec[19]. C_m 1.3% w/v, TCS 5 sec[22]. C_m 5% v/v, TCS 7 sec[23]. C_m 5% w/v, h/u 1.6 sec[24]. C_m 2.5% w/w, h/u 8.4 sec[21].

1-Propanol: C_m 2.6% w/v, TCS 11 sec[22]. C_m 15% v/v, TCS 22 sec[19].

2-Propanol: C_m 0.7 v/v, TCS 18 sec[19]. C_m 1.5 w/w, h/u 14.5 sec[21].

1-Butanol: C_m 4.8% v/v, TCS 31 sec[19]. C_m 0.4% w/w, h/u 24.5 sec[21]. C_m 1% w/v, h/u 6.2 sec[24].

2-Butanol: C_m 0.35% w/w, h/u 20.5 sec[21].

2-Methylpropanol: C_m 0.67% w/v, TCS 12 sec[22]. C_m 0.375% w/w, h/u 22.3 sec[21].

1-Pentanol: C_m 0.97% v/v, TCS 52 sec[19]. C_m 0.12% w/v, h/u 6.6 sec[24].

3-Methylbutanol: C_m 0.33% w/v, τ 53 sec (in an open space)[25]. C_m 0.32% w/v, TCS 17 sec[22]. C_m 0.44% w/v, TCS 25 sec[25]. C_m 0.53% w/v, TCS 17 sec[26]. C_m 1.96% v/v, TCS 2500 sec[19]. C_m 0.20% w/w, h/u 25.9 sec[21].

2,2-Dimethylpropanol: C_m 0.25% w/w, h/u 17.6 sec[21].

1-Hexanol: C_m 0.33% v/v, TCS?[19]. C_m 0.36% w/v, h/u 7.9 sec[24].

1-Heptanol: C_m 0.0081% w/v, TCS 8 sec[22]. C_m 0.051% v/v, TCS ?[19].

1-Octanol: C_m 0.004% w/v, TCS 5 sec[22]. C_m 0.011% v/v, TCS ?[19]. C_m 0.013% w/v, h/u 2.6 sec[24]. C_m 0.00028 M (or 0.0038% w/v), h_5 0.5 cm[27].

1-Decanol: C_m 0.000023 M (or 0.00036% w/v), h_5 0.2 cm[27].

In recalculating the data of Gibb and Htin[21] it was assumed that the "height of froth" of the original included the thickness of the liquid layer. The TCS values of different investigators are not directly comparable, as the definition of collapse was different in different papers.

In spite of the sad discrepancies between the above C_m values for a given alcohol, some qualitative tendencies are unmistakable. The C_m is smaller, the greater the molecular weight of the alcohol. For methanol it is probably near 10 moles/l. and for decanol near 0.00002 mole/l. It was suggested[19] that $\log C_m$ might be a linear function of the number of carbon atoms in the alcohol molecule. This may be so, but the real reason of the rapid decay of C_m when the carbon chain becomes longer is that the solubility also decreases in the homologous series. The maximum persistence (TCS or h/u) seems to be greater for pentanols than for the smaller and the bigger molecules. The differences between the lifetimes of isomeric butanols and pentanols are moderate; it is clear that molecular weight is more significant than structure.

Some authors determined the foaminess of alcohol solutions without looking for the C_m. Thus the Σ of §53 was[28] 7 sec for 1% solutions of 1-butanol, in good agreement with Sasaki[24]. Foams of three isomeric amyl alcohols had similar times of collapse[19]. On the other hand, this time for 3-methylbutanol was said[22] to be significantly greater (17 sec) than for 1,1-dimethylpropanol (10 sec).

The change of foaminess with concentration is very gradual for alcohols of a low molecular weight; this presumably is another cause of the large discrepancy between the C_m values found by different observers for methanol, ethanol, and 1-propanol. The foaminess of higher alcohols has a steeper dependence on concentration. Thus the h_5 for 1-octanol is immeasurably small at all concentrations below 0.0001 M and above 0.0005 M, and for 1-decanol the concentration range in which the foam height can be measured is even narrower[27].

The time of foam collapse of ternary mixtures of water, methanol, and ethanol was measured by Nakagaki[20]. In Bartsch's tests[22], electrolytes slightly reduced the time of collapse of foams of 3-methylbutanol, and the electrolyte concentration causing the greatest reduction depended above all on the valency of the cation. Thus the most active range was 0.001 to 0.01 N for sodium chloride and potassium thiocyanate, 0.0008 to 0.012 N for sodium sulfate, and 0.0018 to 0.003 N for sodium citrate, that is, was independent of the valency of the anions; but for barium chloride and magnesium sulfate it was 0.000015 to 0.0005 N, and about 100 times as small for aluminum chloride and ferric chloride. It would be interesting to repeat these tests at different acidities, e.g., at pH 4 and 10.

§67. *Other hydroxy compounds.*   The time t of collapse after bubbling and the ratio h/u (see §66) of aqueous solutions of ethylene glycol both had maxima near the composition "glycol 3, water 7 parts," when glycol was not thoroughly purified, and t was approximately 25 sec. After careful purification, both t and h/u decreased and C_m moved to 5% glycol solution[29]. This example shows how strong the influence of impurities on C_m and t may be. The highest TCS was only 1.5 sec and was observed in 6 M solution (i.e., 37 g glycol in 100 cm^3 of solution)[22].

Also aqueous glycerol is a poorly foaming liquid. The highest TCS was 3 sec and C_m again was 6 M (i.e., 55 g glycerol in 100 cm^3 of solution)[22]. The time t_e for the first bare spot to appear (see §61) had the greatest value (7 sec) at 15% w/w, while the lifetime of the most persistent bubble (t_b) had a weak maximum (14 sec) at this concentration but also a much taller maximum (128 sec) at 88% w/w[30]. Since the foam was produced by shaking, the results must have been greatly influenced by the viscosity of water-glycerol mixtures and are thus unreliable indicators of the foaming ability proper.

Both phenol and benzyl alcohol had greatest TCS at C_m of 0.1 M; the collapse times were, respectively, 12 and 10 sec[22]. As *m*-cresol is used in flotation, its foaming was studied in several laboratories; unfortunately, there is no agreement between their results. Aleinikov[31] introduced 100 cm^3 of air through a sintered glass plate and then measured the time (t) of collapse of the foam obtained. The t had a maximum (22 sec) at $C_m = 0.02$ M, but the decrease at higher concentrations was insignificant and t still was about 19 sec in a 0.18 M solution. Bartsch[22] shook 10 cm^3 of solution with 30 cm^3 of air and obtained TCS values about one-third as long as the above t; C_m was near 0.025 M, the corresponding TCS was roughly 9 sec, and there was no measurable foam at concentrations above 0.2 M. Lubman[26] also used shaking (25 cm^3 of solution and 25 cm^3 of air) but obtained a much more exciting curve; there was a maximum of TCS of 88 sec near 0.12 M, while TCS was less than 10 sec at all concentrations below 0.01 M and above 0.20 M. Thus for 0.12 M solutions Lubman observed collapse times about 20 times as long as Bartsch did. No explanation is possible at present.

The time t of collapse of *o*-cresol foams produced by bubbling had an uncertain maximum (20 sec) in 0.016 M solution, while t of *p*-cresol was almost independent of concentration (22 to 26 sec) between 0.003 and 0.05 M but reached 32 sec in 0.18 M solution[31].

As cresols are weak acids, their foaminess depends on the acidity of the liquid. Hydrochloric acid accelerates foam collapse after shaking[26]. The effect of sodium hydroxide was less simple. It is shown in Table 4.2 for the highest TCS but certainly needs a re-examination.

Table 4.2.   Effect of NaOH on TCS of 0.11 M *m*-cresol

Mole ratio NaOH: cresol	0	0.04	0.08	0.39	3.9
Time of collapse (sec)	88	108	108	22	6

The effect of acidity on the foaming capacity of chlorinated phenols has been observed[32] in a qualitative fashion only. "Stable" foams were obtained with pentachlorophenol between pH 1.5 and 5, and between 8 and 8.5; with 2,4,5-trichlorophenol at pH 7 and 8; with 2,3,4,6-tetra-chlorophenol at pH 9 and 10, and so on.

Triangular diagrams for the foam height of ternary mixtures of water, ethanol, and glycerol show three shallow maxima (0.8, 1.2, and 2.0 cm) after a "standard" shaking[30]. Their significance is not clear.

§68. Fatty acids.   The solubility of fatty acids in water is not very different from that of the corresponding aliphatic alcohols; consequently the concentration range in which foaminess can be measured also for the acids is narrower, the greater the molecular weight. Thus, the solubility of 1-octanol at room temperature is near 0.05% (or 0.004 M) and of octanoic (caprylic) acid about 0.08% (or 0.006 M); hence the C_m (see §66) of their aqueous solutions would be expected to be near 0.001 M. Experimental values are not too far from those predicted. Data for C_m and the greatest values of TCS and h/u are arranged here as in §66.

Formic acid: C_m 2.1% *w/v*, TCS 4 sec[22]. C_m 5.8% *w/v*, h/u 2.9 sec[24]. C_m 8.8% *w/w*, h/u 14.7 sec[21].

Acetic acid: C_m 1.2% *w/v*, TCS 8 sec[22]. C_m 15% *v/v*, TCS 11 sec[33]. C_m 2.5% *w/w*, h/u 21.8 sec[21]. C_m 4.9% *w/w*, h/u 3.4 sec[24].

Propionic acid: C_m 1.9% *w/v*, TCS 11 sec[22]. C_m 3.5% *w/w*, h/u 26.4 sec[21].

Butyric acid: C_m 2.5% *w/w*, h/u 31.2 sec[21]. See also below.

Valeric acid: C_m 0.15% *w/v*, TCS 9 sec[22]. C_m 1% *w/w*, h/u 29.7 sec[21].

Hexanoic acid: C_m 0.09% *w/v*, TCS 13 sec[22]. C_m 0.035% *w/w*, h/u 37 sec[21].

Heptanoic acid: C_m 0.019% w/v, TCS 16 sec[22]. C_m 0.11% w/w, TCS 85 sec[34]. C_m 0.18% w/w, persistence τ of single bubbles 19 sec[6].

Octanoic acid: C_m 0.0036% w/v, TCS 12 sec[22]. C_m 0.05% w/w, τ 23 sec[6].

Nonanoic acid: C_m 0.001% w/v, TCS 5 sec[22].

Concentrations for the acids C_7 and C_8 are calculated assuming the solubilities to be 0.22 and 0.062% at the temperature of the experiment.

For butyric acid, Bartsch[22] found the maximum TCS (18 sec) in molar solutions, that is, at 8.8% w/v. Later, on the curve of TCS versus concentration, two maxima were found[35], as is seen in Fig. 4.2, whose abscissa is the concentration in moles/liter and whose ordinate is the time until bare liquid surface appears (i.e., the t_e of §61). The taller maximum corresponds to the weight concentration of 19.8%; it was observed in especially purified samples also. Also the rate of drainage from foams of 2 to 3.5 M butyric acid was unusually slow[36]. A repetition of these experiments would be welcome.

The foaminess of fatty acids would be expected to depend on the acidity of the solution and, at least for the higher members of the series, also on the accidental contamination of the solvent with calcium and similar ions. The effect of pH was determined for decanoic acid. About 0.9 g of this was shaken with 20 cm^3 of a buffer solution (of the ionic strength 0.05) in a 100 cm^3 cylinder[37]; this means that no foam could have an expansion ratio above five. The foam volume immediately after

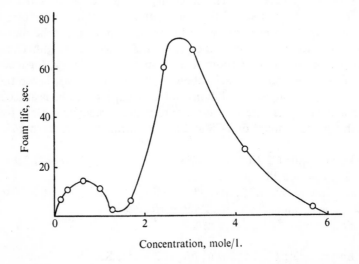

Fig. 4.2. Foam persistence (until bare liquid surface appears) after shaking, in seconds, as a function of the concentration (mole/l.) of butyric acid in water (data of reference [35]).

shaking was small at pH 6 to 7, had a maximum at pH 9, but remained high also at pH 10, 11, and 12; however, the rate of collapse was greater at pH 9 than at other acidities. A mixture of decanoic acid with 3 molecules of 1-decanol also gave the most voluminous foam at pH 9, but its stability was much greater than that of decanoic acid foams; for instance, only about one-fifth of the foam collapsed on standing for 30 hr. It is not known how foam stability was affected by the nondissolved excess of decanoic acid and decanol.

Ferric acetate was listed already by Plateau[2] as a foaming agent. Sodium acetate and its homologs have been studied[38] by the dynamic method of §53. The ratio h/u of the steady-state foam height to the linear rate of air injection had highest values for 0.1 M sodium acetate (4.2 sec), 0.06 M sodium propionate (6.7 sec), and 0.06 M sodium butyrate (7.6 sec). Unfortunately the h/u of water was recorded not as zero but as 1.2 sec.

The C_m of lactic acid is said[22] to be 1.0 M. The time of collapse after shaking was 7.5 sec.

Curves for the dependence of the persistence τ of single bubbles on the concentration of heptanoic and octanoic acids can be found in reference [6]. As, however, the greatest values of τ for these solutions were almost equal to those in 0.4 to 0.8 N potassium chloride and potassium bromide, the results do not appear trustworthy.

The τ of bubbles under unimolecular films of fatty acids is reviewed in §81. For the foaminess of salts of long chain fatty acids (i.e., soaps) see §70.

§69.   In this section data are collected on the foaminess of several organic compounds which do not belong to the groups reviewed in other sections; water was the solvent in all instances. First, the longest time of collapse after shaking (TCS) and the corresponding concentration C_m, as reported by Bartsch, are listed in Table 4.3.

Table 4.3. TCS and C_m of Miscellaneous Aqueous Solutions

Solute	C_m, mole/l.	TCS, sec
Acetaldoxime	0.37	10
Acetone	0.5	2.5
Aniline	0.1	11
Ethyl acetate	0.01	2.5
Ethylamine	0.4	12
Nitrobenzene	0.005	6
Paraldehyde	0.03	9
2-Pentanone	0.05	3
Sucrose	0.5	0.5
p-Toluidine	0.04	6

The C_m and TCS for α-naphthylamine are reported[26] as 0.0017 M and 18 sec. The TCS of 0.0003 and 0.0006% solutions of methyl salicylate was, respectively, 2 and 1 sec; the C_m was not ascertained[39]. For the following ethers the C_m was a little greater than their critical micelle concentration: octyl glycol ether $C_8H_{17}OCH_2CH_2OH$ (TCS > 10 min), octyl glycerol ether $C_8H_{17}O\cdot C_3H_7O_2$ (TCS 20 min), and β-d-octyl glycoside (TCS 600 min); the long collapse times contrasted with the moderate foam heights achieved by shaking, namely 2, 5, and 11 cm for the three ethers tested[40].

The time of collapse after bubbling air through solutions of *p*-toluidine, methylaniline, and dimethylaniline was determined by Aleinikov[31]. The persistence of single bubbles had a maximum (19 sec) at the aniline concentration $C_m = 0.24$ M[6]. Foaming of 0.2% solutions of 9-fluorenylsuccinic acid, 1-acenaphthylsuccinic acid, and 1,2,3,4-tetrahydro-1-naphthylsuccinic acid was observed[41] between pH 3 and 8. The TCS of solutions of methyl violet 5B, brilliant green, malachite green, and crystal violet was determined by Nakagaki[11].

Strongly Foaming Solutions

§70. Soap solutions. As would be expected, numerous data are available for salts of fatty acids having a higher molecular weight than those of §68. First, measurements on individual compounds and then on homologous series are reviewed.

A 3.7% sodium laurate solution, pH 8.6, gave films which persisted for 120 to 400 sec when their diameter was 1.0 cm, and for 13 to 32 sec when the diameter was 3.5 cm[42]. The lifetime of the most persistent bubble of a foam layer on 0.1% sodium laurate, pH 7.6, was 61 sec; and at pH 8.6 it was 150 sec[43]. The critical micelle concentration (CMC) of sodium laurate at pH 11.7 is near 0.023 mole/l. (i.e., 0.51% *w/v*) at room temperature; and the ratio h/u, presumably almost identical with the Σ of §53, was about 1400 sec for a 0.020 M solution. It was greater (perhaps twice as great) at the CMC and higher concentrations but could not be measured at the linear velocity u (of nitrogen) employed[44].

Fig. 4.3 combines results on the foaminess of sodium laurate obtained in three different arrangements[45]. The abscissa of the graph is the soap concentration in moles/l. Curve *1* shows that the persistence of single bubbles, about 1 cm across, reached the greatest value (about 350 sec) in 0.02 M solution (or 0.44% *w/v*). The time (curve *2*) for a foam produced by pouring to subside to one-half the original volume was much longer; its maximum value was near 15,000 sec and observed at 0.7% *w/v*. The initial foam volume after pouring (curve *3*) was almost independent of concentration between 0.01 and 0.06 moles/l., although the CMC of the salt is situated in this region. Since the acidity of the solution varies with concentration, the curves of Fig. 4.3 may have been affected by pH.

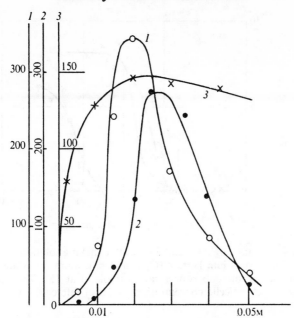

Fig. 4.3. Dependence of the persistence of single bubbles (scale 1 in sec curve *1*), the half-life of foams after pouring (scale 2 in min, curve *2*), and the, initial volume after pouring (scale 3 in cm³, curve *3*) of aqueous sodium laurate solutions at 20° on their concentration (mole/l.) (data of reference [45]).

Single bubbles on a 0.15% solution (0.005 м) of sodium oleate, 0.1 cm in diameter, persisted for about 1400 sec, but the lifetime was only 270 sec for bubbles of 0.45 cm[25]. On 0.1% sodium oleate at pH 7.5, bubbles of unknown size lasted up to 49 sec[43]. The difference may have been caused not only by different diameters but also by evaporation: in the earlier series the vapor phase was enclosed while dry air was streaming over the bubbles in the latter series. According to Pankhurst[46], h/u in a pneumatic method increased with concentration from 0.001 to 0.01 м solutions of sodium oleate; the CMC of this salt is said[47] to be 0.0019 м.

In a series of measurements the pH of each solution was adjusted so as to reach the highest foam volume in a pour test[48]. As in Fig. 4.3, the initial foam volume varied but little with the concentration. It increased by only about 10% when the concentration of sodium myristate or sodium oleate was raised from 0.1 to 0.4%. This is perhaps another confirmation of the mediocre sensitivity of the test, which is pointed out in §58.

The volume of foam produced by a vibrating perforated plate usually increases with the soap concentration as long as this is small[49] but, when the concentration was varied between 0.1 and 0.4%, the volume was constant within a few percents both for sodium palmitate and sodium

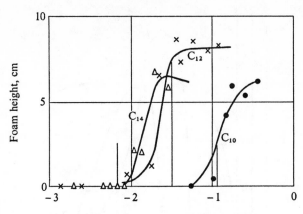

Log of soap concentration, mole/l.

Fig. 4.4. Foam height (in cm) 30 sec after shaking solutions of potassium decanoate (C_{10}), potassium laurate (C_{12}), and potassium myristate (C_{14}) as a function of the log of soap concentration (mole/l.) at 15°. The vertical lines indicate the critical micelle concentrations (data of reference [27]).

oleate. The time t needed for the foam to collapse to one-half generally increased with concentration, but t for sodium palmitate was 50 and 10 min for, respectively, 0.2 and 0.8 % solutions.

Figs. 4.4 and 4.5 show the foam height (h cm) 30 sec after shaking stopped[27]. The abscissa is the logarithm of the soap concentration (in mole/l.). The left-hand curves are for potassium myristate, the middle curves for potassium laurate, and the right-hand curves for potassium decanoate; Fig. 4.4 is for 15° and Fig. 4.5 is for 35°. The vertical lines indicate the CMC, apparently at 15°. All solutions contained a small excess (2 parts for 100 parts soap) of potassium hydroxide. It is seen from Fig. 4.4 that h of the laurate was almost constant between 0.03 and 0.2 M, in a reasonable agreement with Fig. 4.3 (for the sodium salt). The nearly constant stretch started at a concentration equal to, or several times as great as, the CMC. The absolute values of h are, of course, unimportant, as they depend on the size of the vessel and other external circumstances.

The foam volume at 35° (Fig. 4.5) showed maxima rather than plateaus. Maxima at moderate concentrations have been observed also after shaking, in a separatory funnel, solutions of potassium laurate, potassium myristate, sodium oleate, and potassium oleate[50]. An example of the effect of acidity is illustrated[48] in Fig. 4.6. 0.1 % soap solutions at 57° gave, in a pour test, foam heights (h cm) plotted along the ordinate, while the abscissa represents the pH at 50°. The effect of pH is more pronounced than that of the concentration, except at high dilutions. The optimum pH increases from sodium decanoate (10 in the graph) to sodium laurate (12), sodium myristate (14), and sodium palmitate (16). A change of pH from

Results of Foaminess Measurements

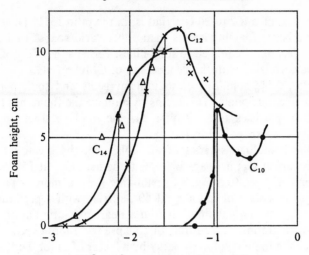

Log of soap concentration, mole/l.

Fig. 4.5. Foam height (in cm) 30 sec after shaking solutions of potassium decanoate (C_{10}), laurate (C_{12}), and myristate (C_{14}) as a function of the log of soap concentration (mole/l.) at 35°. The vertical lines indicate the critical micelle concentrations (data of reference [27]).

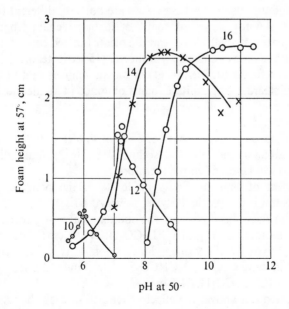

pH at 50·

Fig. 4.6. Foam heights (in cm) after pouring 0.1% solutions of sodium decanoate (C_{10}), laurate (C_{12}), myristate (C_{14}), and palmitate (C_{16}) at 57° as a function of pH (at 50°) (data of reference [48]).

8 to 10 raises h from 2 to 26 cm, that is, in the ratio 13:1. Thus "neutral soaps" are better foaming agents than either "acid soaps" or free acids; also this effect may be related to solubilities. Perhaps the effect of pH on h should be correlated with the dependence of CMC on pH.

At 25°, 0.1% solutions of sodium myristate and sodium palmitate contain a fine precipitate. This seems to stabilize the foam, as is done by many other particles (see Chapter 5); the most persistent bubble on these solutions existed[43] for 150 sec.

In the homologous series of alcohols, §66, the greatest foaminess usually occurs at C_5; in soaps it seems to be near C_{12} at 15 and 35°. At higher temperatures (see Fig. 4.3) palmitate gives a more copious foam than either myristate or laurate. At 60°, stirring with a perforated plate gave more foam in a palmitate than in a stearate solution at an identical weight concentration[49]. The time of collapse (or, more exactly, the time needed for the foam column to decay by a factor of 2) for 0.4% solutions was 20, 600, 1800, and 3000 sec for sodium decanoate, laurate, palmitate, and stearate. The effect of temperature on foaminess undoubtedly is related to its effect on the solubility of the soaps.

The double bond in the chain apparently has no decisive effect. At the most favorable pH and 57°, the foam heights in a pour test were, for 0.1 and 0.4% solutions, 3 and 16 cm for sodium undecylenate and 25 and 27 cm for sodium oleate[48]; these values are not very different from those seen in Fig. 4.6 for saturated soaps of identical chain length. In the perforated plate test at 60°, 0.4% sodium oleate foamed better than sodium stearate, but the time of collapse was shorter (15 sec instead of 3000 sec)[49]. The expansion factor (defined in §1) of oleate foams is said to be smaller than that of stearate or palmitate foams obtained in an identical manner; and the time of half-collapse seems to be longer for sodium than for potassium soaps[51].

Some data on the foaminess of sodium linoleate and sodium ricinoleate[43], sodium ricinoleate and sodium elaidate[48], and sodium ricinoleate[46] also can be found in the literature.

The effect of branching was moderate in the pour test, perhaps because this test is not particularly sensitive. At 60°, 0.25% solutions gave the following foam heights[52] for the sodium soaps of: stearic 23.8 cm, 2-methyleicosanoic $C_{18}H_{37}CH(CH_3)CO_2H$, 14.9 cm; 2-methylstearic, 21.7; 2-butylmyristic, 21.0; 2-heptylundecanoic $C_9H_{19}CH(C_7H_{15})CO_2H$, 21.0; 2-octyldecanoic $(C_8H_{17})_2CHCO_2H$, 23.0; and 2,2-dimethylpalmitic acid $C_{14}H_{29}C(CH_3)_2CO_2H$, 21.5 cm.

As pointed out above, the effect of temperature on the foaminess of soap solutions depends above all on the temperature coefficient of solubility or CMC. However, it is claimed[53] that crystal transformations of soaps can be detected by foam measurements; unfortunately, the original could not be consulted.

The foaming ability of poorly defined soap mixtures was measured by Tschakert[54] and Kohler[55]. Mixtures of sodium laurate and sodium palmitate may produce a greater foam volume in the pour test than either of the pure soaps at an equal concentration[48]. The result, however, greatly depends on the acidity of the two solutions, which so drastically affects the foaming capacity (see Fig. 4.6). Thus 0.1% sodium oleate solutions had the greatest foam volume after shaking when their pH was 9 to 10, and at pH 7.5 the volume was negligibly small. When 5 cm^3 of 2% ricinoleate solution of pH 6.7 were added to 100 cm^3 of 0.1% oleate, pH 8.3, the pH was lowered to 7.7 and the foam volume and the time of collapse were nearly halved. If the ricinoleate solution was first made alkaline (pH 8.4) and then mixed with the oleate solution, the foaminess was not altered[56]. Commercial sodium abietate, which did not affect the pH of the oleate, did not change its foaminess either[56]. When less attention was paid to acidities, rosin (i.e., mainly abietic acid) was found to depress the foam volume of tallow soaps[57].

Calcium soaps of fatty acids are poorly soluble, and their aqueous suspensions do not foam well; utilization of this property for analysis is described in §194. Consequently, addition of calcium ions to soaps of alkali metals strongly reduces foaminess. Cations which form no precipitates have less influence on foaming ability; but it should be remembered that a "neutral" salt can affect this ability by altering the pH of the liquid. Perhaps this consideration accounts for the observation[43] that K_3PO_4 and $K_4P_2O_7$ (whose solutions are alkaline) raise the persistence of single bubbles in the surface of 0.1% sodium oleate more than do sodium chloride or potassium sulfate. The foaminess of mixtures of soaps and "builders" (see §153) in hard water was tested, for instance, by Morrisroe[58] and by Dedrick[59].

Not much is known of the foaminess of quaternary ammonium soaps. A foam of trimethylvinyl ammonium palmitate was stable for weeks[60]. The volume of foam achieved by shaking 0.14% solutions of triethanolamine soaps increased from hexanoate to laurate and then decreased to stearate, while the oleate was almost as active as the laurate[61]. Additional data can be found in reference [62].

Methanol, ethanol, and *n*-butanol lower the foaminess of sodium oleate solutions more the higher the molecular weight of the alcohol and the greater the concentration[63].

§71. *Alkyl sulfates.*

The composition of these surfactants can be represented by the formula $RO \cdot SO_2 \cdot OM$, R being an alkyl radical (e.g., $n\text{-}C_{12}H_{25}$), and M, a univalent cation (such as Na^+). Their foaminess is small when R is C_8H_{17} or smaller. For instance, in a pneumatic method, foam formation was noticeable[46] at 38° when the concentration of sodium octyl sulfate was at least 0.03 M, while the minimum concentration for sodium decyl sulfate, sodium dodecyl sulfate, and sodium tetradecyl

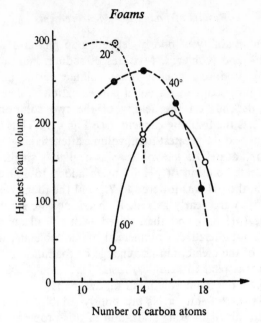

Fig. 4.7. Maximum foam volume (arbitrary units) after shaking sodium alkyl sulfate solutions; the number of carbons in the anion is plotted along the abscissa. The temperature is marked at the curves (data of reference[64]).

sulfate was 0.02, 0.004, and 0.00025 M; thus, addition of six methylene groups lowered the threshold in the ratio 1:120. As would be expected, the chain length corresponding to the greatest foaminess depends on temperature. This is seen in Fig. 4.7, taken from an early publication[64]. The abscissa is the number of the carbon atoms in the radical R, and the ordinate is the foam volume (in arbitrary units) achieved by "standard" shaking. The best foaminess is observed for the sodium dodecyl compound at room temperature, for tetradecyl at 40°, and for hexadecyl at 60°. In every instance the concentration is that corresponding to the greatest foam volume.

When the foam was produced by a vibrating perforated plate and the concentration (weight by weight) kept constant, the foam volume was greater (and the time of collapse was shorter) for dodecyl than for hexadecyl and heptadecyl salts, as long as the concentration was small (e.g., 0.016%); but in 0.4% solutions of sodium salts the volume was almost identical for the saturated C_{12}, C_{13}, C_{14}, C_{16}, and C_{17} compounds, and also for the C_{18} compound derived from oleic acid[49]. (See also §73.)

In a pour test the foam volume increased with concentration as long as this was small but had a nearly constant value at higher concentrations[48,65]. The concentration C_1 at which the limiting value was attained was at 46° about 0.005 M for sodium dodecyl sulfate, about 0.003 M for sodium tetradecyl sulfate, and about 0.001 M for sodium hexadecyl sulfate. So far in this section all radicals were supposed to be primary; for instance, dodecyl means $CH_3(CH_2)_{10}CH_2$—. Among the derivatives of secondary alcohols, C_1 also usually is smaller the longer the chain. This was confirmed for sodium sulfates of 2-hendecanol [i.e., $CH_3(CH_2)_8CH(CH_3)O \cdot SO_2 \cdot ONa$], 2-tridecanol, 2-pentadecanol, 2-heptadecanol, and 2-nonadecanol. There was no definite difference between the C_1 values for the sodium sulfates of 2-pentadecanol and 4-pentadecanol [i.e., $CH_3(CH_2)_{12}CH(CH_3)O \cdot SO_2 \cdot ONa$ and $CH_3(CH_2)_{10}CH(C_3H_7)O \cdot SO_2 \cdot ONa$], but the derivatives of 6-pentadecanol and 8-pentadecanol had a greater C_1 and also a greater limiting foam height than the derivatives of 2-pentadecanol.

In another series of pour tests[66] at the concentration of 0.1% and at 40° the foam height was greatest for sodium tetradecyl sulfate among the straight-chain compounds, and for a sodium eicosyl sulfate when the chain was much branched. In all instances the concentrations C_m (see §66) and CMC were very similar. Still in pour tests, sodium pentadecyl sulfate at 30° gave the most voluminous foam in the series C_{13} to C_{18}, and at 60° the even-numbered chains had a different foaming behavior from that of the odd-numbered compounds; in the first series the greatest foam height occurred at C_{16}, and in the second, at C_{15}[67,68].

Whatever the method of measurement, the foaminess usually exhibits a maximum at some medium concentration (C_m), but the relevant data are in a poor agreement with each other. The persistence of single bubbles, about 1 cm across, had a maximum (90 sec) in 0.01 M sodium dodecyl sulfate solutions[45]. In a pour test at 40° the foam height of these solutions increased with concentration between 0.02 and 0.24% but was almost constant between 0.25 and 0.4%[69]. The concentration at the kink (i.e., 0.24% w/w or 0.008 moles/l.) is near the CMC of the salt, which usually is given as 0.20 to 0.25%. The connection between C_m and CMC is mentioned in the preceding paragraph and is visible also in Figs. 4.4 and 4.5. For additional data see Kashiwagi[70].

The h/u in a pneumatic test (see §53) increased with concentration from 120 sec in 0.001 M sodium dodecyl sulfate to 1800 sec at 0.007 M, and there was still no indication of a maximum or a plateau[44].

When the foaminess of alkyl sulfates is determined in the presence of inorganic salts (sodium sulfate, sodium carbonate, sodium tripolyphosphate, and so on), the C_m shifts so as to remain near the CMC; thus the shift of CMC produced by electrolyte additions can be approximately determined by measuring foaming ability at different concentrations[71]. Large excesses of neutral salts (sodium chloride, magnesium chloride, calcium chloride) and of hydrochloric acid generally increase the foam height (in a pour test) of 0.0023 M solution of the sodium sulfate of 7-tridecanol[65]. Changes in pH between 3 and 9 had but little effect on this height for 0.1 % sodium dodecyl sulfate[48]. The foam volume (apparently produced by shaking) and the stability of this foam depended on the cation (M) in dodecyl sulfate salts $C_{12}H_{25}O \cdot SO_2 \cdot OM$; as a rule they were greater, the smaller the cation in the series M = ammonium, or tetramethyl ammonium, or tetraethylammonium, or tetrabutylammonium, or N,N,N′,N′-tetramethyl-hexamethylenediamine

$$[(CH_3)_2N \cdot CH_2CH_2CH_2—]_2{}^{72}.$$

The foaminess of some "built" solutions of several unsaturated sulfates was estimated by Weil[73]; "built" means mixed with salts chelating calcium ions, and the unsaturated chains included, among others, octadecenyl $CH_3(CH_2)_7CH:CH(CH_2)_7CH_2—$, 2-octadecenyl-oxyethyl $C_{18}H_{35}OCH_2CH_2—$, and 2-octadecenyloxy-1-methyl-ethyl

$$C_{18}H_{35}OCH_2CH(CH_3)—.$$

A temperature rise raised the foam volume (obtained by shaking) but lowered the foam stability and the foam density of solutions of sodium dodecyl sulfate, sodium dodecylbenzene sulfonate, and some other surfactants[74].

The major part of the data presented in this section was obtained on impure samples. As mentioned in §15, sodium dodecyl sulfate is readily hydrolyzed. Its commercial samples contain the free alcohol (dodecanol), which strengthens the walls of the foam films of sodium dodecyl sulfate (see §134) and enhances foam stability.

Other mixtures containing alkyl sulfates have been studied, for instance, by Stirton[75], who used saturated and unsaturated tallow alcohol derivatives. Alkyl sulfates sometimes are incompatible with soaps. For instance, the time of collapse (after shaking) of 0.25% solutions of the sodium 6- or 7-tetradecyl sulfates was lowered from 15 min to a few seconds by addition of 1 part of sodium stearate to 200 parts of sulfate, while other ratios of sulfate to stearate were less deleterious[76]. Calcium salts affect the foaminess of mixtures of some soaps with sodium tetradecyl sulfate[48], but the foam height of sodium tetradecyl sulfate and sodium hexadecyl sulfate (1000 parts) was not altered by addition of 20 parts of sodium palmitate and 2 parts of calcium chloride[77].

§72. Sulfonates. Salts of alkanesulfonic acids $R \cdot SO_2 \cdot OM$ did not receive much attention; R is an alkyl radical and M, an univalent metal. In 0.1% aqueous solutions sodium tetradecane sulfonate $(R = C_{14}H_{29}, M = Na)$ gave a foam volume (after beating with a perforated plate) and a time of collapse very similar to those of sodium tetradecyl sulfate[49]. The foaminess of sodium hexadecane sulfonate was weaker than the foaminess of sodium tetradecane sulfonate or sodium tetradecyl sulfate[49]. The volumes of foam of 0.1% solutions (still obtained with a perforated plate) depended on the position of the acid radical in

the *n*-hexadecane; they were 380, 470, 535, 580, 640, 720, 775, and 800 cm^3 for 1-, 2-, 3-, 4-, 5-, 6-, 7-, and 8-sulfonate, so that the foaming ability was greatest when the —SO$_3$Na group was attached to the middle of the straight chain[78]. When the salt concentration was raised to 0.2%, the foam volume increased by 10 to 20%. For these tests, water pretreated with sodium carbonate was employed[78]. The foaminess of sulfonates was compared with that of disulfonates. The foam volume was twice as great for a mixture of sodium octadecane sulfonates as for a mixture of disodium octadecane disulfonates or a pure disodium octadecane-1,12-disulfonate[79].

On the other hand, an extensive literature exists for the foaminess of alkylbenzene sulfonates R·C$_6$H$_4$·SO$_2$·OM. The metal M was almost always sodium, but the alkyl radical R was changed within wide limits, and also the relative position of the alkyl and the sulfo groups in the ring was altered. Table 4.4 collects data obtained[80] in a pour test at 60° C for sodium salts. The alkyl groups were individual isomers but the sulfonic acids obtained from the hydrocarbons were mixtures, mainly of the ortho- and the para-isomers, that is,

Table 4.4.   Foam Heights of Alkylbenzene Sulfonates

Alkyl	CMC (mmoles/1.)	Concentration (mmoles./l.)					
		0.05	0.2	0.5	1	3	5
n-Propyl	None	—	—	—	—	—	0 mm
n-Butyl	None	—	—	—	0	5	10
n-Pentyl	None	—	—	0	5	20	40
n-Hexyl	37	—	0	5	10	35	62
n-Heptyl	21	0	6	10	15	46	80
n-Octyl	1.4	0	8	13	18	50	88
n-Nonyl	6.5	2	8	15	23	55	96
n-Decyl	3.7	10	23	46	80	133	155
n-Dodecyl	1.2	20	70	128	152	167	175
n-Tetradecyl	0.7	44	110	143	153	174	180
n-Hexadecyl	0.5	36	94	100	110	127	144
n-Octadecyl	0.6	35	48	40	20	22	30
2-Ethylhexyl	25	2	20	38	70	128	135
2-Propylheptyl	8.5	3	20	40	75	140	150
2-Butyloctyl	3.2	10	45	105	125	155	165
2-Pentylnonyl	3.3	18	45	110	150	160	168
Dodecyl-6	3.1	5	30	78	125	160	160
Tetrapropylenyl	3.7	8	30	100	145	160	165

The first column lists the side chain R; the second indicates the critical micelle concentration (mmole/l.) of the sodium salt at 75°, and the foam height (in millimeters) is shown in the following columns for six different concentrations (mmole/l.) of the surfactant. Dodecyl-6 is

$$CH_3(CH_2)_4 \overset{\prime}{C}H(CH_2)_5CH_3$$

and tetrapropylenyl is a mixture of branched radicals $C_{12}H_{25}$; because it contains at least one quaternary carbon atom [e.g., a group $(CH_3)_3C—$], the surfactants containing tetrapropylenyl are poorly biodegradable. The importance of CMC, repeatedly mentioned in the previous section, is clear also in Table 4.4; for instance, the foam heights above 100 mm are obtained at the concentration of 1 mmole/l. only for the salts whose CMC are situated in the range 0.5 to 3.7 mmole/l.

Individual ortho- and para-isomers have been tested by Gray[81]. The foam heights (in mm), again obtained in a pour test at 60°, are shown in Table 4.5 for two concentrations (0.015 and 0.15%) of each isomer of sodium alkylbenzene sulfonate.

In Fig. 4.8 some additional data for a wider concentration range are shown[82]. The ordinate is the foam height (in cm) found in a pour test at 60°, but the abscissa indicates concentrations in mmole/l. rather than in weight percents; for the dodecyl derivatives, 0.015% is equal to 0.43 mmole/l.

It is clear from Table 4.4 that the salt with R $= C_{14}H_{29}$ has the highest foaming ability (at 60°) and from Table 4.5 that there is not much difference between the ortho- and the para-isomers. According to McAteer[83], the greatest foam stability is observed for sodium pentadecane and sodium hexadecane sulfonates. According to Yamane[84], the foaminess was almost constant between C_{10} and C_{13} but decreased from C_{13} to C_{17} as R. The discrepancies are readily accounted for by the different contaminations present in different samples, by the different methods of testing, and so on.

Table 4.5. Foam Heights of *o*- and *p*-Alkylbenzene Sulfonates, mm

Alkyl	Ortho-isomers		Para-isomers	
	0.015%	0.15%	0.015%	0.15%
n-Octyl	27	148	24	134
n-Decyl	12	125	112	186
n-Dodecyl	137	206	81	201
2-Hendecyl	10	165	86	162
2-Tridecyl	115	190	164	210
2-Heptadecyl	65	105	116	129

Tetrapropylenyl (a mixture) 86 (0.015%), 201 (0.15%).

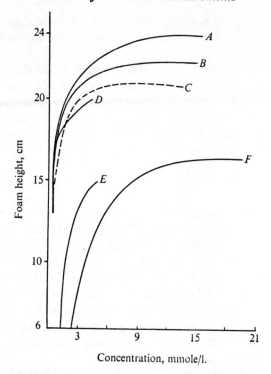

Fig. 4.8. Foam height in a pour test (cm) as a function of surfactant concentration (mM/l.). *A*: sodium *p*-(1-butyloctyl) benzene sulfonate. *B*: sodium *p*-(1-methyldodecyl) benzene sulfonate. *C*: sodium *o*-dodecylbenzene sulfonate. *D*: sodium *p*-dodecylbenzene sulfonate. *E*: sodium *o*-octylbenzene sulfonate. *F*: sodium *p*-octylbenzene sulfonate (data of reference [82]).

The foaminess of 2,5-di-*n*-alkylbenzene sulfonates, the alkyls being methyl, ethyl, butyl, hexyl, and octyl, was weaker than that of the corresponding 4-*n*-alkylbenzene sulfonates[85]. For the foaminess of α-olefin sulfonates see Yamane[84].

The effect of concentration on foam volumes is seen in Tables 4.4 and 4.5 and in Fig. 4.8. Additional observations have been reported by Saito[86], who noticed also that ferric and aluminum ions depressed the foaming ability, and by Vaughn[87], who used a simplified washing machine for his tests.

The foaminess of mixtures of sodium alkylbenzene sulfonates with phosphates and sulfates[88], with proteinaceous materials[89], and with soaps[50] also has been determined. The effect of soaps (sodium laurate, sodium oleate, sodium palmitate, or sodium stearate) on the foam volume (after shaking) of 0.1% sodium dodecylbenzene sulfonate was poorly reproducible in the absence of calcium salts; but when the ratio of sulfonate to soap to calcium chloride was 1000:20:2, the palmitate and the stearate lowered the foam volume more drastically than the laurate and the oleate[77].

An increase in temperature from 15 to 30° raised the ratio h/u in a pneumatic method for 0.1% alkylbenzene sulfonate solutions[90]; no doubt, this effect is related to the temperature coefficient of solubility (or CMC).

§73. *Ethylene oxide adducts.* By condensing a molecule RH containing a reactive hydrogen atom with a molecule of ethylene oxide, a hydroxy compound is formed:

$$RH + \underset{\diagdown O \diagup}{CH_2 - CH_2} = R \cdot CH_2CH_2OH.$$

This can react with another molecule of ethylene oxide and so on, so that finally polyethers of the general type $R(CH_2CH_2O)_nH$ are obtained. These polyethers in which RH is an alkylphenol are common non-ionic surfactants. If a greater solubility is desired, they can be transformed into anionic surfactants by sulfating the terminal hydroxyl; in this manner, compounds of the type $R(CH_2CH_2O)_nCH_2CH_2O \cdot SO_2 \cdot ONa$ are synthesized. The foaminess of many materials of these two classes has been examined, but it should be borne in mind that they were all mixtures, as compounds of different degrees of polymerization (i.e., different n) are produced simultaneously in every reaction vessel.

When the average n in the series $R(CH_2CH_2O)_nH$ was equal to 3, the foam height in a pour test at 50° for 0.5% solutions decreased from $R = C_{10}H_{21}O$ to $R = C_{16}H_{33}O$; and when n increased from 3 to 6, the foam height increased for a given alcohol[91,92]. When R was $C_8H_{17} \cdot C_6H_4O$, the octyl being a $C_5H_{11}C(CH_3)_2$— in the para-position to the hydroxyl, the foam height in a pour test at 49° was[93] about 2 cm as long as the average n was 3, 5, 7, or 8; and 18 to 22 cm when n was 10, 13, 15, 20, or 40. When R was $C_8H_{17} \cdot C_6H_4O$, $C_9H_{19} \cdot C_6H_4O$, or $C_{18}H_{35} \cdot C_6H_4O$ (the octadecenyl was derived from oleic acid), the tallest foam (again in a pour test) was observed[94] when n was 7, 8, or 9.

Usually not only the adducts but also the initial alcohols RH are mixtures. Consequently the foam height h (in a pour test) depends on the provenience of the alcohol used. Thus the h at 50° of 0.1% solutions of adducts from purified alcohols (1 mole) plus 5 moles of ethylene oxide was[95] 8 and 13 cm for, respectively, n-octanol and n-decanol, but only 1 and 10 cm for C_8 and C_{10} alcohols obtained by the "oxo" process. The following data are valid for adducts from purified alcohols. When the number n of ethylene oxide molecules per alcohol molecule varied from 5 to 30, n-decanol always gave the highest, or almost the highest h. There was not much difference between the h values for a given alcohol when n varied between 10 and 20 but, at $n = 5$ and $n = 30$, the h was markedly smaller. When, instead of the straight-chain primary alcohols, 5-ethyl-2-

nonanol (C_{11}), 2,5,8-trimethyl-4-nonanol (C_{12}), 2-methyl-7-ethyl-4-hendecanol (C_{14}), and 3,9-diethyl-6-tridecanol (C_{17}) were employed, the h was smaller; for instance, for C_{12} and $n = 5$, h was 6 or 7 cm for *n*-dodecanol and 2 cm for the branched secondary alcohol. When n was 5, the highest h was observed for C_{14}; when n was 10, 15, or 20, C_{11} was the best; and at $n = 30$, h was greatest for C_{17}. Hardness in water had but a little effect on h, but accelerated the collapse of the foam column.

The foaminess can be enhanced by substituting a cresol for the phenol, e.g., by making R = $C_8H_{17} \cdot C_6H_3(CH_3)O$, and so on[96].

Instead of an alcohol or a phenol, a fatty acid can be condensed with ethylene oxide to give poly(oxyethylene) esters

$$RCO \cdot O(CH_2CH_2O)_nH,$$

R being this time a hydrocarbon radical. The foaming ability of these compounds is small compared with that of the alcohol derivatives $RO(CH_2CH_2O)_nH$. Wrigley[97] determined the h (in a pour test) for 0.25% solutions at 60°; the n values reached 30 and the acids ranged from lauric to stearic and oleic. A comparison of compounds

$$RCO \cdot O(CH_2CH_2O)_nH \text{ and } RO(CH_2CH_2O)_nH$$

was performed by Gadaskina[98]. Additional data have been published by Popescu[99], Geipel[100], and Satkowski[101]. The low foaminess of the acetate $C_{18}H_{35}(OCH_2CH_2)_nO \cdot CO \cdot CH_3$ is mentioned in a patent[102].

Sulfates $R(OCH_2CH_2)_nO \cdot SO_2 \cdot ONa$ are powerful foaming agents. Some foam volume data obtained (apparently by a perforated plate method) at 70° for 0.015% solutions are collected in Table 4.6 from Götte[103]. It is seen that foaminess generally increases with the number n.

The lowest concentrations at which foam became noticeable were[103] for alkyl sulfates (i.e., at $n = 0$) 0.0058%, 0.0026%, 0.00065%, 0.00030%, and 0.00020% when R was *n*-dodecyl, *n*-tetradecyl, *n*-hexadecyl, *n*-octadecyl, and *n*-octadecenyl. When n was 2, the minimum concentrations usually were smaller; namely 0.0026%, 0.0011%, 0.00043%, 0.00026%, and 0.00032%. On the average, the concentrations of polyether sulfates at

Table 4.6. Foam Volumes (cm³) of Sulfates $R(OCH_2CH_2)_nOSO_3Na$

Alkyl	$n = 0$	1	2	3
Dodecyl	80	150	205	220
Tetradecyl	190	280	340	370
Hexadecyl	330	370	390	410
Octadecyl	—————————approximately 400—————			
Octadecenyl	300	520	550	550

$n = 2$ were less than a half those of alkyl sulfates at equal foam volumes. Additional data on the foaminess of ether sulfates have been published by Gohlke[104], who compared the pour test results (in a hard water) for alkyl sulfates, alkylbenzene sulfonates, and poly(ethylene oxide) sulfates. A similar comparison for 41 commercial surfactants was performed by Ramayya[241]. The foaminess of compounds $ROCH_2CH(R_1)$ $(OCH_2CH_2)_nOH$ was studied by Kuwamura[105].

Propylene oxide CH_3CH—CH_2 can be used instead of ethylene oxide

to synthesize surfactants, but the foaminess of these materials often is weaker than that of ethylene oxide derivatives[106]. Data for the h (pour test) of copolymers of ethylene oxide and propylene oxide are presented by Satkowski[101].

The h (pour test) of a solution of p-octylphenyl poly(ethylene ether) $C_8H_{17}{\cdot}C_6H_4(OCH_2CH_2)_8OH$ was almost independent of temperature between 27 and 40° (and equal to 19 cm), but at 46° it was[95] only 5 cm and at 65° only 2 cm[93]. As also in other compounds of this type, a lessening of the foaming ability occurred at the cloud point (which is independent of concentration between 0.5 and 10 % and increases from 20° to 100° when the number n of ethylene oxide units rises from 7 to 15). Thus, copious foaming is observed below the cloud point only.

The foaminess of mixtures of various soaps with non-ionic ethylene oxide adducts was studied, for instance, by Yano[107]. Addition of 0.2 mole of laurates, stearates, or ricinoleates of lithium, sodium, potassium, etc., to 1 mole of the hexadecyl ether $C_{16}H_{33}O(CH_2CH_2O)_{18}H$ had almost no effect on the foam volume (apparently in a pneumatic method) of 0.001 molar solutions[108].

§74. *Cationic surfactants.*   An early observation was that of Sabetay[109], who emphasized the ability of triethanolamine laurate to foam in fairly concentrated hydrochloric acid. For more recent studies of similar surfactants see Sturm[110] and Kusano[111]. Also esters of betaine chloride $RCO{\cdot}O{\cdot}CH_2NCl(CH_3)_3$ produce foam in acid, neutral, and alkaline solutions[112].

Cetyltrimethylammonium bromide, also known as trimethylhexadecyl ammonium bromide $C_{16}H_{33}N(CH_3)_3Br$, is a popular cationic surfactant. The mass of foam produced in its solutions (see §54) increased with concentration to a shallow maximum near 0.0003 mole/l. and decreased when the concentration was greater (up to 0.0015 M)[113].

Foaminess of alkyl-β-amines from long-chain olefins was reported upon by Ishizuka[114]. Condensates of ethylene diamine with ethylene oxide and propylene oxide of the general formula $[(C_2H_4O)_x(C_3H_6O)_yNCH_2{-}]_2$ showed dynamic foam heights h (see §53) much lower than those obtained with sodium dodecyl sulfate[115]; an adduct of the average molecular weight MW of 4750 had a greater h than those of MW 5750 or 6750. For the foaminess of solutions of N-alkyl-N'-gluconyl-ethylene diamines see Okahara[116].

Narasaki studied the foaming ability of various cationic surfactants, for instance, dimethyl dodecyl amine oxide $C_{12}H_{25}N(CH_3)_2O$, condensation products of dodecyl acrylate and 2-vinylpyridine N-oxide[117], N-alkyl pyridinium bromides, α-alkyl-N-dodecyl pyridinium bromides[118], α-methyl-β-ethyl-N-dodecyl pyridinium bromide, and N-dodecyl derivative of poly(α-vinyl-β-ethyl pyridinium bromide)[119]. Surfactants from pyridine and trimethylamine have been investigated by Izawa[120].

Cationic surfactants containing silane groups, such as

$$(C_4H_9)_3SiCH_2CH_2CH_2N(CH_3)_3Cl,$$

gave large foam volumes at 0.1 and 1 % concentrations at 30°[121].

As mentioned in §71, addition of long-chain alcohols to sodium alkyl sulfates enhances foaminess; this addition has a similar effect also on the foam stability of triethanolamine salts of fatty acids[122]. Extensive measurements of the foam volume v (apparently after air injection) on aqueous solutions of N-dodecyl pyridinium chloride plus metal soaps at 30° have been published by Suzuki[123]. For instance, 0.05 molar solution of the chloride had v of 80, 200, and 250 cm³ when the concentration of barium ricinoleate was 0.0, 0.001 mole/l., and 0.01 mole/l.; and sodium ricinoleate was almost as helpful as the barium salt. When the chloride concentration was 0.02 mole/l., its v was 110 cm³, but in the mixtures with 0.004 mole/l. of manganese laurate, manganese ricinoleate, or analogous zinc, cadmium, cobalt, and copper soaps the v was near 250 cm³.[124]

§75. Miscellaneous surfactants.   Disodium α-tocopheryl phosphate (a salt of vitamin E) is not used as a surfactant but has very similar surface properties. A 0.43% solution of its monohydrate gave films which lasted[42] for 30 to 39 sec when the film diameter was 3.5 cm, and for about 400 sec when the diameter was 1.0 cm. Schwartz[125] measured the foaming of sucrose esters in a washing machine.

The time t_e until bare liquid surface appears (see §61) was determined for solutions of sodium dioctyl sulfosuccinate

$$C_8H_{17}O_2C\cdot CH(SO_3Na)CH_2CO_2C_8H_{17};$$

the octyls presumably were branched[126]. When each octyl is 2-ethylhexyl $C_4H_9CH(C_2H_5)CH_2-$, the compound is marketed as Aerosol OT (see §60). No foam was obtained in 0.015% solutions, but t_e in 0.125% solution was about 2 hr. When agar was added (0.08 to 0.1%), t_e values of over an hour were observed for the 0.015%, and about 15 hr for the 0.125% solution. Presumably the foam stability in this instance was caused by high viscosity, as pure agar solutions do not foam appreciably. The foam of sodium dioctyl sulfosuccinate (free of agar) is readily destroyed by ordinary soaps[76].

Many other sulfo derivatives of fatty acids and esters have been used for pour tests of foaminess at 60°. Monosodium α-sulfopelargonic acid $C_7H_{15}CH(SO_3Na)CO_2H$ (in a hard water) gave a foam height h of 0.5 cm at the 0.25% concentration, while its octyl ester afforded $h = 18.5$ cm. The monosodium and the disodium salts of an α-sulfophenylstearic acid gave $h = 16$ cm and $h = 21$ cm. The corresponding salts of 9,10-dichloro-2-sulfostearic acid (from elaidic acid) had h of 21.5 and 23 cm. The two stereomeric disodium salts of 9,10-dihydroxy-2-sulfostearic acid had $h = 17$ cm (the low-melting salt) and $h = 20$ cm (the high-melting salt). It is seen that the foam heights (in the pour test!) were not greatly different from each other, although the concentration employed (i.e., 0.25%) was in some instances greater, and in the others, smaller than the CMC of the salt; this CMC was, in the above order, 1%, 0.08%, 0.05%, 0.05%, 0.056%, 0.017%, 0.15%, 0.59%, and 0.38%[127].

Solutions in distilled water (still at 60° and 0.25%) of disodium salts $R \cdot CH(SO_3Na)CH_2 \cdot OSO_3Na$ (of sulfate-sulfonic acids) gave $h = 12.5$ cm when R was $C_{15}H_{31}$, and $h = 15.5$ when R was $C_{17}H_{35}$; under identical conditions sodium dodecyl sulfate showed $h = 20.5$ cm. The disulfonates $R \cdot CH(SO_3Na) \cdot CO \cdot OCH_2CH_2SO_3Na$ had h of 13.5 and 17.5 cm when R was[128] $C_{15}H_{31}$ and $C_{17}H_{35}$.

In the series $R \cdot CH(SO_3Na)CO_2R_1$ both radicals (R and R_1) were systematically changed[129]. When R was $C_{14}H_{29}$, the h was 20.5, 19.5, 19.5, 20.5, 20.0, 19.0, and 20.5 for R_1 representing methyl, ethyl, n-propyl, isopropyl, n-butyl, isobutyl, and sec. butyl. When R was $C_{16}H_{33}$, the corresponding h values were 19.0, 18.0, 16.0, 17.5, 17.5, 15.5, and 16.0 cm. Also salts of even more complicated sulfonates gave[130] foam heights between 16.5 and 21.0 cm. "Built" solutions of sodium salts of sulfo-fatty acid esters, e.g.,

$$C_7H_{15} \cdot CH(SO_3Na)CO \cdot OC_6H_{13} \quad \text{and} \quad C_{14}H_{29} \cdot CH(SO_3Na)CO \cdot OC_3H_7,$$

afforded h values not higher than 19.5 cm[131]. It would be interesting to test several of the above compounds by other methods to see whether the insensitivity of foaminess to the composition is a true effect or is caused by the crudeness of the pour test used.

The effect of the cations was investigated, still by a pour test, in distilled water at 60°, for 0.1% solutions[132]. Monoammonium salt of α-sulfopalmitic acid $C_{14}H_{29}CH(COOH) \cdot SO_3NH_4$ had $h = 17.5$ cm, and the diammonium salt of the same acid, $h = 19$ cm. Acid and neutral triethanolammonium salts of this acid had h of 16 and 10 cm, and of α-sulfostearic acid $C_{17}H_{35} \cdot CH(COOH) \cdot SO_3H$, $h = 16$ and $h = 19$ cm. The h of the corresponding sodium salts was 17.5 and 2.5 cm (α-sulfopalmitic) and 7.5 and 1.5 cm (α-sulfostearic). The four corresponding magnesium salts showed h of 22, 21.5, 16, and 22 cm. The acid and neutral potassium salts of α-sulfopalmitic acid had an h equal to 20 and 12.5 cm; and the acid calcium salts of α-sulfopalmitic and α-sulfostearic acids had $h = 22$ cm and $h = 12.5$ cm.

Cyclohexyl α-sulfolaurate (or its sodium salt) produced an abundant and persistent foam[133]. The foaminess of sulfonates of glycidyl esters of branched carboxylic acids and several other sulfonates was praised by Kortland[134]. Jedlinski[135] determined the foaming ability of mono-glyceride sulfates $R \cdot CO \cdot O \cdot CH_2CHOHCH_2O \cdot SO_3Na$, R being an alkyl radical between C_{11} and C_{13}.

The foaming capacity of potassium salts of copolymers of dodecyl vinyl ether $C_{12}H_{25}OCH:CH_2$ and maleic anhydride or dimethyl maleate was greater, the smaller the molecular weight of the polymer[136]. The foam volume of 0.2% aqueous solution of a poly(vinyl alcohol) after shaking was one-quarter to one-half that of a comparable sodium dodecyl sulfate solution[137]. The effect of hydrolysis on the foaminess of poly(ethyl acrylate) was studied by Ito[138]. The maximum persistence of single bubbles in aqueous methyl cellulose solutions rapidly increased with the molecular weight of the sample[139]. The foaminess of methylethyl cellulose was depressed by sodium carboxymethyl cellulose[140].

§76. Saponins.   Saponins obtained from different plants or by different procedures have different foaming abilities[141,142]; for instance, the time needed for a saponin foam to reduce its volume (after shaking) by 80% was, for 0.05% solutions, 480 hr when the sample was obtained from England, and only 33 hr when a Czech sample was tested[143]. Unfortunately, many investigators used saponins "as received," so that different data often refer to different compositions.

One of the plants containing significant amounts of saponins is *Polygala senega* (Seneka snakeroot), and the main saponin extracted from it is called senegin. This was purified until further recrystallization ceased to affect its molecular weight. When nitrogen was bubbled through its solutions, no foam formed as long as the concentration was below 0.004 to 0.005 g/l.[144]. For a commercial *saponinum purum* this limiting concentration was[145] 0.0026 g/l. For senegin the ratio h/u in a pneumatic test at (apparently) a concentration of 0.1 g/l. was too high for measurement between pH 5 and 8, but amounted to 2900, 4300, and 4400 sec at, respectively, pH 3.2, 4.5, and 8.8[144].

The foaminess of crude saponin was studied by Plateau, who also measured the surface viscosity of saponin solutions.[2] The minimum concentration at which foaming became noticeable was between 0.0002 and 0.0009%; that is, smaller than for purified saponin[146]. Foam stability in a pneumatic method seemed to have a maximum value at a moderate concentration. The time of collapse of foams produced by shaking was 170 sec when the concentration was 0.5 to 1.0%, but it amounted to only 120 sec in 0.008% and 160 sec in 0.25% solution[25]. Schütz[139] noticed an increase of collapse time with concentration and believed the logarithm of the former to be a linear function of the latter. Joos[144] could not measure foam volumes for concentrations above 0.02 g/l. at some pH. The time of collapse of the "English" saponin increased from 31 to 480 hr when the concentration decreased from 1.0 to 0.05%[143]. Obviously, additional experiments are needed.

Electrolytes may lower the ratio h/u in a pneumatic test. This ratio was > 0.5 sec in a solution of 0.006 g of senegin per liter, but only 0.03 sec when 1.9 g of calcium chloride was introduced in this liquid; sodium chloride and aluminum chloride had an insignificant effect on h/u[144].

Destruction of saponin foam by ethanol was known to Plateau[2]. The ethanol concentration needed to suppress foaming of an (apparently) 0.01% solution of senegin was estimated as 25%, while it was only 15% for 1-propanol[144]. The foam volume after shaking 0.0009% saponin solution was slightly higher in 1.2% ethanol than in water, but 50 g ethanol in a liter reduced the time of collapse without significantly affecting the initial foam volume[146]. For 0.0009% saponin solutions in aqueous isoamyl alcohol, the time of collapse divided by the initial foam height was greater than for purely aqueous solutions whenever the alcohol concentration was above 0.0026 g/l., and the difference was particularly great at 0.026 g/l.[147] The time of collapse for 0.008% saponin foams was markedly shorter in an isoamyl alcohol solution of 3.3 g/l. than in water[25]; perhaps there really is a maximum of persistence at small alcohol amounts.

Addition of soap to saponin or of saponin to soap in aqueous solutions generally lowers the foam volume[148]. Some soap-saponin mixtures give no noticeable foam at all; their composition depends on the nature of the saponin and the soap used[149]. The mutual quenching of soap and saponin foams appear strange at first glance; there ought to be marked foaminess whether saponin displaces soap or soap displaces saponin. Apparently in these systems the acidity of the solution is more important than the organic contamination[150]. The pH of usual saponin solutions is near 5. As Fig. 4.6 shows, the foaminess of soaps at this acidity is practically zero. Apparently, "acid soaps," which form at pH 5 and do not foam, displace saponin and thus substitute a surface layer incapable of foaming for a saponin-rich layer. Experimentally[150], 0.1 g sodium oleate in 100 cm^3 water produced 525 cm^3 of foam on shaking; on addition of 0.1 g saponin the foam volume decreased to zero, and the pH of the liquid decreased from 8.1 to 6.9. Saponins are chemically affected by alkalis but apparently achieve a steady state. When an alkaline saponin solution was aged until its pH stabilized at 8.4, it ceased to depress the foaminess of soaps. As indicated above, senegin gives copious foams at pH as high as 8.8, and a commercial saponin foamed equally well at any pH between 6.2 and 9.6. The time t of collapse of saponin foams was more sensitive to hydroxyl ions (and to soap) than the foam volume after shaking. Thus t at pH 8.1 was nine times as great as at pH 9.6, and 0.01 g saponin added to 0.1 g sodium oleate reduced t by a factor of 50.

§77. Persistent films and bubbles.   Foam rubber, cellular plastics, and analogous industrial products are foams and possess an indefinitely long life. Their skeletons are undoubtedly solid. When the major component of a film or a bubble wall is liquid, then their persistence is limited and it may be asked at what compositions is this persistence particularly long. Plateau had great difficulties in obtaining bubbles and foams which lasted long enough to be suitable for his experiments. When he finally procured odorless and colorless glycerol, mixtures of this glycerol and water were the vehicle in which commercial soap (mainly sodium oleate) was dissolved to give stable foams[151]. Presumably glycerol fulfilled two functions in these experiments: it raised the viscosity of the films and retarded the evaporation of water from them. Both these services can be rendered also by vegetable gums (gum karaya, gum tragacanth)[43]. Dewar[152] preferred 1% aqueous glycerol as the solvent and ammonium oleate (0.1%), triethylamine oleate, tetramethylammonium oleate, and tetraethylammonium oleate as foaming agents.

In more recent times, the following recipes have been advocated.

1. Dissolve 30 g castile soap in 1 l. distilled water without heating. After 24 hr of sedimentation siphon off the clear liquid and add to it three-fourths of its volume of glycerol. Immediately before using add 0.025 to 0.1 g tannin to each 25 cm^3 of the solution[153].

2. Stir 10.5 g triethanolamine with 19.5 g oleic acid and let the mixture react for 24 hr. Dissolve it in 1 l. distilled water without heating. Let settle for 24 hr. Siphon off the lower layer and mix 4 volumes of it with 3 volumes of glycerol[153].

3. Mix 15.2 cm^3 oleic acid with 73 cm^3 of 10% triethanolamine solution, fill with distilled water to 200 cm^3, and add 164 g glycerol. Use the clear bottom layer[154].

4. Shake 2 g sodium dioctyl sulfosuccinate (marketed as "Aerosol OT") with 98 g distilled water. Dissolve, with gentle warming, 20 g sucrose and 13.5 cm^3 glycerol in 25 cm^3 of the above mixture[155].

5. A patented solute[156] consists of alkane sulfonates 42–44, borax 34–35, gelatin 17–18, tragacanth 3 parts, the rest being preservatives or dispersing agents.

§78. Plant extracts.   Saponins of §76 could have been discussed here but are so prominent in the history of foam that their separate treatment seems justified.

The foaminess of fresh green alfalfa extracts was raised by traces of nickel (as nickel sulfate), while zinc was less active and alkali metal ions had no effect[157]. Malt residue (e.g., 2 g in 80 cm^3 of an aqueous solution) proved to be an efficient foaming agent[158], but its suspensions more properly are discussed in §87. Many plant extracts showed measurable foaminess in Yankulov's tests[159].

Pine oil is used as frother in flotation. Sun[160] compared the volumes of foam produced (by a pneumatic method) on aqueous solutions of various samples of pine oil and of α- and β-terpineol (which are its main constituents) with that of hexanol. According to Desalbres[161], alcohols $C_{10}H_{18}O$ (of pine oil) containing a tertiary hydroxyl are better frothers than their isomers. He observed foaming of allegedly saturated solutions of a menthanol and a terpineol. The time of collapse of foam produced by shaking an aqueous solution of pine oil was extended by kerosene; for instance, addition of 1 g of kerosene to 11 mg of pine oil in 1 l. of water raised this time from 5.5 to 10 sec[162].

Ostwald[163] measured the foaming ability of humic acid at different acidities. Solutions of gum arabic have been studied by Pattle[164].

Tannin foams are short-lived, but tannin may enhance the stability of other foams. Thus 0.2% aqueous tannin solution had a time of collapse of only few seconds, but it raised[34] the persistence of heptanoic acid foams from 85 sec, as mentioned in §68, to 1500 sec. The utilization of small additions of tannin to enhance the lifetime of soap films is referred to in §77. However, large amounts of tannin quench soap foams, perhaps because they lower the pH of the solution. Thus 0.02 g of tannin can destroy the foaming ability of 0.1 g sodium oleate in 100 cm^3 of water.

§79. Animal fluids. According to Holden[165], foaming of milk depends on its phospholipid content, which causes the difference between the foaming abilities of summer and winter milk. The foams of aged eggs had a short life[166]. Foaming of pathological lung exudates was studied by Dubovik[167], Pattle[168], and Balagot[169]. Foaming of extracts of ground beef is not a reliable indication of changes produced by preservation[170].

Human bile does not foam at dilutions exceeding 5:1,[171] but bile acids and their salts give rise to relatively stable foams. Some liver disturbances can be recognized from an abnormally high foaminess of the urine; the effect of urine amount, dilution, and density on the pneumatic ratio h/u was ascertained[172]. The collapse of foams produced by shaking aqueous solutions of commercial sodium glycocholate and its mixtures with sodium hexanoate, sodium hendecylenate, and other soaps was described by Holwerda[173]. The persistence of the longest-lived bubbles in a foam of commercial sodium cholate and sodium tauroglycocholate has been measured, but the samples were shown to be mixtures by foam fractionation[174].

§80. Proteins. Proteins generally give rise to stable foams; in many instances, after the evaporation of water, solid skeletons remain which are indefinitely persistent. They are used, for instance, in foods. Liquid protein foams are suitable for fire-fighting. The foaminess usually depends on the acidity of the solution.

A gelatin, whose isoelectric point was between pH 4.75 and 5.00, produced[175] the most copious foam (presumably by shaking) at pH 5; the foam volume did not change greatly between pH 5.5 and 8.5 but was much smaller at pH 9 and 10. When a pneumatic method was used, three samples of photographic gelatin gave greatest foam heights near pH 6, but the foaminess of a fourth sample varied with acidity in a more complicated manner[46]. Again in a pneumatic method, 0.05% solutions of electrodialyzed gelatin foamed best near the isoelectric point (pH 4.8), but 0.5% solution of a similar sample (containing, however, also sodium chloride) had a minimum foam height at this acidity[176]. Low foaming ability of alkaline gelatin solutions was reported by Mischke[177], who also employed a pneumatic method. Sodium chloride lowered but calcium chloride raised the foam volume (after shaking) of gelatin solutions[178].

Their pneumatic foam height h was lowered by heavy cations (thorium and zirconium)[179]. Surfactants may greatly affect h, but their influence may be due, at least partly, to the concomitant changes of pH. The h of a mixture of 10 cm^3 of 0.1 M surfactant solution and 990 cm^3 of 6% gelatin solution at 35° was drastically lowered, for instance, when the surfactant was

$$C_7H_{15}PO(OK)(OC_4H_9) \text{ or } C_8H_{15}CH(COO \cdot C_4H_9)CH_2CO_2K,$$

but the foaminess was enhanced by

$$C_{12}H_{25}CH(COO \cdot CH_2CH_2OH)CH_2CO_2K[180].$$

A deaminated gelatin, whose nitrogen content was 17.38% instead of the initial 17.96% and whose isoelectric point was at pH 4, gave the most voluminous foam (after shaking) at this acidity[181].

Several proteins can be isolated from egg white. At the 0.2% concentration the foaminess of ovomucins A and B exceeded that of any other constituent protein[182]. The h (in a pneumatic method) of a 0.05% egg albumin whose isoelectric point was at pH 4.6 to 4.7 had[183] a maximum value near pH 4.7, but in strongly acid (pH about 0) and strongly alkaline (pH 12) solutions h was almost as great, and the two minima were observed near pH 2 and between pH 8 and 10. Also a 0.1% albumin solution foamed best at the isoelectric point of the protein[176]. Naturally, hydrogen ion is not the only one affecting foaminess. When the pH of a 0.05% albumin solution was kept at 10.5, the foam volume was about twice as great when this alkalinity was achieved by adding potassium hydroxide than when barium hydroxide was employed[183].

Cumper[184] determined the foaming ability of some protein solutions by two methods. The lifetime τ of large bubbles in the surface of these liquids increased with the age of the solution but usually stabilized after an hour or a few hours. This steady value of τ was, for instance, 20 sec for pepsin at pH 3.7 and 50 sec for insulin at pH 5.0, both in a buffer solution of ionic strength 0.03 and at an unknown protein concentration. The τ increased with the buffer concentration; apparently the protein was "salted-out." The Σ (see §53) depended on concentration and on pH. For instance, for egg albumin at pH 4.8, Σ was about 45 sec and 75 sec when the concentration was 0.005% and 0.02%; and at identical concentrations it was 97 sec and 135 sec for β-globulin at pH 10. Fig. 4.9 shows the effect of pH on Σ; the ionic strength of the buffer was 0.03, and the concentrations were 0.005%, 0.01%, and 0.015% for, respectively, insulin, β-globulin, and pepsin. There is no clear maximum of Σ for the globulin, but the other two proteins exhibit the usual maxima; the foaming ability of the insulin specimen is particularly striking.

Fig. 4.10 shows the dependence of h/u (almost identical with the above Σ) of 0.011% β-lactoglobulin in 0.1 N aqueous sodium chloride on the pH of the liquid[185]. A clear minimum is visible at pH 5; the isoelectric point of this protein was at pH 5.2. When the protein concentration was raised from 0.003 to 0.038%, the h/u increased 10-fold, but it was almost independent of concentration between 0.038 and 0.25%; unfortunately it is not clear whether the pH varied at the same time. The h/u of a 0.01%

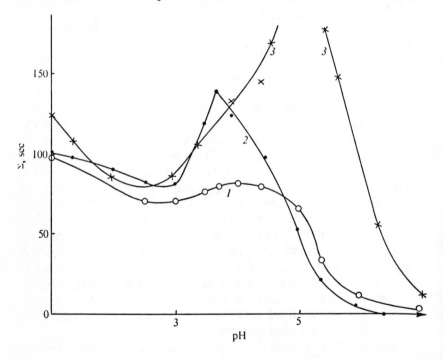

Fig. 4.9. Foam persistence (Σ sec) as a function of pH for aqueous solutions of β-globulin (0.01%, curve *1*), pepsin (0.015%, curve *2*), and insulin (0.005%, curve *3*) (data of reference [184]).

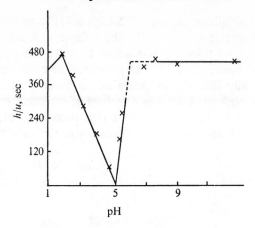

Fig. 4.10. Foam persistence (h/u sec) as a function of pH for 1.1% solution of β-lactoglobulin in 0.1 N sodium chloride (data of reference [185]).

β-lactoglobulin solution in 0.5 M sodium chloride was 70 sec, but 0.05 g or 0.2 g ethanol per liter lowered[186] it to, respectively, 25 sec and 11 sec. The h/u of 0.01% solution in 0.1 M NaCl at pH 3.6 was above 700 sec, but it was reduced to 500 sec and 130 sec when hexanoic acid was present in the concentration of 0.016 or 0.056 M and when pH became 3.2 or 3.0.

The effect of pH on the foam height of horse hemoglobin solutions is caused, partly, by the protein decomposition in acid solutions[176]. Génin[187] described a method for determining the foaminess of casein. The isoelectric point of salmine sulfate is at pH 12, but the average lifetime of single bubbles in 0.5% solution had a maximum near pH 10, and also the foam volume (pneumatic) of 0.05% solutions had the greatest value at this alkalinity[183].

Like egg albumin, the alpha soybean protein showed three regions of high, and two of low foaming ability at different acidities. The ratio v_l/m (of the liquid volume transferred into foam to the amount of solute in the initial liquid) was very high at pH 1.5 and 12.5, moderately high at pH 4 (which is near the isoelectric point, pH 4.1), and had minima near pH 3 and 9. This ratio was raised by electrolytes more the higher the valency of the cation. Thus it was, in 0.0001 molar solutions of sodium chloride, calcium chloride, and lanthanum chloride, 2.5, 13, and 24 times as high as in water in analogous circumstances[188]; but the difference between NaCl and LaCl$_3$ was small when their concentrations were 0.1 molar; presumably, lanthanum ion caused charge reversal. Other examples of foaminess enhancement by polyvalent cations (which perhaps give rise to mechanically strong surface films) also are known[189,190].

Many observations have been reported on the foaming behavior of mixtures of proteins with other organic substances. Water-in-benzene emulsions suppress albumin foams[191]. Stearic acid acts as a foam inhibitor in protein hydrolyzate solutions[192]. The volume of foam produced by shaking animal glue solutions is lowered by olive oil, oleic acid, and sodium and calcium oleates[193]. The foaminess of casein is reduced by milk fat, especially after the oxidation of the latter[194]. Foam volumes produced by a perforated plate in solutions of sodium tetrapropylenyl-benzene sulfonate $C_{12}H_{25}{\cdot}C_6H_4{\cdot}SO_3Na$ plus pepton (or blood albumin or hide glue, or potato protein) are functions of the oxygen demand of the liquid[89].

Additional observations are known on the foaming ability of gelatin[195], albumin[196], tannery proteins[197], and protein mixtures[178,46].

Other Systems

§81. Unimolecular layers.   A bubble released in an aqueous medium covered with a unimolecular film of an insoluble substance has a longer persistence τ than when the "monolayer" is absent. When τ is plotted versus surface concentration (expressed, for instance, as the number of foreign molecules on one square centimeter), curves with a sharp maximum result[198,199]. Naturally, the values of τ are very sensitive to impurities, and different samples of distilled water afforded different lifetimes for supposedly identical films[6]. However, reproducible results were obtained[6] for 0.15-cm large bubbles under ethyl oleate films; when the surface concentration was 0.6×10^{-10} mole/cm^2 (or 3.6×10^{14} molecules/cm^2, or 280 square Å per molecule), the τ was 1.2 sec, and it was 9 sec near 10^{-10} mole/cm^2 and 0.8 sec at 6×10^{-10} mole/cm^2. A complete unimolecular layer is achieved at about 7×10^{-10} mole/cm^2 so that the maximum of τ occurs when the surface film is still very dilute. The highest τ of 1-hexadecanol was 2.3 sec at 6.4×10^{-10} mole/cm^2, that is, near the complete unimolecular coverage.

Trapeznikov[200] used bubbles 0.3 cm across. To obviate the effect of irreproducible ionic impurities in water, films were spread on 0.01 N hydrochloric acid. When this liquid was covered with a "monolayer" of 1-hexadecanol, the greatest τ (3 sec) occurred at 20 square Å per molecule (or 8×10^{-10} mole/cm^2), in a reasonable agreement with the earlier data. Lauric acid (whose solubility in dilute hydrochloric acid is not negligible) showed no maximum of lifetime; its τ was near 2 sec between 60 and 160 Å^2/molecule. For myristic acid the greatest τ was 6 sec at 43 Å^2/molecule; for palmitic acid 6 sec at 24 Å^2/molecule, and for hexadecyl stearate, 3 sec between 22 and 38 Å^2/molecule. The spread of experimental values (see §46) was greater at very small coverages than beyond the maximum of τ; if the bubbles whose τ was zero are disregarded, no alteration of the curve "τ versus surface concentration" occurs at areas less than, say, 100 Å^2/molecule, but this curve is shifted to higher τ values at greater areas. The thus "corrected" curve for ethyl palmitate had a maximum τ (18 sec) at 125 Å^2/molecule. In every instance, reduction of the available area to 20 Å^2/molecule reduced τ to zero. The above values of τ were obtained at temperatures between 18 and 24°, but at 35° the maximum τ for palmitic acid became 10 sec and was situated at 35 Å^2/molecule.

Presumably, at areas greater than, say, 200 Å^2 per molecule the Marangoni effect explained in §11 is still too weak. At 180 to 100 Å^2 it is strong enough to retard the extension of the "dome," necessary for the rise of the bubble. At areas near the complete unimolecular layer or smaller, the mutual attraction of the molecules oriented on the liquid surface causes them to form islands or clusters. When the "dome" expands, these islands do not dissolve rapidly enough, and thus render the Marangoni effect inoperative. No experimental confirmation of this explanation is available however; additional experiments would be welcome.

Bubbles under unimolecular films differ from the other bubbles in that their interior surfaces have a surface tension different from that of the external surface (i.e., the roof of the dome); the former is that of the pure solvent, and the latter is that of the contaminated surface. Consequently bubbles under a "monolayer" ought to be less flat than equally large bubbles of the ordinary kind. Unfortunately the shape of the former systems has not been studied yet, as far as the present author knows.

§82. Organic liquids.   As stated in §64, pure liquids are supposed to be incapable of foaming. Several observations contradicting this view have been described. They can be accounted for either by impurities in the liquids used, see §64, or by the criticism presented in §46; when the liquid is very viscous, the bubble needs considerable time to traverse the last micron (or a few microns) of the surface layer, and this time may easily be mistaken for the lifetime of the bubble.

Gleim[201], who observed long lifetimes τ also under a pure water surface, found measurable values of τ in hydrocarbons; they increased from decane to pentadecane. Terres[202] stated that 2-methylhexadecane, 2-methyloctadecane, and 3-methylpentadecane foamed on shaking within a narrow temperature range (e.g., 11 to 16°); no explanation for this behavior is possible at present. Froth was produced by injecting air into liquid paraffin[203], but this certainly was a mixture. Medicinal paraffin oil gave[204] average lifetimes of gas in the foam (i.e., L_g of §60) between 1 and 7 minutes at 26°; the L_g of a beer was 70 min under identical conditions.

Hydrocarbon mixtures of more definite compositions also have been studied. Concentrated solutions of diphenyl in benzene and of naphthalene in gasoline showed bubble persistences as long as those under unimolecular films[205]; see §81. The time TCS of collapse after shaking single hydrocarbons was less than 1 sec, but mixtures had higher TCS values; for instance, a solution of decane in 4 parts of butylbenzene afforded a TCS of 6 sec[206].

Single bubbles in 0.04 N stearic acid solution in heptane lasted as long as 180 sec, perhaps because the solvent evaporated and a solid film of the solute remained[25].

Single bubbles of air in cyclohexane, *n*-octane, or nitrobenzene burst immediately as expected, but their persistence τ in binary mixtures of these liquids was considerable[204]. When the bubble diameter was 0.3 cm and the gas phase was the saturated vapor of the mixture, the τ had greatest values for the mixtures of 1 mole cyclohexane + 1 mole nitrobenzene (12 sec at 0°, 7 sec at 10°) and 2 moles octane + 3 moles nitrobenzene (60 sec at 21°, 3 sec at 50°). The τ of the latter liquid was poorly reproducible as long as the vapor was not saturated. If the gas phase initially contained nitrobenzene only, τ exceeded an hour; and when it initially was made up of octane (or diethyl ether), the τ was quite small (a few seconds). The authors conclude that volatile liquids destroy the

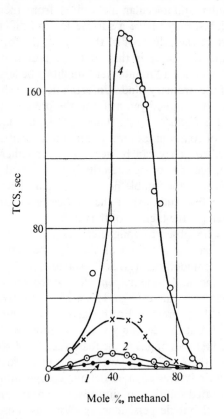

Fig. 4.11. Time of foam collapse after shaking (TCS sec) until only two bubbles remain as a function of the composition (mole % of methanol) of binary solutions of methanol with benzene (curve *1*), toluene (curve *2*), *m*-xylene (curve *3*), and mesitylene (curve *4*) at 0° (from reference [207]).

"emulsoid structure" of the surface layer. Perhaps the Marangoni effect (§11) was the real culprit. A foam film is relatively stable if, on its expansion, the underlying liquid having a greater surface tension γ is exposed. If the gas phase contains large amounts of a compound lowering the γ of the liquid phase, this compound is adsorbed on the new surface as soon as this is formed and precludes any surface elasticity.

Homogeneous mixtures of alcohols and hydrocarbons have been studied by Teitelbaum[207]. Fig. 4.11 shows that, in methanol solutions, the time TCS of collapse after shaking had maxima at methanol concentrations which increased from benzene (about 30 mole % methanol) to toluene, *m*-xylene, and mesitylene (about 50 mole % methanol), and that this TCS simultaneously rose from 3 sec for benzene to 192 sec for mesitylene. All these curves were determined at 0°. The maximum TCS was 142 sec for methanol solutions of *p*-cymene and 20 sec for ethanol solutions of mesitylene. The system ethanol plus *p*-cymene was investigated at 20°, and its greatest TCS was 6 sec. Mixtures of mesitylene and propanol afforded unstable foams, so that TCS decreased from methanol to propanol as long as the other component was mesitylene. Solutions of acetone or diethyl ether in mesitylene gave rise to no measurable foam.

Some substances which act as foaming agents in aqueous solutions may promote foaminess in organic media also. Several surfactants have been semi-quantitatively tested by King[208] and, more thoroughly, by Tarakanov[209]. The solubility of a commercial sodium alkylbenzene sulfonate (in which the side chain was a mixture of various $C_{12}H_{25}$—, $C_{13}H_{27}$—, and $C_{14}H_{29}$— radicals) in dioctyl phthalate was 0.05%. At this and lower concentrations no foam could be obtained, but suspensions of the above sulfonate in dioctyl phthalate gave rise to stable foams when moisture-free carbon dioxide was bubbled in[210]. The time t until only 40 cm^3 out of the initial 80 cm^3 of foam remained was 7 to 10 min when the sulfonate concentration was 0.2, 0.5, or 5%, but it reached 14 to 18 min when this was 1 or 3%; this is another example of maximum foaminess at a moderate concentration. When a phosphotide from sunflower oil was used as the foaming agent, the above t increased with its concentration from several seconds in 0.2% solution to 10 min in 5% solution.

0.4% solutions of various surfactants in tetrachloroethylene CCl_2: CCl_2 were stirred with a sieve, and the height of the foam made was measured 10 sec after the stopping of the movement[211]. This height was negligible in the systems with alkylpyridinium chlorides, the alkyl group being octyl to octadecenyl, with sorbitol esters

$$C_nH_{2n+1}CO \cdot O \cdot CH_2(CHOH)_4CH_2OH$$

when *n* varied from 12 to 18; with alkyl poly(ethylenoxides)

$$C_nH_{2n+1}(OCH_2CH_2)_xOH$$

when *n* was 10 to 18 and *x* was 2 to 40; and with triethanolamine soaps of fatty acids. Measurable foams were given by solutions of sodium dibutyl succinate sulfonate $C_4H_9O\cdot CO\cdot CH_2CH(SO_3Na)\cdot CO\cdot O\cdot C_4H_9$, but no boundary could be seen between the foam and the liquid layer when pentyl, hexyl, heptyl, or octyl radicals were substituted for the butyl. Higher foam columns were obtained for ethanolamine alkylbenzene sulfonates $R\cdot C_6H_4\cdot SO_3\cdot NH_3CH_2CH_2OH$, especially when R was decyl or dodecyl, while there was almost no foam for R = pentyl or R = hexadecyl.

Qualitative observations on foams produced by shaking solutions of copper oleate, magnesium stearate, cadmium palmitate, zinc valerate, sodium dodecyl sulfate, and many other surfactants in benzene and ethanol are described by Manegold[212], who emphasizes the effects of the age of the solution and of the traces of water in it.

1-Amino-octane gave stable foams in some two-phase mixtures of water and *p*-xylene[213].

In ternary solutions of ethylbenzene, polystyrene, and a surfactant, the above time *t* of half-collapse usually was greater, the greater the concentration of polystyrene, that is, also the greater the viscosity of the liquid[214]. This is illustrated in Table 4.7. Its first column lists the concentration (%) of a sodium alkane sulfonate in the mixture, and *t* is shown in the other columns for two solvents: ethylbenzene (viscosity at 25° equal to 0.0064 g/cm·sec) and ethylbenzene +15% polystyrene (viscosity 2.2 g/cm·sec). Data for other compositions and for solutions of an alkylbenzene sulfonate and a sunflower phosphatide will be found in the original paper.

Unusually stable foams have been obtained with 2% solutions of sodium *bis*(2-ethylhexyl) sulfosuccinate in triethanolamine; the L_g of §60 of these mixtures at 100° was 89 minutes[204]. A liquid foaming at 200° was prepared[215] by dissolving 2 g of a commercial soap in 1 l. of paraffin oil.

Table 4.7. Time of Half-Collapse (sec) of Organic Foams

Surfactant concentration (%)	Ethylbenzene	Ethylbenzene + polystyrene
0.01	0	15
0.1	5	200
0.2	30	210
1.0	60	300
3.0	60	300

In mixtures of two siloxane liquids, two maxima of foaminess have been observed by Prigorodov[216].

Foaming which occurs in the distillation of crude glycerol can be eliminated by treatment with sulfuric acid[217].

Every natural fat and oil is a mixture and, consequently, is "permitted" to foam by the theory. Nevertheless, some of them seem to behave like pure compounds in that their foaminess is weaker than that of their mixtures[218,219]. The presence of an "active" oil (e.g., linseed oil) in "inactive" oils (olive, peanut) can be detected from the enhanced foaming ability of the sample[220]. Foaming of butter at high temperatures is attributed to the presence of phosphatides[221]. Cottonseed and some other oils give more foam with steam at 190° the longer they were kept at this temperature, that is, the more extensive their chemical degradation[222].

Hydrogenation raised the foaminess of the methyl esters of several higher unsaturated fatty acids[223]. The foaming ability of hot soybean hydrolyzates depended on the kind of hydrolysis (chemical or biological), on the addition of caramel, changes in pH, and so on[224]. Long heating (e.g., 10 hr at 200°) raises the foaminess of soybean oil; simultaneously it raises the concentration of conjugated dienes and trienes in the oil[225], as well as the hydroxyl number, acid number, epoxy content, and carbonyl content of the latter[226]. Kajimoto[227] noticed a correlation between the foaming ability and the peroxide content of hot cooking oils.

§83. The gas phase. As stated in §1, foams are two-phase systems; in principle, the composition of the gas is as important as that of the liquid. In reality, gases seem to affect the foaminess mainly by altering the chemical composition of the liquid, especially of the foaming agent in it. No more definite statement is possible at present because the experimental material is very meager.

Peper[77] found it advisable to saturate the aqueous solutions (of soap-sulfonate mixtures) with air free of carbon dioxide to prevent changes in pH during shaking. When carbon dioxide rather than air was bubbled through 0.01 % albumin solutions, the ratio h/u of foam height to gas velocity increased from 32 to 36 sec to 190 to 240 sec, and visible coagulation of the protein simultaneously occurred[196]. Unfortunately, the change in acidity caused by carbon dioxide is not known. It is difficult to account for the observation[195] that the foaminess of gelatin solutions increases when the gas is changed from air to nitrogen to carbon dioxide to oxygen.

Clearer results have been obtained by Ostwald[228]. When nitrogen is injected in 0.2 % calcium saccharate solution, voluminous foam develops. When carbon dioxide is injected instead, the gas is absorbed by the alkaline solution and no foaming takes place. When introduction of CO_2 continues past the saturation point, foam appears again and its height finally exceeds that achieved with nitrogen; this happens presumably because the suspended calcium carbonate enhances foaming ability; see Chapter 5.

If in a single layer of foam bubbles there are n_1 bubbles of diameter l_1, n_2 bubbles of diameter l_2, and so on, then the sum $n_1 l_1 + n_2 l_2 + \ldots$ is a measure of foam dispersity. When the liquid was an aqueous solution of "aluminum protein hydrolyzate" and the gas was air, this sum was, for instance, 300 (arbitrary units) and decreased to 100 in about 2500 sec. When so much oxygen or carbon dioxide was injected that the above sum again was 300, this decreased to 100 in only 2000 or only 500 sec. The decrease to 50 required 5500 sec in oxygen and only 1000 sec in carbon dioxide[229]. The change in pH caused by CO_2 was not looked into.

The effect of the composition of the gas phase on the foaminess of organic binary mixtures is mentioned (and tentatively explained) in §82.

§84. Temperature effects. It is easy to see that temperature changes must affect the foaming ability of a solution, but the experimental data are not extensive and, apparently, in no instance was the presumed cause of the foaminess dependence on temperature T confirmed by deliberate tests. Probably the most important mechanism of the alteration of foaminess by T is through the solubility. As T greatly affects ordinary solubility and also CMC (see §70), it must affect foaminess also. This effect is visible in the maxima of foaminess reported in §66, etc., and in Table 4.4. Another important effect also referred to above is that of evaporation; at higher temperatures vaporization is more rapid, so that an increase in T should act analogously to an air current over the surface of the liquid. Because the incorporation of air during agitation depends on the viscosity η of the liquid (see §51), it must depend also on T since a temperature rise almost always means decrease of viscosity. The rate of drainage also is a function of η and consequently of T.

Of course, the above general considerations are insufficient to account for experimental results. Thus the time of collapse of a foam achieved by shaking 0.05 M aqueous isoamyl alcohol was[22] at 50° by 32% shorter than at 18°; was this an effect of the rate of drainage, of solubility, or of something else?

The foam volumes after shaking 0.25% solutions of potassium laurate, potassium myristate, or potassium oleate were identical at 30 and 70°[50]. On the other hand, comparison of Figs. 4.4 and 4.5 shows that the foam heights for potassium laurate and potassium myristate were markedly greater at 35° than at 15°. It is mentioned in §70 that the effect of temperature on foaminess is supposed to depend on crystal transformations in soaps.

The temperature dependence of the foam volume of alkyl sulfate solutions is seen in Fig. 4.7. The increase of the foam volume and the ratio h/u with temperature for sulfates and sulfonates is mentioned in §§71 and 72. The foam height (in a pour test) of non-ionic ethylene oxide adducts was[91] at 50° by about 25% greater than at 20°. The cloud point of aqueous solutions of adducts from ethylene oxide and p-octylphenol (the octyl being branched) is independent of concentration, but increases with the number n of ether linkages in the molecule, and considerable foam height h (again in a pour test) was observed only at temperatures below this point[93] (see §73). Thus, at $n = 8$, h was near 19 cm between 27 and 40°, but dropped to 5 cm at 46° and 2 cm at 65°; the cloud point of this sample was 40°. According to Wemelle[67], in a pour test a temperature increase from 30 to 60° altered not only the absolute values of h but also the order of the foaming agents according to efficiency.

The foam height after shaking 1% gelatin solution (which contained peptones) was[178] 30.0, 25.5, 20.0, 15.0, 13.0, 11.0, 5.5, and 3.5 (arbitrary units) when T was, respectively, 20°, 40°, 50°, 60°, 70°, 80°, 90°, and 100°; this trend is likely to be caused by viscosity decrease. The temperature coefficient of foaminess was determined also for other protein solutions[230], milk[231-233], honey[234], eggs[235], beer[236], sulfite liquor[237], and so on.

Particularly complicated systems have been studied by Mokrushin and Zhidkova[158]. Not only the foam formation from bubbles but the gas release itself depended on temperature. Aqueous suspensions of calcium carbonate and malt residue (finely ground and acting as foaming agent), containing also calcium chloride to lower the melting point of the liquid, were poured into sulfuric acid (e.g., 66%). At 18°, 375 cm^3 of foam formed (although the volume of carbon dioxide which could have been liberated in the reaction $H_2SO_4 + CaCO_3 = CO_2 + CaSO_4 + H_2O$ was near 800 cm^3), and at $-17°$ still 350 cm^3 was obtained. One-half of the foam disappeared in about 40 min at 18° but over 3 hr was needed for this collapse at $-17°$. At $-25°$ and $-35°$ the rate of the above reaction was so slow that almost no foam was obtained without stirring. With stirring, some foam formed; its volume was greater at $-10°$ than at $-25°$. Also the rate of collapse at the very low temperatures was small. Strictly speaking, these observations belong to Chapter 5 because all liquids used contained solid particles, namely the unreacted calcium carbonate and the insoluble malt powder.

The effect of T on η presumably accounts for the behavior of some nonaqueous foams. The value of L_g (see §60) of hydrocarbon oils decreased[204] when T increased, for instance, from 32 min at 26° to 1.1 min at 100°. According to Kichkin[238], temperature increase raises the "foaming tendency" but lowers the "foam stability" of lubricating oils. The foaminess of purified hydrocarbon oils was weaker at 150° than at 24° but oils containing detergents gave more foam at the higher temperature[239]. The temperature dependence of the bubble persistence τ in organic liquids is referred to in §82. The foam volume and the foam stability of some silicone fluids decreased when the testing temperature rose from $-60°$ to $+250°$ and foam stability was particularly great when the air injected was warmer (e.g., 60°) than the liquid (e.g., 10°)[240].

Fig. 4.12 illustrates the effect of temperature (plotted along the

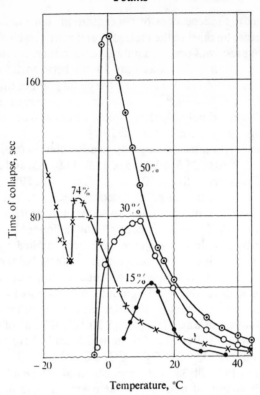

Fig. 4.12. Time of foam collapse after shaking (TCS sec) as a function of temperature (°C). Mixtures of methanol and mesitylene; the numbers at the curves mean mole % of methanol (from reference [207]).

abscissa) on the time of collapse after shaking (TCS) shown along the ordinate[207]. The percentages at the curves mean the mole % of methanol in its mixtures with mesitylene. No doubt, the initial volume of foam produced by shaking varied with T because η was affected by T (see §51); it is not known how this effect influenced the experimental data. Another effect which certainly was present was related to solubility. The 30 and the 50% solutions form two layers a few degrees below zero, that is, near the temperature of zero foaming. The 74 mole % system is heterogeneous below $-12°$ that is, near the minimum of TCS visible in Fig. 4.12.

References

1. Bikerman, J. J. *Foams.* New York: Reinhold. (1953), p. 50.
2. Plateau, J. *Mém. Acad. Roy. Sci. Belg.* 37:52 (1869), 8th ser.
3. Foulk, C. W., and J. E. Barkley. *Ind. Eng. Chem.* 35:1013 (1943).
4. Teitelbaum, B. Ya., and E. E. Sidorova. *Kolloidn. Zh.* 14:372 (1952).

5. Andrew, S. P. S. *Proc. Intern. Symp. Distn., Brighton, Engl., 1960.* 73. *Chem. Abstr.* 56:4558 (1962).
6. Talmud, D., and S. Sukhovolskaya. *Zh. Fiz. Khim.* 2:31 (1931). *Z. Physik Chem.* 154A: 277 (1931).
7. Gleim, V. G. *Zh. Prikl. Khim.* 26:1157 (1953).
8. Gleim, V. G., I. K. Shelomov, and B. R. Shidlovskii. *Zh. Prikl. Khim.* 32: 1046 (1959).
9. Ravinskii, M. B. *Kolloidn. Zh.* 24:58 (1962).
10. Bikerman, J. J. *J. Appl. Chem. (London)* 18:266 (1968).
11. Nakagaki, M. *J. Phys. Chem.* 61:1206 (1957).
12. Ramazanova, E. E., and I. Z. Makinskii. *Chem. Abstr.* 54:19045 (1960).
13. Leaf, W., et al. *Am. Ry. Eng. Assoc.* 45. Bull. 441:58 (1943).
14. Hancock, J. S. *J. Soc. Chem. Ind.* 49:369T (1930).
15. Leaf, W., et al. *Am. Ry. Eng. Assoc.* 46. Bull. 448:59 (1944).
16. Foulk, C. W., and J. W. Ryznar. *Ind. Eng. Chem.* 31:722 (1939).
17. Durov, S. A. *Zh. Prikl. Khim.* 13:693 (1940).
18. Durov, S. A. *Physico-chemical Basis of Priming in Boiler Water.* [Russian] Moscow. (1948).
19. Schütz, F. *Trans. Faraday Soc.* 38:94 (1942).
20. Nakagaki, M. *Bull. Chem. Soc. Japan* 22:21 (1949).
21. Gibb, W., and K. Htin. *J. Roy. Tech. Coll. (Glasgow)* 5:128 (1950).
22. Bartsch, O. *Kolloidchem. Beihefte* 20:1 (1924).
23. Sasaki, T. *Bull. Chem. Soc. Japan* 14:250 (1939).
24. Sasaki, T. *Bull. Chem. Soc. Japan* 13:517 (1938).
25. Rebinder, P., and E. Venstrem. *Kolloid-Z.* 53:145 (1930).
26. Lubman, N. M. *Tsvetn. Metal.* 6:854 (1931). *Zh. Fiz. Khim.* 3:204 (1932).
27. Nakagaki, M., and K. Shinoda. *Bull. Chem. Soc. Japan* 27:367 (1954).
28. Bikerman, J. J. *Trans. Faraday Soc.* 34:634 (1938).
29. Ross, S. *J. Phys. Chem.* 50:391 (1946).
30. Nakagaki, M. *Bull. Chem. Soc. Japan* 21:30 (1948).
31. Aleinikov, N. A. *Tsvetn. Metal.* 6:1546 (1931). *Kolloidchem. Beihefte* 36:82 (1932).
32. Haefele, C. R. *Paper Trade J.* 126, No. 15:66 (1948).
33. Sasaki, T. *Bull. Chem. Soc. Japan* 13:669 (1938).
34. Talmud, D. L. *Zh. Fiz. Khim.* 5:1062 (1934).
35. Sheludko, A. *Compt. Rend. Acad. Bulgare Sci.* 9 No. 1:11 (1956).
36. Sheludko, A., and D. Ekserova. *Izv. Khim. Inst. Bulgar. Akad. Nauk.* 1959:105.
37. Shah, D. O., and C. A. Dysleski. *J. Am. Oil Chem. Soc.* 46:645 (1969).
38. Topper, L., and E. Baer. *J. Colloid Sci.* 9:495 (1964).
39. Taggart, A. F., and A. M. Gaudin. *Trans. Am. Inst. Mining Met. Engr.* 68:479 (1923).
40. Shinoda, K., T. Yamanaka, and K. Kinoshita. *J. Phys. Chem.* 63:648 (1959).
41. Hahn, W. E. *Lodz. Towarz. Nauk., Wydzial III, Acta Chim.* 3:9 (1958). *Chem. Abstr.* 53:13121 (1959).
42. Bikerman, J. J. *J. Phys. Chem.* 56:164 (1952).

43. Merrill, R. C., and F. T. Moffett. *Oil Soap* 21:170 (1941).
44. Ruyssen, R., and P. Joos. *Mededel. Koninck. Vlaam. Acad. Wetenschap. Belg.* 25, No. 3 (1963).
45. Burcik, E. J. *J. Colloid Sci.* 5:421 (1950).
46. Pankhurst, K. G. A. *Trans. Faraday Soc.* 37:496 (1941).
47. Johlin J. M. *J. Biol. Chem.* 84:543 (1929).
48. Miles, G. D., and J. Ross. *J. Phys. Chem.* 48:280 (1944).
49. Götte, E. *Melliand Textilber.* 29:65, 105 (1958).
50. Reutenauer, G., and P. Sicard. *Bull. Mens. ITERG* 4:99 (1950). *Chem. Abstr.* 44:5621 (1950). G. Reutenauer, M. Frelat, and P. Sicard, *Bull. Mens. ITERG* 3:448 (1949). *Chem. Abstr.* 44:2772 (1950).
51. Otero Aenlle, E., and R. Cadorniga Carro. *Congr. Luso-Espan. Farm.*, *2nd Congress* 2: 1ª sect.: 294 (1952). *Chem. Abstr.* 48:14256 (1954).
52. Ault, W. C., et al. *J. Am. Oil Chem. Soc.* 42:233 (1965).
53. Rozhdestvenskii, D. A., M. N. Zaliopo, and G. A. Borodina. *Maslob.-Zhir. Prom.* 25, No. 9:24 (1959). *Chem. Abstr.* 54:8115 (1960).
54. Tschakert, E. *Seifen-Oele-Fette-Wachse* 92:853 (1966).
55. Kohler, S. *Tvaett Ind.* 1969:304. *Chem. Abstr.* 72, No. 91669 (1970).
56. Shkodin, A. M., and G. P. Tikhomirova. *Kolloidn. Zh.* 14:279 (1952).
57. Mysona, M., and K. Rylko. *Przemysl Chem.* 6(29):254 (1950). *Chem. Abstr.* 45:10623 (1951).
58. Morrisroe, J. J., and R. G. Newall. *Ind. Eng. Chem.* 41:423 (1949).
59. Dedrick, C. H., and J. H. Wills. *Ind. Eng. Chem.* 41:575 (1949).
60. Lederer, E. *Angew. Chem.* 47:119 (1934).
61. Fiero, G. W. *J. Am. Pharm. Assoc.* 27:658 (1938).
62. Bolle, J. *Mém. Serv. Chim. État (Paris)* 38, No. 2:171 (1953). *Chem. Abstr.* 49:7874 (1955).
63. Yurzhenko, A. I., and G. F. Storozh. *Nauk. Zap. L'vivsk. Derzh. Univ., Khim. Zb.* 46, No. 5:48 (1958). *Chem. Abstr.* 55:1038 (1961).
64. Götte, E. *Kolloid-Z.* 64:327 (1933).
65. Dreger, E. E., et al. *Ind. Eng. Chem.* 36:610 (1944).
66. Kölbel, H., D. Klamann, and P. Kurzendörfer. *World Congr. Surface Active Agents (3rd)*, Cologne 1960, 1:1 (1962).
67. Wemelle, R. *Ind. chim.* 42:173 (1955). *Chem. Abstr.* 50:588 (1956).
68. Wemelle, R., and J. Bolle. *Congr. Mondial Détergence Prod. Tensio-Actifs*, 1ᵉʳ Congress, Paris, 1954, 1:93. *Chem. Abstr.* 51:10929 (1957).
69. Kashiwagi, M. *Bull. Chem. Soc. Japan* 30:193 (1957).
70. Kashiwagi, M. *Bull. Soc. Chim. Japan* 30:572 (1957).
71. Bespyatov, M. P., and Zh. Ya. Leshchenko. *Maslob.-Zhir. Prom.* 29, No. 9: 19 (1963). *Chem. Abstr.* 60:9487 (1964).
72. Kondo, T., K. Meguro, and S. Sukigara. *Yukagaku* 9:63 (1960). *Chem. Abstr.* 54:21797 (1960).
73. Weil, J. K., A. J. Stirton, and E. B. Leardi. *J. Am. Oil Chem. Soc.* 44:522 (1967).
74. Gara, M., and G. Szatlmayer. *Kolor. Ert.* 12:53 (1970). *Chem. Abstr.* 73, No. 67894 (1970).
75. Stirton, A. J., et al. *J. Am. Oil Chem. Soc.* 44:99 (1967).
76. Winsor, P. A. *Nature* 157:660 (1946).

77. Peper, H. *J. Colloid Sci.* 13:199 (1958).
78. Asinger, F., F. Ebeneder, and G. Richter. *J. Prakt. Chem.* [4] 2:203 (1955).
79. Asinger, F., H. Eckoldt, and G. Richter. *J. Prakt. Chem.* [4] 2:233 (1955).
80. Griess, W. *Fette, Seif. Anstrichmittel* 57:24, 168, 236 (1957).
81. Gray, F. W., et al. *J. Am. Oil Chem. Soc.* 42:998 (1965).
82. Gray, F. W., J. F. Gerecht, and I. J. Krems, *J. Org. Chem.* 20:511 (1955).
83. McAteer, J. H., and L. M. Kinnard. *Chem. Phys. Appl. Surface Active Subst., Proc. Int. Congr., 4th, 1964,* 1:127 (1967).
84. Yamane, I., et al. *Kogyo Kagaku Zasshi* 73:723 (1970); *Chem. Abstr.* 73, No. 78815 (1970).
85. Kölbel, H., D. Klamann, and E. Wagner. *Vorträge Originalfassung Intern. Kongr. Oberflächenakt.* 1:27 (1960/61).
86. Saito, T., et al. *Kogyo Kagaku Zasshi* 67:464 (1964). *Chem. Abstr.* 61: 3309 (1964).
87. Vaughn, T. H., H. R. Suter, and M. G. Kramer. *J. Am. Oil Chem. Soc.* 30:1 (1953).
88. Kooijman, P. L., H. W. Huÿser, and J. J. Tjepkema. *Congr. Mondial Détergence Prod. Tensio-Actifs, Paris, 1954,* 1:93. *Chem. Abstr.* 51:10929 (1957).
89. Götte, E. *Fette, Seif., Anstrichmittel* 66:528 (1964).
90. Patterson, S. J., C. C. Scott, and K. B. E. Tucker. *J. Am. Oil Chem. Soc.* 44:407 (1967); 45:528 (1968).
91. Baldacci, R. *Ann. Chim.* (*Rome*) 40:358, 372 (1950).
92. Rossi, C., and R. Baldacci. *Ann. Chim.* (*Rome*) 41:534 (1951).
93. Fineman, M. N., G. L. Brown, and R. J. Myers. *J. Phys. Chem.* 56:963 (1952).
94. Tagawa, T., T. Sonoda, and N. Ohba. *Kogyo Kagaku Zasshi* 65:1844 (1962). *Chem. Abstr.* 58:10409 (1963).
95. Satkowski, W. B., S. K. Huang, and R. L. Liss. *Nonionic Surfactants.* New York: Dekker. (1967), p. 108.
96. Popescu, F. S., and A. G. Gobjila. *Izv. Vysshikh Uchebn. Zavedenii, Neft i Gaz* 6, No. 3:55 (1963). *Chem. Abstr.* 59:11155 (1963).
97. Wrigley, A. N., F. D. Smith, and A. J. Stirton. *J. Am. Oil Chem. Soc.* 34:39 (1957).
98. Gadaskina, N. D., D. M. Rudkovskii, and E. Ya. Radzemskaya. *Zh. Prikl. Khim.* 30:148 (1957).
99. Popescu, F., and T. Basarabescu. *Petrol Gaze* (*Bucharest*) 16:344 (1965). *Chem. Abstr.* 63:15111 (1965).
100. Geipel, G., F. Wolf, and K. Loeffler. *Tenside* 5, No. 5–6:132 (1968).
101. Satkowski, W. B., S. K. Huang, and R. L. Liss. *Nonionic Surfactants.* New York: Dekker. (1967), p. 164.
102. Roehm & Haas G.m.b.H., German 1 243 312; *Chem. Abstr.* 67, No. 65784 (1967).
103. Götte, E. *World Congr. Surface Active Agents, 3rd Congress, Cologne, 1960,* 1:45 (1962).
104. Gohlke, F. J., and H. Bergerhausen. *Seifen-Oele-Fette-Wachse* 93:519 (1967). *Soap Chem. Spec.* 44, No. 3:60 (1968).

105. Kuwamura, T., and E. Kameyama. *Kogyo Kagaku Zasshi* 65:1265 (1962). *Chem. Abstr.* 58:1648 (1963).
106. Matsuda, M., et al. *Yukagaku* 18:80 (1969). *Chem. Abstr.* 70 No. 88316 (1969).
107. Yano, W., T. Takeda, and W. Kimura. *Yukagaku* 11:304 (1962). *Chem. Abstr.* 58:4743 (1963).
108. Suzuki, H. *Yukagaku* 17:566 (1968).
109. Sabetay, S. *Bull. Soc. Chim. France* [5] 3:1918 (1936).
110. Sturm, H., et al. *Ind. Eng. Chem.* 45:186 (1953).
111. Kusano, T., and J. Mikumo. *Kogyo Kagaku Zasshi* 59:458 (1956). *Chem. Abstr.* 52:3371 (1958).
112. Dean, R. S., and A. B. Hersberger. *Trans. Am. Inst. Mining Met. Engr.* 134:81 (1939).
113. Nash, T. *J. Appl. Chem. (London)* 8:440 (1958).
114. Ishizuka, T., et al. *Chim. Phys. Appl. Prat. Ag. Surface, C.R. Congr. Int. Déterg., 5th, 1968,* 3:169 (1969); *Chem. Abstr.* 74, No. 23818 (1971).
115. Schmolka, I. R., and R. K. Seizinger. *J. Am. Oil Chem. Soc.* 45:715 (1968).
116. Okahara, M., K. Nishino, and S. Komoni. *Yukagaku* 15:155 (1966). *Chem. Abstr.* 64, No. 20004 (1966).
117. Narasaki, H., and Y. Kurihara. *Tokyo Kogyo Shikensho Hokoku* 60, No. 2, 65 (1965). *Chem. Abstr.* 63:803 (1965).
118. Narasaki, H., and H. Suzuki. *Kogyo Kagaku Zasshi* 64:1406, 1441 (1961). *Chem. Abstr.* 57:4077 (1962).
119. Narasaki, H., and N. Ito. *Kogyo Kagaku Zasshi* 64:1790 (1961). *Chem. Abstr.* 57:3580 (1962).
120. Izawa, Y., T. Fujita, and Y. Ogata. *Kogyo Kagaku Zasshi* 71:943 (1968). *Chem. Abstr.* 69, No. 68486 (1968).
121. Maki, H., et al. *Yukagaku* 19:1029 (1970).
122. Sanders, P. A. *J. Soc. Cosmet. Chem.* 17:801 (1966). *Chem. Abstr.* 66, No. 49214 (1967).
123. Suzuki, H. *Yukagaku* 16:667 (1967).
124. Suzuki, H. *Yukagaku* 19:328 (1970).
125. Schwartz, A. M., and C. A. Rader. *J. Am. Oil Chem. Soc.* 42:800 (1965).
126. Savitskaya, E. M., and P. A. Rebinder. *Kolloidn. Zh.* 13:200 (1951).
127. Weil, J. K., et al. *J. Am. Oil Chem. Soc.* 37:679 (1960).
128. Stirton, A. J., F. D. Smith, and J. K. Weil. *J. Am. Oil Chem. Soc.* 42:114 (1965).
129. Stirton, A. J., et al. *J. Am. Oil Chem. Soc.* 42:1078 (1965).
130. Smith, F. D., A. J. Stirton, and M. V. Nuñez-Ponzoa. *J. Am. Oil Chem. Soc.* 44:413 (1967).
131. Bistline, R. G., and A. J. Stirton. *J. Am. Oil Chem. Soc.* 45:78 (1968).
132. Weil, J. K., R. G. Bistline, and A. J. Stirton. *J. Am. Oil Chem. Soc.* 34:100 (1957).
133. Kale, W. M. *Text. Dyer Printer* 1:60 (1968). *Chem. Abstr.* 73, No. 26908 (1970).
134. Kortland, C., and C. Borstlap. *Tenside* 5, No. 2:43 (1968). *Chem. Abstr.* 69, No. 44731 (1968).

135. Jedlinski, Z. *World Congr. Surface Active Agents, 3rd Congress, Cologne, 1960*, 1:51 (1962).
136. Narasaki, H. *Tokyo Kogyo Shikensho Hokoku* 59:241 (1964). *Chem. Abstr.* 62:2919 (1965).
137. Konishi, H., K. Takehara, and T. Ishizuka. *World Congr. Surface Active Agents, 3rd Congress, Cologne, 1960*, **1**:66 (1962).
138. Ito, H., Y. Ono, and S. Suzuki. *Kogyo Kagaku Zasshi* 60:1054 (1957). *Chem. Abstr.* 53:10829 (1959).
139. Schütz, F. *Trans. Faraday Soc.* 38:85 (1942).
140. Morrison, R. I., and B. Campbell. *J. Soc. Chem. Ind.* 68:333 (1949).
141. Ruyssen, R. G., and E. O. K. Verstraete. *Mededel. Koninck. Vlaam. Acad. Wetenschap. Belg., Kl. Wetenschap.* 4(6) (1942). *Chem. Abstr.* 38:3889 (1944).
142. Zotova, K. V., and A. A. Trapeznikov. *Kolloidn. Zh.* 26:190 (1964).
143. Trapeznikov, A. A., K. V. Zotova, and N. V. Shamrova. *Kolloidn. Zh.* 32:437 (1970).
144. Joos, P., E. Moerman, and R. Ruyssen, *C.r. Congr. Sci. Farm., Conf. Comun. 21ᵉ, Pisa, 1961*, 615 (1962).
145. Lauwers, A., and R. Ruyssen. *Conf. Orig., 3ᵉ Congr. Int. Détergence, Cologne, 1960*, 3:146 (1962).
146. Sasaki, T. *Bull. Chem. Soc. Japan* 15:449 (1940).
147. Mitsui, S., and T. Sasaki. *Bull. Chem. Soc. Japan* 18:259 (1943). *Chem. Abstr.* 41:4352 (1947).
148. Kind, W., and F. H. Zschacke. *Z. Deut. Oel-u. Fett-Ind.* 43:520 (1923).
149. Tyutyunikov, B., N. Kasyanova, and N. Gvirtsman. *Maslob.-Zhir. Delo* 1930, No. 7–8: 48. *Chem. Abstr.* 25:2317 (1931).
150. Shkodin, A. M., and G. P. Tikhomirova. *Kolloidn. Zh.* 13:134 (1951).
151. Plateau, J. *Mém. Acad. Roy. Sci. Belg.* 36 (1867); 7th ser.
152. Dewar, J. *Proc. Roy. Inst. Gt. Brit.* 24:197 (1925).
153. Cook, G. A. *J. Chem. Educ.* 15:161 (1938).
154. Bragg, L., and J. F. Nye. *Proc. Roy. Soc. (London)* 190A:474 (1947).
155. Kuehner, A. L. *J. Chem. Educ.* 25:211 (1948).
156. Aktiebolaget Alfred Ullman & Co., German pat. 1 049 352 (1959); *Chem. Abstr.* 55:2150 (1961).
157. Harris, P. J., and F. Sebba. *Nature* 208:869 (1965).
158. Mokrushin, S. G., and L. G. Zhidkova. *Kolloidn. Zh.* 21:336 (1959).
159. Yankulov, I. *Herba Hung.* 5:137 (1966). *Chem. Abstr.* 68: No. 41322 (1968).
160. Sun, S.-C. *Mining Eng.* 4:65 (1952).
161. Desalbres, L. *Rev. Ind. Minérale* 38:379 (1956). *Chem. Abstr.* 50:16595 (1956).
162. Aleinikov, N. A. *Zh. Prikl. Khim.* 22:812 (1949).
163. Ostwald, Wo., and A. Steiner. *Kolloid-Z.* 36:342 (1925).
164. Pattle, R. E. *J. Soc. Chem. Ind.* 69:369 (1950).
165. Holden, T. F., et al. *J. Dairy Sci.* 49:346 (1966). *Chem. Abstr.* 64:18307 (1966).
166. Czaika, J., and T. Radomański. *Roczniki Panstwowego Zakladu Hig.* 6:33 (1955). *Chem. Abstr.* 49:10542 (1955).

167. Dubovik, V. I., and V. A. Bandarin. *Sb. Nauchn. Rabot, Minskii Gos. Med. Inst.* 13: 268 (1953). *Chem. Abstr.* 50:4377 (1956).

168. Pattle, R. E., *J. Pathol. Bacteriol.* 72:203 (1956). *Chem. Abstr.* 51:3038 (1957).

169. Balagot, R. C., R. M. Reyes, and M. S. Sadove. *J. Am. Med. Assoc.* 163:630 (1957). *Chem. Abstr.* 51:6844 (1957).

170. Barresi, E. *Boll. Soc. Ital. Biol. Sper.* 30:1145 (1954). *Chem. Abstr.* 49: 12744 (1955).

171. Otero Aenlle, E., and S. Garcia Fernandez. *Anales Real Soc. Españ. Fis. Quim. (Madrid)* 48B:338 (1952). *Chem. Abstr.* 49:1839 (1955).

172. Bikerman, J. J. *Instrument Soc. America*, Paper No. 54-13-4 (1954).

173. Holwerda, K. *Biochem. Z.* 295:11 (1937).

174. Bader, R., and F. Schütz. *Trans. Faraday Soc.* 42:571 (1946).

175. Bogue, R. H. *J. Am. Chem. Soc.* 44:1343 (1922).

176. Peters, D. *Kolloid-Z.* 125:157 (1952).

177. Mischke, W. *Kolloid-Z.* 90:77 (1940).

178. Trotman, S. R., and J. E. Hackford. *J. Soc. Chem. Ind.* 25:104 (1906).

179. Shveikina, R. V., and S. G. Mokrushin. *Kolloidn. Zh.* 20:233 (1958).

180. Levi, S. M., and O. K. Smirnov. *Kolloidn. Zh.* 20:179 (1958).

181. Loebel, Z. C. *J. Phys. Chem.* 32:763 (1928).

182. Nakamura, R., and Y. Sato. *Nippon Nogei Kagaku Kaishi* 37:393 (1963). *Chem. Abstr.* 63:6244 (1965).

183. Thuman, W. C., A. G. Brown, and J. W. McBain. *J. Am. Chem. Soc.* 71:3129 (1949).

184. Cumper, C. W. N. *Trans. Faraday Soc.* 49:1360 (1953).

185. Lauwers, A., *Colloquium over Grenslaagverschijnselen* 1966, 113.

186. Ruyssen, R., A. Lauwers, and P. Joos. *Mededel. Koninck. Vlaam. Acad. Wetenschap., Kl. Wetenschap.* 23, No, 10:1 (1961)

187. Génin, G. *Lait* 34:156 (1954). *Chem. Abstr.* 48:9733 (1954).

188. Perri, J. M., and F. Hazel. *J. Phys. Colloid Chem.* 51:661 (1947).

189. Shtamer, R. Great Britain 476 552. France 823 992. *Chem. Abstr.* 32:4257 (1938).

190. Clark, N. O. *A Study of Mechanically Produced Foam for Combatting Petrol Fires.* London: DSIR (1947), p. 55.

191. Dumanskii, A. V., T. A. Granskaya, and G. R. Vishnevskaya. *Chem. Abstr.* 34:5201 (1940).

192. Hadorn, M. *Mitt. Gebiete Lebensm. Hyg.* 38:46 (1947). *Chem. Abstr.* 41: 5053 (1947).

193. Sauer, E., and W. Aldinger. *Kolloid-Z.* 88:329 (1939).

194. Tarassuk, N. P., and G. A. Richardson. *Proc. 23rd Ann. Meeting, Western Div., Am. Dairy Sci. Assoc.* 1937:70. *Chem. Abstr.* 32:1795 (1938).

195. Viktorin, O. *Chem. Listy* 34:265, 283 (1940). *Chem. Abstr.* 43:8794 (1949).

196. Clark, G. L., and S. Ross. *Ind. Eng. Chem.* 32: 1594 (1940).

197. Ermolenko, N. F., and N. A. Abramchuk. *Zh. Fiz. Khim.* 8:587 (1938).

198. Hardy, W. B. *Proc. Roy. Soc. (London)* 86A:610 (1912).

199. Hardy, W. B. *J. Chem. Soc.* 127:1207 (1925).

200. Trapeznikov, A. A. *Acta physicochim. URSS* 13:265 (1940).

201. Gleim, V. G., T. A. Aleksandrova, and S. G. Tereshchenko. *Khim. Tekhnol. Topliv Masel* 10, No. 11: 23 (1965). *Chem. Abstr.* 64:3261 (1966).
202. Terres, E., et al. *Brennstoff-Chem.* 35:263 (1954).
203. Maminov, O. V., and R. Sh. Latypov. *Tr. Kazansk. Khim.-Tekhnol. Inst.* No. 35:67 (1965). *Chem. Abstr.* 67, No. 34360 (1967).
204. Brady, A. P., and S. Ross. *J. Am. Chem. Soc.* 66:1348 (1944).
205. Bresler, S. E., and D. L. Talmud. *Zh. Fiz. Khim.* 4:796 (1933).
206. Robinson, J. V., and W. W. Woods. *J. Phys. Colloid Chem.* 52:763 (1948)
207. Teitelbaum, B. Ya. *Kolloidn. Zh.* 12:375 (1950).
208. King, E. G. *J. Phys. Chem.* 48:141 (1944).
209. Tarakanov, O. G., and E. G. Eremina. *Kolloidn. Zh.* 25:596 (1963).
210. Tarakanov, O. G., and E. G. Eremina. *Kolloidn. Zh.* 27:274 (1965).
211. Wedell, H. *Melliand Textilber.* 40:798 (1959).
212. Manegold, E. *Schaum.* Heidelberg: Strassenbau, Chemie und Technik. (1953), p. 119.
213. Friberg, S., and S. I. Ahmad. *J. Colloid Interface Sci.* 35:175 (1971).
214. Tarakanov, O. G., and E. G. Dubyaga. *Kolloidn. Zh.* 28:732 (1966).
215. Hartner-Seberich, R. *Ges. Abhandl. Kenntnis Kohle* 11:628 (1934).
216. Prigorodov, V. N. *Kolloidn. Zh.* 32:793 (1970).
217. Irodov, M. V. *Maslo-Zhir. Prom.* 35:21 (1969). *Chem. Abstr.* 72, No. 22819 (1970).
218. Wada, T. *J. Chem. Soc. Japan* 65:369, 440 (1944). *Chem. Abstr.* 41:3309, 3983 (1947).
219. Naudet, M., O. Micaelli, and P. Denuelle. *Bull. Mens. ITERG* 1948, No. 6, 32.
220. Dieterle, R. *Seifensieder-Ztg.* 66:54 (1939). *Chem. Abstr.* 33:9693 (1939).
221. Ritter, W. *Mitt. Lebensm. Hyg.* 29:253 (1938). *Chem. Abstr.* 33:2598 (1939).
222. Robinson, H. E., H. C. Black, and H. S. Mitchell. *Oil Soap* 17:208 (1940).
223. Higashi, H., and T. Kaneko. *Bull. Japan Soc. Sci. Fisheries* 16:505 (1951). *Chem. Abstr.* 47:8279 (1953).
224. Y. Yamamoto. *Kagawa-ken Shoyu Shikensho Hokoku* 49:33 (1957). *Chem. Abstr.* 53:10654 (1959).
225. Toi, B., and K. Oue. *Yukagaku* 6:87 (1957). *Chem. Abstr.* 55:4013 (1961).
226. Ota, S., A. Mukai, and I. Yamamoto. *Yukagaku* 12:409 (1963). *Chem. Abstr.* 59:15863 (1963).
227. Kajimoto, G., and K. Mukai. *Eiyo To Shokuryo* 16:425 (1964). *Chem. Abstr.* 61:12551 (1964).
228. Ostwald, Wo., and A. Siehr. *Kolloid-Z.* 76:33 (1936).
229. Clark, N. O., and M. Blackman. *Trans. Faraday Soc.* 44:1 (1948).
230. Oehman, V. *Tek. Tidskr.* 82: 317 (1952). *Chem. Abstr.* 46:6461 (1952).
231. Mohr, W., and C. Brockmann. *Milchw. Forsch.* 11:48 (1930). *Chem. Abstr.* 25:1916 (1931).
232. Sirks, H. A. *Chem. Abstr.* 30:2649 (1936).
233. Richardson, G. H., and M. S. El-Rafey. *J. Dairy Sci.* 31:223 (1948). *Chem. Abstr.* 45:274 (1951).
234. Milum, V. G. *Am. Bee J.* 79:186 (1939). *Chem. Abstr.* 33:4337 (1939).
235. Hawthorne, J. R., and E. B. Bennion. *J. Soc. Chem. Ind.* 61:153 (1942).

236. Helm, E., and O. C. Richardt. *J. Inst. Brewing* 42:191 (1936).
237. Smirnov, V. A., and A. N. Bondarenko. *Zh. Prikl. Khim.* 20:97 (1947).
238. Kichkin, G. I. *Khim. Tekhnol. Topliv Masel* 11, No. 4:49 (1966). *Chem. Abstr.* 65:3634 (1966).
239. Trautman, C. E. *Lubrication Eng.* 2:143 (1946).
240. Prigorodov, V. N., and L. V. Gornets. *Khim. Tekhnol. Topliv Masel* 15, No. 9:38 (1970). *Chem. Abstr.* 74, No. 14413 (1971).
241. Ramayya, D. A., S. D. T. Rao, and K. S. Murti. *Indian Std. Inst. Bull.* 17:314 (1965).

CHAPTER 5

THREE-PHASE FOAMS

§85.   A few of the foams mentioned in Chapter 4 contained suspended solid particles in addition to the two essential phases, that is, a dispersed gas and a liquid medium. Foams, in which a third (solid) phase is highly important, are used on a large scale in froth flotation. This procedure is employed to separate valuable (say, A) from unneeded minerals (say, B). A comminuted mixture of A and B is suspended in a solution of a foaming agent and (usually) air is injected. The foam rising above the liquid contains a high proportion of A and only very little of B. When the foam is mechanically removed, extraction of A from the mixture has been achieved. See also §156.

The physical property which above all determines whether the powder of a particular mineral (or ore) is, or is not, concentrated in the foam is the wettability of the powder particles. If these particles are perfectly wetted by the liquid, they are stirred up by the air current but do not reach the gas-liquid interface and, consequently, are not carried into the foam. When the receding contact angle θ_R (in the liquid!) is large, for instance, 60° (or over a radian), then the particle becomes attached to a bubble and rises with it because of buoyancy. Fig. 5.1 may help in understanding the mechanism of this effect. The solid particle is supposed to hang at the bottom of the bubble; every other position would be just as suitable but a little more difficult to draw. Surface tension γ of the liquid pulls at the solid along the three-phase line (in which gas, liquid, and solid meet). If the tangent to the gas-liquid interface at this line forms an angle β with the vertical (upward!), then force $L\gamma \cos \beta$ pulls the particle vertically up; L is the length of the three-phase line. When this force exceeds that of gravitation, the particle remains attached to the bubble.

The angle β depends on two other angles. One is the receding contact angle θ_R mentioned above. A receding, rather than the equilibrium or an advancing contact angle, is deciding because the particles are initially immersed in the liquid, and the air in the bubble has to displace some of the liquid from the solid surface. The other angle is that formed by the vertical and the solid surface at the three-phase line. If φ is this (obtuse) angle, then (see Fig. 5.2) $\beta = \varphi - \theta_R$. The value of θ_R is, to an extent, determined by the compositions of the solid and the liquid; in actual flotation processes θ_R is deliberately adjusted, whenever indicated, by addition of "collectors" which become adsorbed on the particle and alter its wettability. The value of φ cannot be predicted. It is seen, however, that β is smaller, that is, cos β and the force $L\gamma$ cos β are greater, when φ decreases (for a given θ_R). This means that a particle is more likely to be immobilized in the interface when the three-phase line follows a nearly horizontal than a nearly vertical plane.

Consider, as an example, a particle of volume 0.001 cm³ and specific

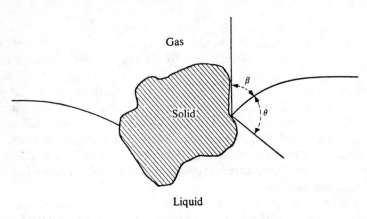

Fig. 5.1. Solid particles suspended in gas-liquid boundary by surface tension. β is the angle formed by the vertical with liquid surface at a point of the 3-phase line, and θ is the contact angle at this point.

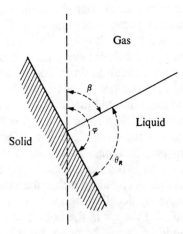

Fig. 5.2. Relation between the three angles β (between the vertical and the liquid surface), φ (between the vertical and the solid profile), and θ_R (the receding contact angle).

gravity 2 g/cm^3. In an aqueous medium the gravitational force on it is
about 1 dyne. The force pulling it up is $L \gamma \cos \beta$. If $\gamma = 60$ g/sec^2 and cos
$\beta = 0.83$ (corresponding to $\beta \approx 34°$), then L needs to be only 0.02 cm
long to achieve balance of forces. The length of the edge of a cube whose
volume is 0.001 cm^3 is 0.1 cm, that is, much greater than this L.

   §86.   Particles attached to a bubble as indicated in Fig. 5.1
would prevent bubble coalescence and, in this manner, would enhance
foam stability. Fig. 5.3 illustrates a solid particle in contact with two
bubbles. The θ_R along both three-phase lines is approximately 70°. The
three-phase lines run along the small faces m because only this position
renders the attachment possible without bad distortion of the bubble
(which would be resisted by surface forces); thus this arrangement is
favored whenever possible. In order to achieve coalescence, it would be
necessary for at least one of the gas-liquid interfaces to slide along the
solid to the vertical faces nn. The position of one of these interfaces during
this sliding is indicated by dashes near the upper left-hand corner of the
figure. It is clear that the direction of this interface near the three-phase
line (which direction is necessary to maintain $\theta_R = 70°$) is such that the
upper bubble is completely distorted. Consequently the motion considered requires work against surface forces. Thus coalescence of bubbles
protected by poorly wetted particles should be a rarer phenomenon than
coalescence of "bare" bubbles. A three-phase foam ought to be more
persistent than a comparable two-phase system.

   The effect presumably should depend on the wettability and the
particle size of the solid material. As mentioned in §85, particles perfectly
wetted by the liquid do not tend to join the bubbles and thus cannot in-

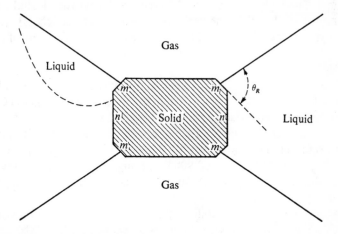

Fig. 5.3.   A solid particle between two bubbles prevents their coalescence
because the profile of the gas-liquid interface becomes distorted (see the broken
line) when the 3-phase line descends to a vertical face *nn*.

fluence foam stability, except by adsorbing the foaming agent or by enhancing the viscosity of the liquid. On the other hand, when the contact angle in the liquid is very high, the particle will be almost completely surrounded by the gas phase and again will have no direct effect on the rate of coalescence. Consequently, a medium wettability would be expected to be most efficient.

Big particles, whatever their wettability, will not rise with the foam because of their weight. If the specific gravity of (bubble + its particles) is greater than that of the liquid (which almost always is near 1 g/cm^3), the bubble will sink rather than rise. Also very small particles will be poor protectors of foam, even if their wettability is right. Let the particle be l cm across. It will be able to exert its protective action only when the liquid film between the two bubbles is considerably thinner than l cm. If the amplitude of waves crossing each film during the bubbling is commensurate with l, contact between the bubbles may soon be established along an unprotected stretch of the film; this is more likely to happen the smaller l. Thus the greatest foam protection would be observed at some middle degree of dispersion.

Unfortunately, the available experimental material on three-phase foams is meager in comparison with their industrial importance and too sketchy for a convincing comparison with the above theoretical expectations.

§87.   As seen in §66, the TCS (time of collapse after shaking) of aqueous solutions of 3-methylbutanol, in one series of experiments, had a maximum of 17 sec at the concentration of 0.32%. An identical shaking treatment was applied[1] to suspensions of chalcopyrite, $CuFeS_2$, in above solutions. The highest TCS was observed again near the alcohol concentration of 0.3% but the time itself was much longer, namely 60 sec; thus chalcopyrite markedly retarded the collapse of a foam. Powders of galena, PbS; sphalerite, ZnS; or molybdenite, MoS_2; raised the TCS to several minutes.

The conclusion of §86 that foam stability should be enhanced most by particles of medium size was qualitatively confirmed[1]. The TCS was not markedly raised by solid particles bigger than 0.03 cm across nor by colloidal particles. The greatest enhancement was achieved with powders which were retained by sieves of 12,000 openings per square centimeter but traversed sieves of 9000 openings. Thus the more favorable average diameter of the dust presumably was near 0.005 cm or somewhat smaller.

Also the effect of wettability predicted in §86 seems to be visible in Bartsch's data. An excess of galena was shaken with 0.25% aqueous 3-methylbutanol, and the wettability of the mineral was gradually altered by adding small amounts of oleic acid. If x is the final concentration of this acid in the mixture (without regard to any possible adsorption) and

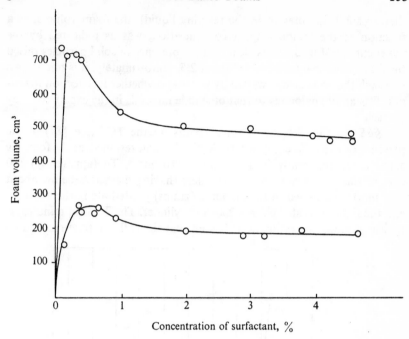

Fig. 5.4. The volume of foam (cm^3) obtained by condensation as a function of the concentration of malt residue (g/100 cm^3). Upper curve: solid residue. Lower curve: extract (data of reference [2]).

H is the foam height after shaking, the dependence of H on x can be seen in the following table:

x	0	0.01	0.05	0.1	0.3	0.5	1.0	mmole/l.
H	11	13	11	8	3	2	1	cm

Presumably this behavior means that the receding contact angle on the galena surface has a value most favorable for foam protection when 1 l. of suspension contains 0.01 mmole of oleic acid. Unfortunately the effect of oleic acid on the foaming of particle-free solutions of 3-methyl-butanol seems to be unknown. Also no contact angles have been actually measured.

   Zhidkova[2] used "malt residue" as the foaming agent. This is a by-product of beer manufacture, containing many organic substances. It was comminuted to particles of about 0.2 mm across. When x g of the powder was added to a NaHCO$_3$ solution and the suspension was poured into strong sulfuric acid so that the total liquid volume was 100 cm^3. the foam volume had the values shown in the upper curve of Fig. 5.4. When an aqueous extract of the malt residue was substituted for the powder (so that no solid was present in the reacting liquid), the foam volume, as a function of the foaming agent concentration, was as indicated by the lower curve of Fig. 5.4. It is clear that the presence of solid particles raised the foam volume by a factor of about 2.5. Unfortunately, it is not known how well the powder was wetted by water and whether its effect on foaminess was at all analogous to that of sulfide minerals in the experiments by Bartsch.

§88. Fig. 5.5 shows some results on the TCS (see §87) in the presence of five different powders. Again the malt residue was the foaming agent used, apparently in 1.2% concentration[3,4]. To (apparently) 100 cm³ of this suspension various powders (having diameters below about 0.05 mm) were added in the amounts (grams) plotted along the abscissa, and the liquid was shaken in a 250-cm³ cylinder. The TCS (in minutes) is indicated along the ordinate. It is seen that the collapse time reached 40

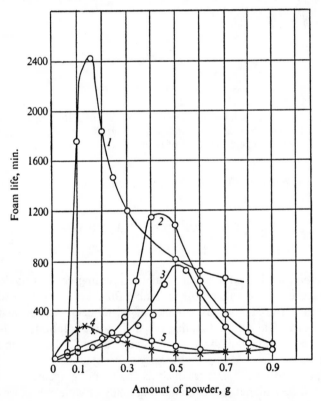

Fig. 5.5. Time of collapse after shaking (minutes) malt residue suspensions containing also oxide powders as a function of the powder amount (grams). Curve *1*: cadmium hydroxide. Curve *2*: cobalt hydroxide. Curve *3*: nickel hydroxide. Curve *4*: copper hydroxide. Curve *5*: barium hydroxide (from reference [4]).

hr in the presence of 0.15 g of cadmium hydroxide; in the absence of mineral powders the TCS was near 35 min. The enhancement of TCS by the other hydroxides (of cobalt, nickel, copper, and barium) was less spectacular but, as proved by Fig. 5.5, still very strong. Much smaller increases in TCS were achieved by small amounts of zinc hydroxide, magnesium hydroxide, zinc oxide, asbestos, mica, and starch; and when the concentration of any of these powders was above 0.4%, the TCS was less than 35 min, and continued to decrease when this concentration was raised. The members of a third group of solids lowered the TCS at almost all concentrations. Thus at 0.1% content of the powder, TCS was 35 min for chromium hydroxide, 20 min for iron hydroxide, 26 min for aluminum hydroxide, 14 min for calcium carbonate, and 4 min for bentonite. Also some zinc sulfide strongly suppressed foam stability.

Evidently, these observations cannot be accounted for by wettability. The stability of 3-methylbutanol foams was greatly raised by hydrophobic zinc sulfide (see §87), but this mineral destroyed the foam of malt residue. Perhaps the data of the preceding paragraph are characteristic for the unusual foaming agent employed. It would be instructive to repeat Mokrushin's experiments with a more familiar frother (a synthetic surfactant or soap).

The effect of finely dispersed metal hydroxides on gelatin foams[5] may have been caused by acidity changes, as foaminess of gelatin is very sensitive to pH. Bubbling of air through 0.1% gelatin solution produced a foam column 17 cm tall, which collapsed to 3 cm within 40 min. A contraction of equal magnitude (that is from 17 to 3 cm) required 60, 57, and 48 min when the gelatin solution contained, respectively, 0.03% ferric hydroxide, 0.03% aluminum hydroxide, or 0.018% aluminum hydroxide. These compounds had been prepared by blowing ammonia gas over the surfaces of iron or aluminum salt solutions. Contrary to the expectations of §86, a greater protection was afforded by colloidal hydroxides. The time of collapse was 90 min after the addition of 0.001% of colloidal ferric hydroxide, 60 min for 0.0008% chromium hydroxide sol, and so on. However, larger amounts of colloidal hydroxides accelerated collapse, in analogy to the data of Fig. 5.5. Thin films of copper sulfide, lead sulfide, or zinc sulfide shortened, and colloidal sulfides extended, the time of collapse of gelatin foams.

§89.   Several observations of a less systematic kind also are known. It is mentioned in §83 that calcium carbonate apparently raises the foaming ability of saccharate solutions. This compound enhances also the foaminess of sugar beet juice[6]. On the other hand, spheroidal silica (of 30 μ diameter), added in the 0.0025% amount to aqueous solutions of lower soaps, had almost no effect on the maximum foam volume achieved in a pneumatic test or on the time of collapse of these foams[7]. Colloidal silica seems to raise the stability of foams of pickling liquor, as reported in §147.

Some sea foams in the Aegean Sea were particularly persistent. This stability was attributed[8] to a colloidal alumosilicate whose chemical composition is claimed to be $Na_2Al_{40}Si_{40}O_{141}$, 67 H_2O.

Gypsum, talcum, and clay in concentrations up to 8% did not alter the height of foam produced in the reaction of calcium carbonate and sulfuric acid in the presence of malt residue[9]. After drying, the foam of sugar molasses sometimes contains 40% of rust, sand, and other insoluble solids[10]; perhaps they cause its marked persistence.

Foam is a nuisance during several stages of pulp and paper manufacture, and it is always a three-phase foam, as it contains paper fibers. In sulfite spent liquor the greatest foam stability is said to occur when the fiber concentration is 4.4 g/l.; but the concentration corresponding to the maximum persistence decreases during aging[11]. According to a review[12], foams in paper stock suspensions usually contain 12 to 15% of fibers and less than 1% of materials which may act as foaming agents. Cellulose fibers in the amounts up to 3% raised the pneumatic foam height (and the Σ of §53) of the "black liquor"[13]. It is unlikely that this effect is caused by poor wettability. Do the fibers retard the drainage of liquid from the foam?

Three-phase foams of solutions in organic solvents have been studied by Tarakanov[13]. An alkylbenzene sulfonate (the side chain being a mixture of dodecyl to tetradecyl) or a sunflower phosphatide were dissolved in diethyl phthalate at 150° and the liquid was cooled; at room temperature it was a suspension or a colloidal solution. It was foamed by injecting carbon dioxide. When the concentration of the sulfonate was varied between 0.2 and 5%, the half-life of the foam was particularly long (18 min) at the concentration of 1%.

§90.   Foaming of boiling aqueous salt solutions, related to the priming and carry-over of §145, often is enhanced by solid particles. This effect presumably has little connection with wettability; it may derive from the fact that powders contain occluded air, which acts as nuclei for water vapor. This explanation certainly fits the observation[14] that almost any powder dropped into a boiling solution of 0.5% sodium chloride and 0.005% sodium hydroxide increased priming, while no powder maintained its activity after being immersed in the liquid for some time. The rate of losing activity was different for different solids; it was small for galena, greater for limestone, still greater for dolomite, and so on. Since gelana is much less hydrophilic than the carbonates, it may be argued that the hot alkaline solution displaces air from its surface more slowly than from the surfaces of limestone and dolomite; and this process determines the rate of disactivation of the solid. The results of several (later) experimental studies[15-19] are less definite.

On the other hand, the air release mechanism cannot be operative in the numerous experiments by Villar[20] and Durov. An aqueous mixture of calcium chloride (10%) and sodium bicarbonate (5%), in which obviously calcium carbonate was present, readily boiled over but foaming was prevented by small amounts (e.g., 0.001%) of sodium dodecylbenzene sulfonate or a sodium alkyl sulfate[20]; see also §145. The foam height h of boiling 0.5 N sodium chloride solution was raised by ferric hydroxide or basic magnesium carbonate (for instance, 0.002 g-equiv./l.); and it was particularly great when both colloids were present in equal amounts[21]. The h of boiling suspensions of basic magnesium carbonate depended on the soluble salts present in the suspension; if this salt was sodium sulfate, the h was greater than when sodium chloride or, especially, calcium chloride was employed[22]. Both basic magnesium carbonate and calcium carbonate accumulated in the foam; when the concentration of the former salt in the suspension increased, the degree of enrichment (i.e., the lesser accumulation ratio of §157) also increased and reached the value of 3.2; this means that the carbonate concentration in the collapsed foam was 3.2 times that of the initial solution[23]. Occluded air could not be responsible for these results, but there is no proof that wettability was.

References

1. Bartsch, O. *Kolloidchem. Beihefte* 20:1 (1924).
2. Zhidkova, L. G. *Tr. Ural'sko. Politekhn. Inst.* 94:16 (1960).
3. Mokrushin, S. G., and V. I. Borisikhina. *Zh. Prikl. Khim.* 28:109 (1955).
4. Mokrushin, S. G., and V. I. Borisikhina. *Tr. Ural'sko. Politekhn. Inst.* 94:4 (1960).
5. Mokrushin, S. G. *Kolloidn. Zh.* 12:448 (1950).
6. Szarejko, B., S. Zagrodski, and J. Dobrzycki. *Roczniki Technol. Chem. Zywności* 2:101 (1957). *Chem. Abstr.* 52:13293 (1958).
7. Topper, L., and E. Baer. *J. Colloid Sci.* 9:495 (1954).
8. Akcetin, D., and F. H. Constable. *Istanbul Univ. Fen. Fak. Mecmuasi Rev.*, Fac. sci. Univ. Istanbul, Ser. C, Astron.-Fiz.-Kimya, Ser. CA30: 125 (1965). *Chem. Abstr.* 67, No. 67894 (1967).
9. Mit'kevich, G. P. *Zh. Prikl. Khim.* 26:873 (1953).
10. Vavrinecz, G. *Cukoripar* 10:128 (1957). *Chem. Abstr.* 52:15938 (1958).
11. Kleinert, T., and M. Ruck-Floriantschitsch. *Papier* 8:13 (1954). *Chem. Abstr.* 48:5494 (1954).
12. Petersen, K. *Wochbl. Papierfabrik* 91:771 (1963). *Chem. Abstr.* 59:13005 (1963).
13. Tarakanov, O. G., and E. G. Eremina. *Kolloidn. Zh.* 27:274 (1965).
14. Foulk, C. W., and S. F. Whirl. *Ind. Eng. Chem.* 26:263 (1934).
15. Foulk, C. W., and H. C. Brill. *Ind. Eng. Chem.* 27:1430 (1935).
16. Foulk, C. W., and R. Ulmer. *Ind. Eng. Chem.* 30:158 (1938).
17. Durov, S. A. *Zh. Prikl. Khim.* 13:693 (1940).
18. Leaf, W. B., et al. *Am. Ry. Eng. Assoc.* 45. Bull 441:58 (1943).
19. Harvey, E. N., W. D. McElroy, and A. H. Whiteley. *J. Appl. Phys.* 18:162 (1947).

20. Villar, G. E. *Quim. Ind.* (*Uruguay*) 2, No. 3:5 (1952). *Chem. Abstr.* 47:3081 (1953).
21. Durov, S. A., and N. G. Chen. *Zh. Prikl. Khim.* 30:1096 (1957).
22. Durov, S. A., and N. G. Ch'eng. *Tr. Novocherk. Politekhn. Inst.* 65:23 (1959). *Chem. Abstr.* 55:2214 (1961).
23. Durov, S. A., et al. *Tr. Novocherk. Politekhn. Inst.* 65:33 (1959). *Chem. Abstr.* 55:2214 (1961).

CHAPTER 6

FOAM DRAINAGE

§91. As pointed out in §43, drainage is one of the three processes occurring in every liquid foam. In principle, foam drainage and foam collapse (discussed in Chapter 4) are independent of each other; thus the numerous attempts to estimate the stability of a foam from the rate of its drainage are fundamentally incorrect. As long as the films surrounding gas volumes exist, the foam also exists; drainage renders the films thinner but does not cause their rupture.

In reality, there is a considerable interplay between drainage and collapse. First, when a foam bubble bursts, the liquid initially present in its walls joins the liquid flowing down in the other walls and the neighboring Plateau borders; thus the draining liquid originates partly in defunct bubbles. This part is greater, the smaller the foam stability. If the foam is not very stable, both fractions (i.e., that due to drainage proper and that created by bubble rupture) have comparable magnitudes, and the rate of drainage is a complex quantity. Only when the persistence is very high, can pure drainage be measured.

Secondly, the probability of bursting generally depends on wall thickness and thus is influenced by drainage. This dependence is mentioned in §18 for single films and in §59 for "three-dimensional" foams, and an explanation for it is presented in §121.

Thirdly, the rate of drainage and the probability of collapse may be influenced by the same property of foam lamellae. Surface viscosity and rigidity of the surface layers belong to these properties; see §99 and 134.

§92. Methods of measurement. Usually these methods are very simple. A foam is formed by whatever means and then left alone in a vessel. Gradually some liquid collects at the bottom of the latter; its volume is measured after a prescribed time (or after several time intervals) and the rate of drainage is calculated.

Fig. 6.1 is an example of instruments employed for drainage measurements[1]; once upon a time this apparatus was very popular in Germany. It is a flask containing a large (1900 cm^3) and a small (50 cm^3) bulb connected with each other by means of a graduated tube of 50 cm^3. The liquid to be tested (100 cm^3) is shaken in it for 30 sec; then the flask is placed in the vertical position indicated in Fig. 6.1, and the volume of the

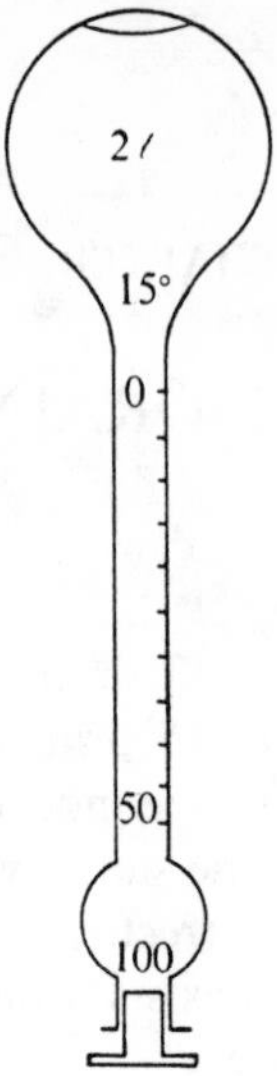

Fig. 6.1. A flask for determining the rate of drainage ("foam number") (from reference [1]).

liquid underneath the foam is read 3 min later. If this volume is x cm^3, then the "foam number" of the sample is $100 - x$. In a miniature variant of this method[2], the total volume of the flask is only 100 cm^3 and the volume drained is read after 1 min. It should be noted that the expansion factor of the foam in Stiepel's flask cannot exceed 20.

In the Carlsberg procedure for estimating the "foaming capacity" of beer[3], "the foam is formed in a natural way, viz., by the free fall of the beer." The liquid which oozed out of the foam is collected and removed after 2 min (volume a) and after 10 min (volume b); then the remaining foam is destroyed and the resulting liquid measured (volume c). The ratio $100(b+c)/(a+b+c)$ is said to be the total foam or the total head; obviously it is the relative amount (in percent) of the liquid remaining in the foam after 2 min aging. The ratio $100c/(a+b+c)$ is denoted as residual foam or residual head.

In a similar method[4], the beer is first saturated with carbon dioxide under pressure, the pressure is released, and a part of the foam obtained is used to fill a vertical cylinder of 100 cm^3. The volume of the liquid at the bottom is read after 1, 4, 5, and 8 min (volumes a, b, c, and d), the rest of the foam is destroyed by 2-propanol, and the total liquid volume (A) is determined. The "foam index" of the beer is defined as $50(b+c-a-d)/A$.

The "percentage leakage" was defined[5] as $100\ M_2/M_1$. Egg white was whipped and weighed; M_1 was its mass. The foam was permitted to drain for one hour when M_2 g of liquid was collected.

The "foam power" of soap solutions is another measure of drainage rate[6]. Air is introduced into V_1 cm^3 of the liquid through a horizontal tube rotating about a vertical axis. Presumably the air stream is broken to small bubbles by this agitation and the liquid is completely transferred into foam. Then the system is left alone for 1 min. If the volume of the liquid collected during this interval is V_2, then $100(V_1 - V_2)/V_2$ is the foam power.

To determine the "half-life," t_1, of a foam, 100 cm^3 of air is bubbled through 25 cm^3 of an animal or vegetable oil (see §54). Then the time necessary to collect 12.5 cm^3 of the oil at the bottom of the vessel is measured and cited as the half-life[7].

The methods indicated above have been devised strictly for testing. In Fig. 6.2 an apparatus used for research[8] is illustrated; obviously, it is not greatly different from the former instruments. The solution falls drop-wise from burette D onto the sintered glass filter C and is transformed into foam because vessel A is being evacuated by a pump attached to E. As soon as the vessel is filled with the foam, both D and E are closed, and the volume of the liquid drained out of it is observed in the graduated capillary B. Recently[9], the rate of drainage was measured for identical solutions by three methods differing in the technique of foam formation; the foam was obtained by shaking, or by agitation in a beaker, or in an apparatus similar to that of Fig. 6.2; and there was only a qualitative agreement between the three sets of results.

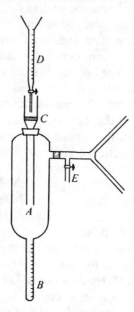

Fig. 6.2. Measurement of the rate of drainage. A is the orifice at which bubbles form. B: the calibrated tube in which liquid is collected. C: sintered glass filter. D: burette. E: connection to vacuum pump (from reference [8]).

The volume oozed out can be measured also by electric means. On the porous bottom of a vertical tube, two electrodes were placed and a small amount of liquid was poured. The whole liquid was transferred into foam by air injected through the porous plate. Then the air stream was stopped and the electric conductance of the medium between the electrodes was measured at suitable intervals. The resistance gradually decreased because initially only foam was present between the electrodes, while later more and more liquid and less and less foam filled the space[10]. It is clear that the curves of conductance versus time would be steeper the smaller the electrode plates in the vertical direction, but this effect has not been investigated.

§93.   None of the test methods indicated in §92 can be considered satisfactory. For instance, the volume V of liquid collected after an arbitrary time interval (1 min, 3 min, etc.) cannot have any general validity. The rate of drainage is not constant, and a liquid which afforded a greater V after 1 min may give rise to a smaller V after 10.

Probably the effect of the experimental details on the test results would prove even more disconcerting; Kruglyakov's observations[9] seem to demonstrate this danger. Thus in the example of Fig. 6.1, the rate of increase of V must depend on the diameter of the graduated capillary, and "foam numbers" obtained with different capillaries would not be consistent with each other. The foam volume obtained by pouring depends on the distance of fall, the rate of pouring, and so on, and the rate of drainage naturally depends on this volume. The amount of carbon dioxide dissolved in beer depends on pressure, time, and, in all probability, on the composition of the beer; thus, the "foam number" would indicate properties not related to true foaminess. The volume of foam produced by whipping or agitation is affected by the details of the procedure used, the size and shape of the vessel, etc.; and again the rate of drainage is a function of this volume and of the shape of the foam layer.

As emphasized in §51, the volume of gas incorporated in a liquid by agitation varies not only with the kind and intensity of mixing but also with some properties (mainly viscosity and density) of the liquid. This incorporation is not standardized in any of the tests referred to so far in this section. However, in estimating the "half-life," an attempt is made to introduce 100 cm³ of air into 25 cm³ of liquid. This is an advance, but the bubble diameter also should be standardized because the rate of drainage depends on the cross-section of the liquid channels, and this is related to the bubble size. Approximate, for the sake of simplicity, bubbles as cubes and disregard the flow in Plateau's borders. Let the edge of the gas cube be l_g cm and the edge of the cube comprising this gas and the liquid belonging to it be l_1 cm. As long as the ratio of the air volume to the liquid volume is constant (for instant, 100:25 in the "half-life" method), also the ratio $l_1:l_g$ remains constant. But the volume rate of percolation through the idealized system depicted here would be approximately proportional to $(l_1 - l_g)^3 / l_1{}^2$ per unit cross-section of the tube. Thus it is not constant when l_1 / l_g is constant.

The rest of this chapter deals with research on drainage rather than with tests on foams.

§94. Rate of drainage.   If the volume V of liquid exuded is measured after several time intervals, the rate dV/dt of drainage can be found; t is time. It is clear that neither V nor dV/dt have general meaning because they depend on the initial amount of foam; the greater this amount, the greater the rate dV/dt. Consequently, in many instances V/V_0 rather than V was plotted against t; V_0 is the initial volume of the liquid in the foam. The dependence of V, V/V_0, or dV/dt on t can be expressed graphically or algebraically. Two geometrical representations are reproduced in this section.

Fig. 6.3 shows the results of an early series of measurements on saponin foams[8]; the instrument of Fig. 6.2 was used. The initial volume of liquid (i.e., V_0) was 10 cm^3, and time t is expressed in minutes. It is seen that the curves of V/V_0 (the ordinate) versus t are very similar for 0.2 and 2% saponin solutions, but the drainage in the latter foam takes eight times as many minutes as in the former. The rate of drainage (i.e., dV/dt or $dV/V_0 \cdot dt$) increases with time up to about $V/V_0 = 0.2$, is roughly constant between $V/V_0 = 0.2$ and $V/V_0 = 0.5$ and gradually decreases at greater ratios V/V_0. The authors pointed out that the behavior was similar to the time dependence of the synersis of gels and, consequently,

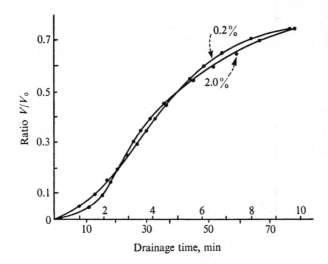

Fig. 6.3.   Drainage of saponin foams. Ordinate: ratio of liquid exuded to initial liquid (V/V_0). Abscissa: time in minutes. Upper scale for 0.2%, lower scale for 2% solutions (data of reference [8]).

defined drainage as syneresis of foams. As explained in §91, the liquid collected under a foam usually originates partly from broken and partly from unbroken bubbles. The authors believed that in their experiments bubble collapse was more important than drainage proper, and the collapse started from the top of the foam column and advanced downward. This was supposed to explain the observed increase of dV/dt on the earliest stages of drainage, as the liquid liberated near the top required considerable time to reach capillary B. The subsequent decrease of dV/dt presumably was caused by (1) insufficient amount of liquid in the foam at later stages of the process, and (2) stabilization of the lower foam layers by the excess saponin coming from the collapsed upper layers. The liquid oozed out contained less saponin than the initial solution, so that accumulation in foam (see Chapter 12) took place. If the authors' impression (that bubble rupture was the main cause of drainage) was correct, then these data cannot be used for testing the explanations of drainage proper.

In another study[11], an effort was made to determine drainage not complicated by bubble bursting. The results are presented in Fig. 6.4, analogous to Fig. 6.3. The longest time shown is 649 sec, while the first rupture was noticed at about 1000 sec. The initial liquid volume V_0 in the foam was 45.2 cm^3. A commercial sulfonate was the foaming agent, and its concentration was 1%. The rate dV/dt of drainage monotonously decreased in time and no inflection point was visible. Curves similar to

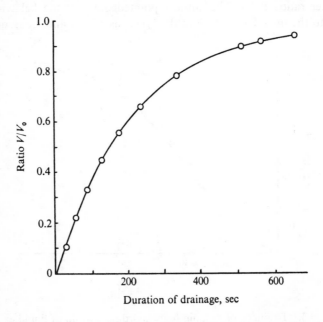

Fig. 6.4. Drainage of a sulfonate foam. Ordinate: ratio of liquid exuded to initial liquid (V/V_0). Abscissa: time in seconds (data of reference [11]).

that of Fig. 6.4 have been obtained[9] also with several surfactant foams produced by shaking or in a laboratory blender, but maxima of dV/dt were observed with identical solutions when foams were obtained by a pneumatic method and the expansion factor of the foam (see §1) was greater than 15.

A maximum of dV/dt at some middle value of V/V_0 is visible on the curves published by Barmore[12] for egg white foams prepared in a kitchen mixer.

§95. *Equations for drainage.*   Several more or less empirical equations for the volume V exuded during time interval t and for the rate dV/dt are collected in this section. Because the absolute values of V and dV/dt may be expected to be greater, the greater the horizontal cross-section of the foam column, the symbol V may mean also the volume exuded per unit cross-section; this alteration would not affect the form of the equations. Also V_0 may mean either the absolute volume (cm^3) of liquid initially present in the foam or the ratio (cm) of this volume to the cross-sectional area (cm^2). In all instances, k is an empirical constant.

$$V = V_0(1 - e^{-kt}) \tag{6.1}$$

$$\frac{dV}{dt} = kV_0 e^{-kt} = k(V_0 - V) \tag{6.2}$$

This formula was valid for the second half of the drainage of saponin foams[8] and in some other systems[13-15]. According to it, the rate of drainage is proportional to the volume of liquid still present in the foam, so that drainage appears as a first-order reaction.

$$V = V_0 - (kt)^{-1} \tag{6.3}$$

This equation[16] cannot be correct because it predicts negative values for V at very small times t. Its differentiation leads, however, to expression

$$\frac{dV}{dt} = \frac{1}{kt^2} = k(V_0 - V)^2 \tag{6.4}$$

which is characteristic for second-order chemical reactions; the rate is proportional to the square of the liquid volume still available for drainage. Hence, instead of Equation 6.3, the following equation

$$V = \frac{V_0^2 kt}{1 + V_0 kt} \tag{6.5}$$

should be employed. Equations closely related to Equations 6.3 and 6.5 have been derived later using the foam density (§107) or the expansion factor as variables. According to Haas[17] and Wace[18],

$$\frac{1}{\varphi} - \frac{1}{\varphi_0} = \frac{k_1 l^2}{h} t; \tag{6.6}$$

$1/\varphi$ and $1/\varphi_0$ are the expansion factors at time t and at zero time, l is the average bubble diameter, and h is the height of the foam duct which is supposed to be almost horizontal. The total volume $[V]$ cm^3 of foam is assumed to remain constant during the drainage (i.e., no bubble bursting occurs). Consequently, $1/\varphi_0 = [V]/V_0$, and $1/\varphi = [V]/(V_0-V)$ (Equation 6.7). Introduction of these expressions into Equation (6.6) results in Equation 6.5, if k in this formula is set equal to $k_1 l^2/h[V]$.

If drainage is measured not in a stationary foam treated so far but in a foam continually rising with the linear velocity u cm/sec, then φ is a function of u. Empirically,

$$\frac{1}{\varphi} = \frac{l^2 D^{0.2}}{k_2 u} \; ; \tag{6.8}$$

k_2 is another constant (of an unusual dimension) and D is the diameter of the vertical tube in which the foam rises[19]. Substituting φ from Equation 6.7, neglecting $D^{0.2}$ (which is almost constant as long as similar tubes are compared), and writing H/t for u (H is the height of the tube and t is the time needed for the foam to reach the top, that is, the time available for drainage), formula

$$V = V_0 - \frac{[V]k_2 H}{l^2 t}$$

is obtained. If the ratio $k_2 H[V]/l^2$ which is constant for any given foam (which does not collapse) is abbreviated as $1/k$, Equation 6.3 results.

The equation derived by Kruglyakov[11] is equivalent to

$$V = \frac{V_0 t}{t + k V_0} \; , \tag{6.9}$$

giving the rate of drainage of stationary foams

$$\frac{dV}{dt} = \frac{k V_0^2}{(t + k V_0)^2} = \frac{k V^2}{t^2} \; ; \tag{6.10}$$

k is supposed to mean the inverse drainage rate at $t = 0$ (that is, dV/dt at $t = 0$ is $1/k$) but is treated as an empirical constant. The dimension of this k is sec/cm^3, of the constant in Equations 6.1 and 6.2: sec^{-1}, Equations 6.3, 6.4 and 6.5: cm^{-3} sec^{-1}, in Equation 6.6: cm^{-1} sec^{-1}, and in Equations 6.11 and 6.12: cm^{-6} sec^{-1}.

$$\frac{1}{(V_0 - V)^2} - \frac{1}{V_0^2} = 2kt \tag{6.11}$$

is an equation[20] for the third-order reactions, as becomes clear after differentiation:

$$\frac{dV}{dt} = k(V_0 - V)^3. \tag{6.12}$$

$$V = V_0(1 - e^{-k\sqrt{t}}) \tag{6.13}$$

was proposed by Lederer[21], but is erroneous, as is manifest in its differentiated form

$$\frac{dV}{dt} = \frac{kV_0}{2\sqrt{t}} e^{-k\sqrt{t}} = \frac{k^2(V_0-V)}{2\ln[V_0/(V_0-V)]} ; \qquad (6.14)$$

the rate of drainage would be infinitely great at $t = 0$ or at $V = 0$. The dimension of k this time is $\sec^{-0.5}$.

When none of the above equations is valid and a maximum of dV/dt is observed, the cause may be, according to Kruglyakov[11], that the experimental rate of drainage is, more or less, a result of different flows, namely that from lamellae into the Plateau borders and that downward in these veins.

There is only one arbitrary constant in each of the above equations. An equation with three constants was advocated by Miles[22]; it is

$$V_0 - V - a\log(V_0 - V) + b = kt. \qquad (6.15)$$

However it is easy to eliminate b from it and to obtain an equivalent expression with two constants, namely

$$-V + a\log\frac{V_0}{V_0 - V} = kt. \qquad (6.16)$$

If the logarithmic term is much greater than V, then Equation 6.16 is nearly identical with Equation 6.1; that is, drainage is a first-order reaction; and if the logarithmic term is smaller than V, then the volume oozed out decreases in time, which is impossible.

Another two-constant equation was derived by Jacobi[23]. It is

$$V = \frac{2b}{a}[1 - (at+1)^{-0.5}] \qquad (6.17)$$

or

$$\frac{dV}{dt} = \frac{b}{(at+1)^{1.5}} ; \qquad (6.18)$$

the dimensions of a and b are, respectively, $\sec^{-1}$ and $cm^3/\sec$ (or cm/sec if V is referred to unit cross-section). As long as t is very small, dV/dt is almost constant and almost equal to b, but it continuously decreases when t increases further; no inflection point is predicted. Such an inflection point would be expected, however, if in the initial foam the lamellae were thicker near the top than near the bottom. Also in this equation, constant b can be dispensed with. At $t = \infty$, V must be equal to V_0. Hence, $2b/a = V_0$ and Equation 6.17 can be written as

$$V = V_0[1 - (at+1)^{-0.5}]. \qquad (6.17')$$

Experiments on eight pneumatic foams of commercial detergents were in a reasonable agreement with Equation 6.17 having two arbitrary constants, but did not confirm Equation 6.17[1]. When b varied from 6.55 to 31.5 cm^3/sec and a from 0.20 to 1.00 sec^{-1}, the ratio $2b/a$ varied between 55 and 86 cm^3, that is, was not constant and not equal to 50 cm^3, although V_0 was 50 cm^3 in all tests.

An equation with three empirical constants

$$\log(V_0 - V) = a - b \log(t + t_0) \qquad (6.19)$$

was applied[24] to the drainage of foams produced by shaking a dilute aqueous solution of hexadecyltrimethyl ammonium chloride and ethanol but was found invalid for a solution of this chloride plus α-naphthol and ethanol.

Even more complicated is the equation suggested by Gleim[10], namely

$$V = (V_0 - V_r)e^m + V_r, \qquad (6.20)$$

in which V_r is the liquid volume remaining in the foam at the completion of drainage, and $m = k/[1 + (t/t_0)^n]$; k, t_0, and n being constants to be determined by the experiment.

None of the above equations has a firm theoretical foundation, and it is not known why the behavior of a particular system agrees with one equation better than with the others. In every instance it should be tested whether the measurable drainage is, or is not, affected by bubble collapse and/or by evaporation.

The "average lifetime of liquid in foam" was defined[25] as

$$L_l = \frac{1}{V_0} \int_0^V t \cdot dV \qquad (6.21)$$

If Equation 6.2 is valid, the product $t \cdot dV$ is equal to $k t V_0 e^{-kt} dt$, and $L_l = 1/k$; the dimension of L_l is, of course, seconds. The half lifetime, τ_1 of the foam, that is, the time needed for V to become equal to 0.5 V_0, also is related to the various constants k; if again Equation 6.1 is employed, then $\tau_1 = 0.693/k$ sec.

§96.   Almost all equations of §95 referred to liquid draining out of a stationary foam, that is, a foam which was left alone after formation. As mentioned in §59, it is possible to add the solution (capable of foaming) to the top of the foam column at the same time as a gas is introduced at its base. If the two rates of flow (of the liquid and the gas) are kept constant, a steady state is soon reached. In this steady state the density of the foam is greater the greater the rate dV_1/dt of liquid input, assuming the gas input not to vary. This increase in density was measured[26], for instance, for solutions of commercial sodium dodecyl sulfate but not expressed in mathematical form. Perhaps the following reasoning will prove helpful.

As long as evaporation is negligible, $(dV_l/dt) = (dV_f/dt) + (dV/dt)$, that is, the liquid flowing in is distributed between the foam (this portion is dV_f/dt) and the "bottoms" (dV/dt). The volume transferred into foam in unit time apparently little depends on foam density. Thus, when dV_1/dt increases, also dV/dt is greater. But dV/dt increases with the volume V_0 of the liquid in the foam, that is, also with its density. Thus a great value of dV_1/dt means great foam density. Additional experimental data on, and mathematical treatment of, the steady-state drainage were published later by Leonard[27], Lemlich[28], Haas[17], Shih[29,30], and Hofer[31].

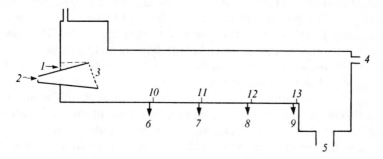

Fig. 6.5. A horizontal foam tower. *1* is the liquid pool; *2* is the air inlet; *3* is the perforated plate; *4* is the air outlet; *5* is the outlet for the foam; *6* to *9* are outlets for exuded liquid; and *10* to *13* are baffles (after reference [32]).

An increase of dV/dt with dV_1/dt was observed[32] also in a horizontal "foam tower," whose principle is shown in Fig. 6.5. Liquid which overflows pool *1* and runs over the perforated plate *3* is transferred into foam by air stream *2*. After filling the horizontal tube, the foam is broken under outlet *5*, while air escapes through tube *4*. The liquid oozed out is collected through the narrow tubes *6* through *9*; *10* through *13* are baffles. The rate of drainage increased from tube *6* to tube *9*; some liquid escaping through tube *6* perhaps was never incorporated in a foam. The flow in each of the narrow tubes was more rapid the greater the liquid influx dV_1/dt (at *1*) but almost independent of the air volume injected. The solutions contained 0.2% saponin. When their viscosity η was raised by adding varying amounts of hydroxyethyl cellulose, the rate of drainage was a linear function of log η. When the saponin concentration was varied between 0.1 and 0.75% and in this manner the surface tension of the liquid was changed from 44.5 to 36.6 g/sec^2, the rate of drainage was not significantly altered. The effect of the apparatus dimensions on this rate was not investigated.

Evaporation has been mentioned above in this section. Numerous data on the relative importance of vaporization and drainage in foams of some synthetic surfactants can be found in Manegold's book[33].

§97. *The mechanism of drainage. Films.* Drainage in foams has two causes and proceeds by two paths. The more obvious cause is gravitation: in every foam, gas is present below liquid, and the center of gravity of the system descends when liquid flows down and gas rises up. The second cause was discovered by Plateau[34]. The pressure in Plateau borders (see §7) is smaller than in the central parts of the lamella because the lamella faces are almost plane and the surfaces of Plateau borders are strongly curved. Consequently the liquid content of the bubble walls continually flows into the borders; and this flow can cease only if the mutual repulsion of the two electric double layers (see §13) becomes equal to the capillary pressure P_c or if the bubble wall becomes solid. For this flow, there is only one path: in the wall toward the border. The flow in the gravitational field takes place in both borders and lamellae.

When the liquid flows from the center of a film (where pressure P is almost equal to the atmospheric) toward a border where the pressure is $P - P_c$, the flow pattern is analogous to that calculated by Stefan[35] (see also Bikerman[36]) for the squeezing of a liquid out of the space between two parallel plane disks. If the disks are circles of radius a cm, the time t required for the distance between them to be reduced from δ_1 to δ_2 cm is given by the equation

$$\frac{1}{\delta_2{}^2} - \frac{1}{\delta_1{}^2} = \frac{4P_c}{3\eta a^2}\, t; \tag{6.22}$$

η is the viscosity (g/cm·sec) of the liquid. The value of P_c is assumed to be independent of time; that is, the shape of the gas-liquid interfaces is supposed not to vary while drainage proceeds.

In every foam the value of P_c no doubt varies within wide limits, but the shorter radius R_1 of curvature of a Plateau border in many instances will have the order of 10^{-4} cm. Hence, $P_c (= \gamma/R_1)$ of 10^5 g/cm·sec^2 seems to be a reasonable value. If $\eta = 0.01$ and $a = 0.1$, then the ratio $4P_c/3\eta a^2$ is near 10^9. With this value, the decrease of film thickness from a much greater value to $\delta_2 = 10^{-5}$ cm would take about only 10 sec.

The flow represented by Equation 6.22 is the only one occurring in horizontal films. In a vertical lamella the flow caused by gravitation enhances or reduces the motion due to capillary pressure. The lamella may be crudely approximated as a slit between two plane parallel and immobile walls. The mean linear velocity u cm/sec in such a channel is

$$u = \frac{g\rho\delta^2}{12\eta}; \tag{6.23}$$

g is the acceleration (cm/sec^2) due to gravity, ρ is the density difference, and δ is the distance between the walls.

However, the "crusts" of a usual soap film need not be immobile, parallel, and plane. The thickness δ of a lamella tends to be smaller near the top than near the bottom because the hydrostatic pressure increases with depth. In simplest instances[37,38],

$$\delta = \left(\frac{4\eta h}{g\rho t}\right)^{0.5};$$ (6.24)

h is the distance of the level considered from the top of the film. For instance, for a typical aqueous solution, 10 sec after the start of drainage the thickness δ at the level $h = 10$ cm below the top would be about 60 μ. The experimental thickness for some surfactant solutions was a little smaller than predicted[38].

Equation 6.24 represents a case opposite to that discussed in §7. There it is supposed that the film is fully supported by surface tension and no flow occurs at all; and here only the flow is taken into account and surface phenomena are neglected. No doubt, the reality generally is nearer the present extreme, but the neglect of surface tension ought to be justified separately for every system under consideration.

§98. The mechanism of drainage. Foams.
Equations 6.22 and 6.24 refer to the thickness of a film. The rate of drainage in foams is expressed as dV/dt (see §95); that is, as the volume of liquid leaving the foam for the bulk liquid in unit time. It is necessary to find the connection between these two approaches.

The flow in foams takes place partly through the films and partly in Plateau's borders. The cross-section of the latter looks like a three-way star. The hydrodynamics of flow through a channel of this shape has been subjected to mathematical treatment[27], but it is simpler to approximate Plateau borders as cylindrical capillaries having an equivalent radius r. With the approximations used to derive Equation 6.23, the mean linear rate of flow in a cylindrical capillary (of radius r) in gravitational field is

$$u = \frac{g\rho r^2}{8\eta}$$ (6.25)

and the volume flow is $\pi g\rho r^4/8\eta$ cm^3/sec.

The volume rate of flow in a vertical slit of width w cm (w is horizontal and perpendicular to δ) from Equation 6.23 is $g\rho w\, \delta^3/12\eta$ cm^3/sec. The ratio of the two volume rates of flow is $2w\delta^3/3\pi r^4$. If this is greater than unity, the flow in the lamellae is more important than that in the Plateau borders. However, in many (probably in the majority of) foams, the ratio is much smaller than 1. A reasonable value for w would be 0.1 cm, for δ 0.1 μ, and for r 10 μ. In such a foam, $2w\delta^3/3\pi r^4 = 0.022$. Some investigators believe that the flow in bubble walls, as a rule, may be neglected.

If the major part of the liquid volume $V_0 - V$ in the foam (see §95) really is present in the Plateau borders, then $V_0 - V \approx \pi n l r^2$, if there are n borders in the foam and l is their mean length (so that the volume of an average vein is $\pi r^2 l$). This l depends on n. Approximate bubbles as cubes, all equal. If l is the length of a cube edge, the volume of each cube is l^3 and the number of cubes in the whole foam volume $[V]$ is $[V]/l^3$. Each cube has 12 edges but each edge is common to 4 cubes, so that the number of edges is $n = 3[V]/l^3$. The numerical coefficient will be different from 3 in real foams which are not composed of cubes. Consequently,

$$V_0 - V = \frac{m[V]r^2}{l^2},\tag{6.26}$$

m being a numerical constant not much different from 3π. In other words, the expansion factor $1/\varphi = [V]/(V_0 - V) = l^2/mr^2$.

The rate of drainage in cylindrical capillaries is

$$\frac{dV}{dt} = \frac{n}{3} \cdot \frac{\pi g \rho r^4}{8\eta};$$

(see Equation 6.25); the value $n/3$ rather than n is used because, on the average, only one out of three veins is in vertical position. When r is substituted from Equation 6.26 and all numerical coefficients are combined to a number m_1, the equation

$$\frac{dV}{dt} = \frac{m_1 g \rho}{\eta} \cdot \frac{l^4}{[V]^2} (V_0 - V)^2 \tag{6.27}$$

results. The quantity $g\rho/\eta$ depends on the liquid only; the ratio $l^4/[V]^2$ is a property of a given foam and, ideally, does not vary during the drainage; thus dV/dt appears to be proportional to $(V_0 - V)^2$, in agreement with Equation 6.4, §95.

A different relation is obtained for those foams in which the liquid is present preponderantly in the bubble walls. Approximate the bubbles again as cubes; let w be the cube edge. The total number of cubes is $[V]/w^3$, and the number of lamellae, $3[V]/w^3$. Hence, $V_0 - V = (3[V]/w^3) w^2 \delta = 3[V]\delta/w$, so that $\delta = w(V_0 - V)/3[V]$. Substituting this δ into the above expression for the volume flow in slits and remembering that not all cube faces are vertical (and the bubbles are not cubes), equation

$$\frac{dV}{dt} = \frac{g \rho w^4}{K \eta [V]^3} (V_0 - V)^3 \tag{6.28}$$

is obtained; K is a numerical coefficient not very different from 324 ($= 3^3 \times 12$). This relation is the theoretical justification[20] for Equations 6.11 and 6.12, §95. However, according to Equation 6.24, the lamella "crusts" are neither plane nor parallel to each other. When δ is permitted to vary not only in time but also with the level of the horizontal section considered, then Equation 6.17 rather than Equation 6.11 can be obtained[23]. Straightforward application of Equation 6.24 leads[37] to the flow rate

$$\frac{dV}{dt} = \frac{2w}{3}\left(\frac{\eta}{g\rho}\right)^{0.5}\left(\frac{h}{t}\right)^{1.5} \tag{6.29}$$

so that this rate is proportional to $t^{-1.5}$ rather than to $(at+1)^{-1.5}$. However, the difference is not significant because the film shape prescribed by Equation 6.24 generally forms as a result of flow and thus, as a rule, is absent in fresh foams; consequently, t in Equation 6.29 is shorter than t in Equation 6.17.

Equations 6.27 to 6.29 refer to gravitational flow. The flow caused by capillary pressure P_c and expressed by Equation 6.22 also can be described by an analogous equation. As long as drainage renders the lamellae thinner without affecting their total area, the volume $V_0 - V$ is proportional to δ. Let δ_1 be the thickness before drainage, so that $V_0 = k_1\delta_1$, k_1 being the proportionality constant. Then Equation 6.22 is equivalent to

$$\frac{k_1}{(V_0-V)^2} - \frac{k_1}{V_0{}^2} = \frac{4P_c}{3\eta a^2}t$$

or, writing k_2 for $4P_c/3\eta a^2 k_1$,

$$\frac{1}{(V_0-V)^2} = k_2 t + \frac{1}{V_0{}^2},$$

in agreement with Equation 6.11, §95. Hence,

$$\frac{dV}{dt} = \frac{k_2 V_0{}^{1.5}}{2(1+k_2 V_0 t)^{1.5}} \tag{6.30}$$

The dimension of k_2 is $cm^{-3}\ sec^{-1}$. Of course, the flow dV/dt of Equation 6.30 is different from the experimental flow rate; it is only the volume of liquid supplied in unit time to Plateau borders.

Experimental Data

§99. Drainage and Viscosity. According to the theory[22] of drainage which leads to Equation 6.15 of §95, the descent of the liquid in the capillary spaces of a foam is retarded by the tendency of the liquid to rise in these spaces, as water does in a clean glass tube. If this hypothesis were correct, it would mean that solutions having a high surface tension γ would, on the whole, drain more slowly than those of a smaller γ. However the hypothesis is not applicable to ordinary foams because bubble walls filled with air (rather than with liquid) cannot be detected in them. Thus it appears at present that surface tension does not influence drainage. The importance of the Marangoni effect is mentioned below.

A property on which drainage undoubtedly depends is viscosity η. It is represented in each of the Equations 6.22 through 6.25 and Equations 6.27 through 6.30 of §§97 and 98, and its importance is confirmed by experiments. For instance, the empirical constant k of Equation 6.1, §95, was[20], as expected, inversely proportional to the η of seven lubricating oils in which foam drainage was measured at several temperatures. When the viscosity of 0.1% sodium dodecyl sulfate solution (containing also sodium sulfate) was raised by a factor 1.6 (approximately) by addition of glycerol, the rate of drainage was lowered[22] by a factor near 1.5 or 1.6.

The maximum value of dV/dt in pneumatic foams was[11], as an approximation, inversely proportional to the kinematic viscosity η/ρ cm^2/sec, ρ meaning the density of the liquid, when this quantity was altered by adding sucrose (200 g to 1 l.) to a solution of a commercial sulfonate. The effect of η was[9] (qualitatively) confirmed also when η was raised by 0.5 to 1% of the sodium salt of carboxymethyl cellulose.

The viscosity of solutions of hexadecyltrimethyl ammonium chloride in aqueous ethanol is raised by α-naphthol, especially when the mole ratio of naphthol to chloride is 1:1. When these solutions were shaken, the time of drainage was, for instance, 10, 400, and 30 sec for the above ratios 4:1, 1:1, and 1:4; thus the slowest drainage corresponded to the highest viscosity[24].

Viscosity changes cannot account for the observation[39] that single films of very dilute solutions (0.001 to 0.005 M) of dodecanol in aniline drain at a higher rate than those in more concentrated solutions (0.005 to 0.1 M). The authors believe that the Marangoni effect (§11) is responsible for the difference. When the surface layer of a film flows toward the bulk liquid, the concentration of the surfactant on the central area of the film-gas interface presumably decreases; hence γ increases there and the difference between the two surface tensions works against the flow. This effect is supposed to be strong enough only if the surfactant concentration c is relatively high; thus the drainage is retarded at a high but not at a low concentration.

An alternative explanation of the effect is based on surface viscosity η_s (see §10). The surface layer of a dilute solution has a lower η_s than those at higher c values. In §97, equations for two extremes can be found: in one instance the liquid flows between two rigid walls so that its layer immediately adjacent to the wall is stagnant; and in the opposite case there is no wall at all and the motion along the boundary with air is less restricted than in the center of the lamella. If surface viscosity is present, an intermediate flow pattern must occur; the flow near the interface is retarded, but not as much as when the interface is between a liquid and a solid. A theory of drainage for this intermediate case has been derived[27], but in the following only an experimental confirmation of the importance of η_s is reviewed. Unknown amounts of 1-dodecanol, too small to affect the bulk viscosity, lowered the rate of drainage of sodium dodecyl sulfate foams[22], and it is known that dodecanol stiffens the surface films of this sulfate. Thus 0.15% aqueous solution of sodium dodecyl sulfate 92% and dodecanol 8% affords foams with practically solid film walls[28]; for the most recent study of this effect see Trapeznikov[40]. However, the clearest proof of the relation between the rate of drainage and η_s was supplied by the measurements at varying temperatures, reported in the next section.

§100. Drainage and temperature.

Temperature would be expected to affect the rate of drainage by altering the bulk viscosity η of the liquid. The η of dilute aqueous solutions at 40° is approximately 0.66 that at 20°. The rate of drainage of 0.1% sodium dodecyl sulfate solutions at 20° was 0.63 to 0.75 that at 40°, and the ratio was near 0.7 for 0.25% solution of a commercial surfactant ("Igepon T"), whose main component is said to be $C_{17}H_{33}CO\cdot N(CH_3)CH_2CH_2\cdot SO_3Na$; thus dV/dt was roughly proportional[22] to $1/\eta$. (See also §101).

The temperature coefficient of η cannot account for sharp changes of dV/dt with temperature T in some systems. Thus, when liquid was supplied to the top of the foam column at the same rate as it drained out of the foam (see §96), then the volume V_0 of liquid in the column was at 40° almost twice as great as at or above 42°; the rate of feed and drainage was maintained constant and the liquid was 0.25% aqueous solution of commercial dodecyl sulfate (SDS). When this surfactant was extracted with ether, the dV/dt of the purified sulfate did not show any kink near 41°. When a small amount of 1-dodecanol was incorporated in the purified surfactant, the kink appeared again, and the T of the kink depended on the percentage of dodecanol introduced[26]. An analogous effect was observed[41] for single films of SDS + dodecanol and of sodium tetradecyl sulfate + tetradecanol. For instance, the kink (or the transition) occurred at 40° in the solution of 0.1% SDS + 0.01% dodecanol and at 10° when the concentrations were SDS 2%, dodecanol 0.02%.

It has been known for many years[42] that viscosity η_s of insoluble surface films manifested a sharp drop when T rose, in a specific interval, by one or two degrees; something like melting of the "crust" takes place. Ross[43] showed that these "melting temperatures" (or better, transition temperatures) were very near to those of the above kinks. Thus a sharp change of η_s was observed at 41° for 0.1% SDS plus 0.001% dodecanol, and at 21° for 0.2% SDS plus 0.002% dodecanol, According to Trapeznikov[40], the transition temperature of 0.1% SDS increases from 31° to 38° when the concentration of dodecanol increases from 0.001 to 0.1%.

In all probability, drainage is slow below the kink temperature because the liquid has to flow between two nearly solid walls; and it is rapid above the transition temperature because then the walls are nearly liquid.

§101. Drainage and composition.

Several series of experiments on the drainage of single surfactant films have been performed. When the films were horizontal, the rate of the decrease in their thickness (i.e., $-d\delta/dt$) was similar to that predicted by Stefan's equation (6.22), §97, but often not equal to it[39,44,45]. Presumably the lamellae were too small (i.e., δ/a was not a very small number), or the mutual repulsion of the electric double layers retarded the approach of the opposite surfaces (i.e., the ceiling and the floor of the lamella), or the surface viscosity of the "crusts" was too small, and so on.

The rate of thinning of vertical lamellae has been studied using optical methods, by Dewar[46], Miles[47], Mysels[38], and others. Thompson[48] measured the thickness of the horizontal middle section of vertical films of 2% sodium alkyl sulfate and 3% "alkylarylsulfonate" solutions. For the latter, the time needed to thin from 1.4 to 0.2 μ was 130 sec at 22° and 220 sec at 10°. Thus the rate of drainage increased with temperature, presumably because the bulk viscosity simultaneously decreased. When a calcium salt (e.g., 0.07% calcium carbonate) was added, that is, the hardness of water was raised, then the rate of thinning was not altered for about 400 sec but later was more rapid than in soft water.

The rate of drainage of "three-dimensional" foams was never measured for foams of short duration, such as those of aqueous alcohols, although the present high-speed photography ought to be perfectly adequate for such determinations. The rate of drainage of a foam produced by shaking aqueous mixtures of decanoic acid and decanol had a minimum at the molecular ratio 1:3 of the components[48a] (see also §99).

In the foams of individual soaps the "foam number" after 3 min (see §92) was found to increase in the following order: sodium linoleate < sodium stearate < sodium palmitate < sodium oleate < sodium laurate < sodium myristate; thus the rate of drainage was lowest for sodium myristate[49]. Potassium soaps afforded greater "foam numbers" than the corresponding sodium soaps. Sodium chloride and glycerol retarded the drainage[49]. Interpolation of the data published by Merrill[50] shows that the time t_1 for half the liquid in the foam to drain out was about 90 sec for 0.1% impure sodium oleate and about 20 sec for 0.1% sodium laurate; thus the drainage was more rapid in laurate solutions, contrary to the findings by Godbole[49]. Equation 6.1, §95, was invalid for this drainage.

For a commercial soap, t_1 was greater the greater the concentration. When this was about 0.1, 0.2, and 0.4%, t_1 was, respectively, 11, 30, and 105 sec[51]. The "foam power" (§92) of three commercial soaps in 0.1% solutions was 70 to 80 when the solutions were tested fresh, and only 10 to 30 after an aging in air for a day or two[6]. This change may have been caused by absorption of carbon dioxide from the air, and this absorption perhaps affected also the following experiments. When 50 cm^3 of approximately 0.13% soap solution was shaken and the foam permitted to drain, the volume V_0 in the foam after 1 min was 24 to 33 cm^3 when the solution was kept in a closed vessel, and only 10 to 12 cm^3 after keeping the liquid in air for 12 hr; the corresponding liquid volumes after 5 min of draining were[52] 15 cm^3 and 4 cm^3.

§102.   Equation 6.1, §95, was not confirmed by the drainage of foams of some synthetic surfactants[50]. In some instances, however, when a pneumatic foam is produced slowly (so that a considerable fraction of drainage takes place during foam formation), the major part of the observable drainage curve (V versus t) agrees with this equation; on the earlier stages dV/dt is too rapid, and on the latest stages, too slow[25]. This was true for 0.025 to 0.1% solutions of sodium *bis*(α-ethylhexyl) sulfosuccinate ("Aerosol OT") and 0.1 to 0.4% solutions of dodecanesulfonic acid. The values of the empirical constant k were, for instance, 0.005 sec^{-1} for 0.1% sulfosuccinate and 0.004 sec^{-1} for 0.1% sulfonic acid.

A pneumatic foam of sodium dodecyl sulfate drained more slowly, and the k for 0.2% solution appeared to be approximately 0.001 sec^{-1}. The rate of drainage increased from 1-dodecyl sulfate to 2-dodecyl sulfate and was even greater for the 3-, 4-, 5-, and 6-derivatives which showed almost identical rates. In 0.2% solutions, also foams of dodecylamine hydrochloride, *N,N*-diethyl-dodecylamine hydrochloride, and *N,N*-diethanol-dodecylamine hydrochloride $(HOCH_2CH_2)_2N \cdot C_{12}H_{25}$, HCl, drained more rapidly than those of sodium dodecyl sulfate, while there was but little difference between the individual rates[53].

The effect of concentration on the rate of drainage was measured[10] for a surfactant described as sodium hydroxylbutyl diphenylsulfonate. One-half of the initial liquid in the foam has collected at the bottom in about 3, 8, 18, 46, and 60 sec, respectively, when the solution contained 0.1, 0.2, 0.5, 1.6, and 2.0 g of surfactant per liter, so that t_1 of §101 was almost proportional to the concentration c. However, the initial liquid volume V_0 of these foams (produced by air injection) also was greater the higher c, so that more liquid had to drain out during the above 60 sec than during the t_1 of more dilute solutions. If Equation 6.1 is valid, then $kt_1 = 0.693$ and the value of k can be obtained from the above values of t_1, but it is not known which of the drainage equations was best satisfied in Gleim's experiments.

For several synthetic surfactants, dV/dt was not greatly different from those of typical soaps[50]. The time required for 40 cm^3 of liquid to drain from a foam containing 70 cm^3 of 0.1% solution was 27 sec for sodium laurate, 96 sec for commercial sodium octadecenyl sulfate, 122 sec for impure sodium oleate, 131 sec for sodium *bis*(α-ethylhexyl)sulfosuccinate, and 138 sec for a sodium alkylarene sulfonate.

According to Spitzer[54], sodium salts of straight alkyl sulfates and fatty acids give rise to slow-draining foams in the presence of "aliphatic foam promoters" to which presumably 1-dodecanol belongs (see §99). Fast-draining foam films are produced by some alkylarene sulfonates and branched alkyl sulfates. In many instances the transition temperature described in §100 was observed to increase with the number of carbons in the molecule of the surfactant and the "foam promoter."

§103.   The drainage of single saponin films has been studied by Scheludko[55,56]. The lamellae were horizontal and approximately circular, with a radius of approximately 100 μ and a starting thickness of about 0.1 μ. The solution contained 5 parts per million of saponin and also variable concentrations of salts (potassium chloride, barium chloride, and so on). The validity of Equation 6.22, §97, was assumed, and the experimental deviations were attributed to the mutual repulsion of electric double layers and similar causes (see §101).

The curves of Fig. 6.3, §94, indicate a strong effect of saponin concentration on the rate of drainage. This effect was not confirmed by Gray[57], who found k in Equation 6.1, §95, to be about 0.004 sec^{-1} for all solutions between 0.1 and 2%. Concentration had little influence on the rate of drainage also when individual saponins were tested at higher dilutions[2]. Thus the relative volume $(V_0 - V)/V_0$ of liquid remaining in the foam after 1 min of drainage was 0.40 in 0.1% and 0.19 in 0.01% digitonin solution. In 0.1% solutions this ratio little depended on the nature of saponin, but in 0.02% solutions it decreased from guaiac saponin to digitonin to commercial saponin to cyclamin to quillaja saponin.

According to Equation 6.1, §95, the ratio V/V_0 at any given time would be independent of the absolute value of V_0. In reality, this ratio increased with V_0 in foams of 0.1% saponin solutions[58]. When the initial expansion factor (§1), was 31, then V was equal to 0.5 V_0 after 30 min, but the equality $V = 0.5 V_0$ was achieved after less than 9 min when the initial expansion factor was about 8. Avdeev[59] compared the rate of drainage of aqueous saponin foams with those of alkylbenzene sulfonate solutions in dioctyl phthalate.

Siehr[60] submitted a foam of 0.4% saponin solution to the pressure of about 0.13 bar (= 100 mm of mercury) but noticed no alteration in the speed of drainage (see also §114).

§104.   When the concentration of gelatin increased from 0.1 to 2%, the empirical constant k of Equation 6.1 (§95) rose[57] from 0.01 to 0.017 sec^{-1}, while k of egg albumin and of peptone was almost independent of concentration (in the same interval) and equal, respectively, to 0.0055 sec^{-1} and 0.01 sec^{-1}. Drainage of egg white foam produced with a kitchen mixer was slower when the viscosity η of the liquid collected was greater[12]. However, the product of η and dV/dt was not constant; for instance, when η decreased in the ratio 1:0.56, dV/dt was raised in the ratio 5:1. Barmore's observations seem to contradict also the expectation that more succulent foams (that is, those having a smaller expansion factor or, in other words, thicker lamellae) would manifest a greater rate of drainage than the leaner foams. The expansion factor was gradually raised by continued whipping; the density of the foam was, for instance,

0.14 g/cm³ (that is, the expansion factor was about 7) after 1 min, and 0.10 g/cm³ (expansion factor about 10) after 5 min of beating. The drainage was more rapid in the latter foam. This differed from the heavier foam, not only in density but also in the average bubble diameter. Maybe this was the cause of the more rapid drainage.

Equation 6.1 (§95) was valid for the drainage of foams obtained on fermenting molasses and yeast[15], and for some beer foams[13] but other beers afforded[4] a curve (of V versus t) consisting of an early linear part, attributed to pure drainage, and a later exponential part, attributed to coalescence and bursting of bubbles.

The gas in the usual beer foams (obtained by pouring) is chiefly carbon dioxide; in fact, the "total head" of beer (defined in §92) is a linear function of the concentration of carbon dioxide in the sample tested[3]. When foam is produced by bubbling a gas through beer, the rate of drainage depends on the gas used. Thus the constant k of Equation 6.1 was for nitrogen only about one-fourth that for carbon dioxide[13]. Several gases were tested earlier[3], and the values of k for carbon dioxide, hydrogen, oxygen, nitrogen, and air were found to be in the ratio 1:0.41:0.34:028: 0.27. The authors point out that the order of gases in this series is almost identical with that of the product "coefficient of diffusion × solubility" of the gas in water. Thus the rate of drainage would be greater the more rapid the transfer of gas across the foam lamellae. However, this transfer is decisive for the bubble rearrangement reviewed in §112 but is unlikely to affect pure drainage. Perhaps different gases act differently on the foaming agents present in beer (see §151). If, for instance, the gas raises the rigidity of the "crust," this would retard drainage proper; and it is not difficult to imagine other mechanisms.

The effect of surface viscosity, treated in §99, presumably is responsible for the fact that gum arabic and albumin retard, and filtration accelerates, the drainage in beer[61]. Bulk viscosity probably is the cause of the drainage acceleration between 10 and 20°.

The constant k of beer foams varies[3,57] from 0.003 to 0.010 sec⁻¹ and thus is smaller than that of a wine for which $k = 0.03$ sec⁻¹ is given[62].

Data are available also for the rate of drainage of foams of alkaline lignin solutions[16] and of soaps made from the sulfate liquor of wood pulp manufacture[63].

§105.   The drainage of lubricant foams was studied by Brady and Ross[20].

The half-life t_1 of various vegetable and animal oils was determined[7] for pneumatic foams. It ranged from 3 min for hayfish oil to 53 min for a lecithin-containing soybean oil. Purification of the oil generally lowered t_1, and addition of lecithin extended it; presumably this was an effect of surface viscosity. When different gases were bubbled through a sample of linseed oil, t_1 was 36, 36, 31, and 22 min, respectively, for air, nitrogen, hydrogen, and carbon dioxide. Strangely enough, this order of gases is identical with that observed on beer foams (§104); it should be remembered (see §102) that kt_1 is a constant. Obviously, additional experiments are needed to detect the cause of this regularity.

The drainage of the three-phase organic foams described in §89 could be observed undisturbed for a short time only because, later, bubble-bursting interfered[64]. Thus in 0.2% suspension of sodium alkylbenzene sulfonate in dioctyl phthalate, the foam density decreased for about 12 min because drainage was more rapid than collapse and the amount of liquid in unit volume of foam was diminishing; but 2 min later the density was higher than ever because of the collapse of numerous bubbles, which then gave their liquid to those remaining. The first, that is, the drainage part of the curve "V/V_0 versus t" was similar to those shown in Fig. 6.3, §94. It could be represented by a combination of Equations 6.1, 6.3, and 6.15 of §95. The effect of the concentration was unexpected. The half-time of drainage t_1 was about 3 min for 0.2% and 5% suspensions, about 5 min in 0.5% suspension, and about 6 min in 1% and 3% suspensions.

§106. Rate of drainage and rate of collapse. As pointed out in §91, drainage proper is, in principle, completely independent of bubble rupture. It is interesting to see whether, experimentally, there is any correlation between the rate of drainage and the rate of collapse. Unfortunately, the data available for this comparison are rather meager.

Lederer[51] compared the time t_e (§61) with the half-time t_1 of drainage for foams obtained by bubbling air through solutions of commercial soaps and surfactants. Naturally, t_e was much longer than t_1 but the ratio $t_e:t_1$ was not constant for different foams. For instance, t_1 was 12 sec for both 0.05% and 1% solutions of "Igepon A" (which is said to be mainly the sodium salt of the oleyl ester of 2-hydroxyethane sulfonic acid,

$$C_{17}H_{33}CO \cdot O \cdot CH_2CH_2SO_3Na),$$

but the corresponding t_e values were, respectively, 0.5 and 2 hr.

Another comparison was performed by Merrill[50]. Unfortunately, the foam volume used to determine the persistence of the most stubborn bubble (see §61) was not only much smaller (presumably less than 1/100th) than the volume employed in the drainage tests, but, it seems, was also not in a constant ratio to the latter. Hence it is not clear what significance should be attached to the following numbers. When only one bubble remained in the smaller sample, the fraction V/V_0 of the liquid collected under the voluminous foam was 0.75 for 0.1% solution of sodium laurate, 0.23 for 0.1% sodium *bis*(α-ethylhexyl) sulfosuccinate, 0.57 for 0.1% technical sodium octadecenyl sulfate, 0.53 for 0.1% commercial sodium alkylarene sulfonate, and so on. Obviously, if the volumes were equal, the ratio V/V_0, after the bursting of the penultimate bubble, would have been in all systems equal to 0.999. . . .

In solutions of commercial surfactants, no clear correlation was found[65] between Stiepel's "foam number" and the height of foam produced by shaking or by vibrating a perforated plate.

The L_g of §60 was compared with L_l of §95 for several liquids. In only one instance, for a lubricating oil which also otherwise manifested an unusual behavior, was L_g shorter than L_l, so that liquid seemed to be present in the foam when there was no gas left in it. In all other systems L_g was longer than L_l; that is, bubbles persisted after the major part of the liquid in the bubble walls had oozed out. The ratio L_g/L_l usually was less than 2 in lubricating oils, but reached 7.5 in a sample of beer. The value of L_l systematically varied with the viscosity η of lubricating oils but, as expected, no correlation was present between η and L_g; at a given η, the values of L_g were different for different oils.

Introduction of an additional compound differently affects the L_g and L_l of the foaming solution[66]. The initial liquid consisted of water, a commercial surfactant ("Nacconol NRSF," a sodium alkylbenzene sulfonate), and sodium silicate. Tributyl phosphate was the second best foam inhibitor if L_l was considered, but was sixth in its effect on L_g. On the other hand, 4-methyl-2-pentanol was the second best inhibitor in respect to L_g but occupied the fourth place in respect to L_l. In other words, hexanol destroyed foam when this was still very wet, and the phosphate accelerated drainage more than collapse so that bubble-bursting took place chiefly in fairly dry foams.

References

1. Stiepel, C. *Seifensieder-Ztg.* 41:347 (1914).
2. Sieburg, E., and F. Bachmann. *Biochem. Z.* 126:130 (1922).
3. Helm, E., and O. C. Richardt. *J. Inst. Brewing* 42:191 (1936).
4. Brenner, M. W., R. E. McCully, and S. Laufer. *Am. Brewer* 84, No. 2:39, 49 (1951).
5. Bailey, M. I. *Ind. Eng. Chem.* 27:973 (1935).
6. Preston, W. C., and A. S. Richardson. *J. Phys. Chem.* 33:1142 (1929).
7. Kaufmann, H. P., and P. Kirsch. *Fette Seifen* 47:196 (1940).
8. Arbuzov, K. N., and B. N. Grebenshchikov. *Zh. Fiz. Khim.* 10:32 (1937).
9. Kruglyakov, P. M., and P. R. Taube. *Zh. Prikl. Khim.* 39:1499 (1966).
10. Gleim, V. G., V. Ya. Khentov, and V. M. Vilenskii. *Kolloidn. Zh.* 28:648 (1966).
11. Kruglyakov, P. M., and P. R. Taube. *Zh. Prikl. Khim.* 38:1514 (1965).
12. Barmore, M. A. *Colo. Agr. Expt. Sta., Tech. Bull.* 9 (1934).
13. Blom, J. *J. Inst. Brewing* 43:251 (1937).
14. Clark, G. L., and S. Ross. *Ind. Eng. Chem.* 32:1594 (1940).
15. Ramaiah, N. A., and S. K. D. Agarwall. *J. Proc. Oil Technologists' Assoc. India, Kanpur* 12:92 (1958). *Chem. Abstr.* 53:14385 (1959).
16. Erbring, C., and H. Peter. *Kolloid-Z.* 96:47 (1941).
17. Haas, P. A., and H. F. Johnson. *A.I.Ch.E.J.* 11:319 (1965). *Ind. Eng. Chem., Fundamentals* 6:225 (1967).

18. Wace, P. F., P. J. Alder, and D. L. Banfield. *Chem. Eng. Progr., Symp. Ser.* 65, No. 91:19 (1969).
19. Rubin, E., C. R. LaMantia, and E. L. Gaden. *Chem. Eng. Sci.* 22:1117 (1967).
20. Brady, A. P., and S. Ross. *J. Am. Chem. Soc.* 66:1348 (1944).
21. Lederer, E. L. *Seifensieder-Ztg.* 63:331 (1936).
22. Miles, G. D., L. Shedlovsky, and J. Ross. *J. Phys. Chem.* 49:93 (1945).
23. Jacobi, W. M., K. E. Woodcock, and C. S. Grove. *Ind. Eng. Chem.* 48:2046 (1956).
24. Nash, T. *J. Appl. Chem. (London)* 7:392 (1957).
25. Ross, S. *J. Phys. Chem.* 47:266 (1943).
26. Epstein, M. B., J. Ross, and C. W. Jakob. *J. Colloid Sci.* 9:50 (1954).
27. Leonard, R. A., and R. Lemlich. *A.I.Ch.E.J.* 11:18 (1965).
28. Lemlich, R. *Ind. Eng. Chem.* 60:16 (1968).
29. Shih, F. S., and R. Lemlich, *A.I.Ch.E.J.* 13:751 (1967).
30. Shih, F. S., and R. Lemlich. *Ind. Eng. Chem., Fundamentals* 10:254 (1971).
31. Hofer, M. S., and E. Rubin. *Ind. Eng. Chem., Fundamentals* 8:483 (1969).
32. Weissman, E. Y., and S. Calvert. *A.I.Ch.E.J.* 13:788 (1967).
33. Manegold, E. *Schaum.* Heidelberg: Strassenbau, Chemie und Technik. (1953), p. 220.
34. Plateau, J. *Mém. Acad. Roy. Sci. Belg.* 37 (1869), 8th ser.
35. Stefan, J. *Sitzber. Akad. Wiss. Wien, Math.-Naturw. Kl., Abt. II,* 69:713 (1874).
36. Bikerman, J. J. *J. Colloid Sci.* 2:163 (1947).
37. Bikerman, J. J. *J. Colloid Sci.* 11:299 (1956).
38. Mysels, K. J., K. Shinoda, and S. Frankel. *Soap Films.* New York: Pergamon. (1959), pp. 27, 7.
39. Radoev, B., E. Manev, and I. Ivanov. *Kolloid-Z. Z. Polymere* 234:1037 (1969).
40. Trapeznikov, A. A., and E. S. Dokukina. *Kolloidn. Zh.* 32:272 (1970).
41. Epstein, M. B., et al. *J. Phys. Chem.* 58:860 (1954).
42. Trapeznikov, A. A. *Acta Physicochim. URSS.* 20:589 (1945).
43. Ross, J. *J. Phys. Chem.* 62:531 (1958).
44. Sheludko, A. *Dokl. Akad. Nauk. SSSR.* 123:1074 (1956).
45. Manev, E., and M. Bouleva. *Abhandl. Deut. Akad. Wiss., Kl. Chem., Geol., Biol.* 6b:557 (1966).
46. Dewar, J. *Proc. Roy. Inst. Gt. Brit.* 22:179, 359 (1917/8).
47. Miles, G. D., J. Ross, and L. Shedlovsky, *J. Am. Oil. Chem. Soc.* 27:268 (1950).
48. Thompson, W. *J. Appl. Chem. (London)* 12:12 (1962).
48a. Shah, D. O. *J. Colloid Interface Sci.* 37:744 (1971).
49. Godbole, N. N., and Sadgopal. *Kolloid-Z.* 75:193 (1936).
50. Merrill, R. C., and T. F. Moffett. *Oil Soap* 21:170 (1944).
51. Lederer, E. L. *Angew. Chem.* 47:119 (1934).
52. Hetzer, *Chem. Ztg.* 72:715 (1932).
53. Padgett, A. R., and E. F. Degering. *Ind. Eng. Chem.* 32:486 (1940).
54. Spitzer, E. L. T. M. *World Congr. Surface Active Agents, 3rd Congress, Cologne,* 1960, 2:556 (1961).

55. Scheludko, A., and D. Exerowa. *Kolloid-Z.* 165:148 (1959).
56. Scheludko, A., and D. Exerowa, *Kolloid-Z.* 168:24 (1960).
57. Gray, P. P., and I. Stone. *Wallerstein Lab. Commun.* 3:159 (1940).
58. Siehr, A. *Kolloid-Z.* 77:27 (1936).
59. Avdeev, N. Ya. *Uch. Zap. Kabardino-Balkarsk. Univ., Ser. Fiz.-Math.,* No. 24, 9 (1965). *Chem. Abstr.* 68, No. 60772 (1968).
60. Siehr, A. *Kolloid-Z.* 78:156 (1937).
61. Schwarz, R., and S. Laufer. *Am. Brewer* 70, No. 12:30 (1937).
62. Amerine, M. A., L. P. Martini, and W. DeMattei. *Ind. Eng. Chem.* 34:152 (1942).
63. Elkonin, N. V. *Bumazh. Prom.* 19, No. 2:21 (1941). *Chem. Abstr.* 38:2819 (1944).
64. Tarakanov, O. G., and E. G. Eremina. *Kolloidn. Zh.* 27:274 (1965).
65. Raison, M., and R. Matalon. *Mém. serv. chim. état (Paris)* 34:353 (1948).
66. Ross, S., and G. J. Young. *Ind. Eng. Chem.* 43:2520 (1951).

CHAPTER 7

MECHANICAL PROPERTIES OF FOAMS

§107. Density and expansion factor. The density $[\rho]$ of a foam is the mass of the foam divided by its volume $[V]$. As the total mass is the sum of the gas and liquid masses (m_1 and m_2), the equation for $[\rho]$ is

$$[\rho] = \frac{m_1 + m_2}{[V]}. \tag{7.1}$$

The reproducibility and the stability of foams are so poor that a precise determination of $[\rho]$ is unnecessary; hence it is permissible in practically all instances to neglect the mass m_1 of the gas phase and to use the equation

$$[\rho] = \frac{m_2}{[V]}. \tag{7.2}$$

For aqueous foams the approximation $\rho_2 = 1.0$ usually is sufficient; ρ_2 is the density of the liquid phase. Thus the mass (grams) of the liquid in the foam is numerically equal to its volume (V_0 cm^3). In general,

$$\frac{[\rho]}{\rho_2} = \frac{V_0}{[V]}; \tag{7.3}$$

especially for nonaqueous foams the correct value for ρ_2 must be used.

It is fortunate that the value of $[\rho]$ need not be known with high precision. Otherwise, it would have been necessary to take account of the vapors in the gas phase, of the capillary pressure in the bubbles (which affects the gas density ρ_1), of the difference between the chemical compositions (and consequently the densities) of the liquid in the foam and in the bulk, of the negative pressure in Plateau's borders, and so on.

Expansion ratio or expansion factor is defined as the ratio of the total volume $[V]$ to the volume V_0 of the liquid phase in the foam; it is equal to $1/\varphi$ of §1 or to the ratio $\rho_2/[\rho]$, if m_1 may be neglected in comparison with m_2. Naturally, φ and $[\rho]/\rho_2$ are related to bubble dimensions. If the major part of the liquid is present in the bubble walls of thickness δ and the lamellae are approximated as squares of edge l, then the volume of a lamella is δl^2. The number of lamellae in 1 cm^3 of foam is approximately $3/l^3$. Thus the total volume of the lamellae in 1 cm^3 is about $3\delta/l$. Hence $\varphi \approx [\rho]/\rho_2 \approx 3\delta/l$. On the other hand, in many foams more liquid is situated in the Plateau borders than in the bubble walls (see §98). Let A be the average cross-section of the borders; then the average volume of one liquid vein is Al. The number of veins in 1 cm^3 of foam again is $3/l^3$, so that the total volume of the veins is $3A/l^2$ per cm^3. In these systems, $\varphi \approx [\rho]/\rho_2 \approx 3A/l^2$. The coefficient 3 is calculated assuming the bubbles to be cubes, but its numerical value will not be greatly different for real foams. If the volume of the lamellae is comparable with that of the Plateau borders, then

$$\varphi \approx \frac{3A}{l^2} + \frac{3\delta}{l} . \tag{7.4}$$

The terms "foaming power" and "the increase in volume on foaming" have been given[1] to the expression $100\left(\dfrac{\rho_2}{[\rho]} - 1\right)$, which can also be written as $100(1 - \varphi)/\varphi$; it is 100 times the ratio of the gas volume to the liquid volume in foam. Another function of the foam volume was used[2] for estimating the efficiency of a foaming agent. If m grams is the mass of the solute present in V_0 cm^3 of foam, then V_0/m is the foam volume produced by 1 g of the solute and is a measure of the foaming ability of the latter (see §55).

Probably the most common method of determining $[\rho]$ is to fill a vessel of a known mass M_0 and a known capacity $[V]$ with the foam and to weigh the system. If its mass is M_1, then

$$[\rho] = \frac{M_1 - M_0}{[V]} . \tag{7.5}$$

In another method the experimenter starts from a known volume V_0 of the liquid. This is completely transferred into foam (for instance, by whipping or gas injection). If the volume of this foam is $[V]$, then $[\rho] = \rho_2 V_0/[V]$, according to Equation 7.3, which is valid whenever $m_2 \gg m_1$.

Because, as a rule, a liquid absorbs radiation more than a gas does, foam density can be determined from the absorption coefficient. This was achieved, for instance, by using ^{137}Cs or a mixture of ^{90}Sr and ^{90}Y as sources of, respectively, γ and β radiation[3,4]. Let l_f cm be the thickness of the foam layer traversed by the radiation; if the absorption is such as would be achieved by a l_2 thick layer of liquid (and if the foam is uniform), then $V_0/[V] = (l_2/l_f)^3$. When the liquid contains heavy atoms (such as iodine), which strongly absorb X-rays, this absorption can be utilized to determine the liquid content of a foam and some related properties[5]. See also §108.

§108. Foam density is characteristic for a given sample and a given moment of time only. Of the four main processes generally occurring in foams, gas diffusion from bubble to bubble presumably has almost no effect on $[\rho]$. Evaporation raises the latter. The competing effects of drainage and bursting influence $[\rho]$ in a less obvious manner. When the bubbles in the upper part of a foam column collapse and the liquid thus set free slowly percolates through the lower strata, then the density of the latter increases in time. Apparently, an example of this effect was observed in concentrated aqueous solutions of heptanoic acid[6].

The variation of density from top to bottom of a foam column was determined by electric means[7]. A pair of electrodes was immersed in the foam at variable heights h cm above the initial level of the liquid through which air was bubbled. It was assumed (see §131) that the ratio I_f/I_0 of the current intensities between the electrodes when the space was filled with the foam and with the solution, respectively, was equal to the ratio $[\rho]/\rho_2$. At a given rate of air injection into a solution of trisodium phosphate, the value of I_f/I_0 was smaller, the greater h; for instance, it was near 0.3 at $h = 2$ cm, 0.1 at $h = 3.5$ cm, and 0.03 near $h = 4$ cm. It is clear that the expansion factor in this foam (containing no surfactant) was small. At a constant h, the ratio I_f/I_0 was greater for smaller bubbles; e.g., at 4 cm it was 0.3 when the bubble diameter was 5 μ, and 0.03 when this was 100 μ. When the rate of air injection increased, more liquid was lifted between the electrodes, and I_f/I_0 rose. When the phosphate concentration was made greater (e.g., from 0.001 to 0.5 M), the amount of liquid in foam rose, for instance, from $I_f/I_0 = 0.2$ to $I_f/I_0 = 0.8$. The latter system would be classified as gas emulsion.

The unusual time variation of the density of organic foams is referred to in §105. More common is a steady decrease of density caused by drainage. If the time dependence of drainage is known, the decrease of $[\rho]$ in time is easily calculated. When, for instance, Equation 6.1 of §95 is valid, then

$$[\rho] = \frac{\rho_2 V_0}{[V]} e^{-kt}. \tag{7.6}$$

As before, ρ_2 is the density of the liquid, V_0 the initial volume of the liquid in the foam, and $[V]$ the constant foam volume (no bubble bursting!). Thus the ratio $\rho_2 V_0/[V]$ is constant during drainage, and foam density decreases exponentially with time. When the system obeys Equation 6.9 §95, then

$$[\rho] = \frac{\rho_2 V_0}{[V]} \left(\frac{kV_0}{t + kV_0} \right); \tag{7.7}$$

as long as t is small, $[\rho]$ decreases very slowly, but it is almost inversely proportional to t when t is large.

Some time passes between the formation of a foam and the measurement of its density. It is clear from Equation 7.6 that the $[\rho]$ obtained will be smaller than the initial density, the greater the rate constant k. Moreover, some drainage occurs also during foam formation, so that $[\rho]$ cannot be measured exactly at $t = 0$. This produces a tendency for rapidly draining foams to afford low density values or high expansion factors, and for slow-draining foams to possess high values of $[\rho]$ and small expansion ratios.

This presumably is the reason for the fact that it is much easier to obtain highly expanded foams from an aqueous than from a nonaqueous solution. The majority of non-aqueous liquids which give a reasonably persistent foam are markedly more viscous than water and consequently have a lower rate of drainage. When their $[\rho]$ is measured, the volume of liquid in the foam still has almost its maximum value.

Because drainage takes place all the time the foam is being formed, foams produced rapidly are likely to have a greater density than those produced slowly from an identical solution. This apparently was noticed by Koizumi[4] and is particularly easy to observe in pneumatic foams. When a gas is bubbled through an aqueous solution of a commercial surfactant, the density of the foam obtained is smaller, the slower the gas injection. This behavior is clearly seen in Table 7.1, based on a study performed for the Office of Saline Water in 1968 but not published (see also §172). The gas used was nitrogen, and "Alipal CO-433"

$$[\text{chiefly } C_9H_{19} \cdot C_6H_4O(CH_2CH_2O)_nSO_3Na]$$

was the foaming agent. It is seen that increasing the rate of nitrogen supply by a factor of two may increase the density 20-fold.

§109. It follows from the statements of §108 that numerical values of $[\rho]$ are not particularly valuable as long as the conditions of foam formation are not specified. Nevertheless, a brief review of these values is presented in this section.

The range of possible foam densities extends from almost zero to an arbitrary limit beyond which gas emulsions rather than foams are encountered. Extremely low values of $[\rho]$ are found for a vessel containing only one or a few aqueous foam films. Suppose that the capacity of the vessel is 100 cm^3, the total area of the lamellae in it is 100 cm^2, and the average thickness of the lamellae is 10^{-5} cm; then $[\rho]$ would be 10^{-5} g/cm^3, and

Table 7.1. Density and Rate of Gas Injection

Solution	Rate (cm^3/sec)	$[\rho]$ (g/cm^3)	Expansion factor
0.1% Alipal CO-433 +0.05% FeCl$_2$, 4H$_2$O	3.5	0.00054	1840
	7	0.0038	260
0.1% Alipal CO-433 +0.0005% FeCl$_2$, 4H$_2$O	3.5	0.00073	1360
	7	0.0146	69

expansion factor would be 100,000. Let the boundary between foams and gas emulsions be placed at equal gas and liquid volumes; then the greatest value of the ratio $[\rho]/\rho_2$ is 0.5 for foams, and $0.5\rho_2$ is also the minimum density of gas emulsions; ρ_2 is the density of the liquid.

A quantity closely related to $[\rho]/\rho_2$, namely I_f/I_0 (see §108) was determined for many aqueous foams containing inorganic salts[8]. Electrodes were placed near the top of a shallow foam column less than 6 cm tall. The ratio I_f/I_0 varied between 0.04 and 0.2. When solutions of equal molarity (not normality!) were compared, this ratio apparently decreased in the series sodium perchlorate, sodium nitrate, sodium bromide, sodium chloride, sodium fluoride, sodium sulfate, sodium carbonate, and trisodium phosphate. Different cations influenced the conductance less than different anions. An increase in temperature lowered I_f/I_0, and the relative lowering was steeper for trisodium phosphate than for sodium sulfate or sodium chloride (all three in 0.04 M solutions)[9].

When carbon dioxide was bubbled through an aqueous solution of a commercial sodium alkyl sulfate, the ratio $[\rho]/\rho_2$ was[10] approximately 0.2 for 0.2 to 2% concentrations but decreased to 0.13 when the concentration was 0.05%. A similar tendency was observed for a saponin; $[\rho]/\rho_2$ was near 0.22 between 0.2 and 2% but only 0.08 at 0.05%. Gum arabic had $[\rho]/\rho_2$ of 0.04 and 0.20 in, respectively, 1% and 2% solutions. In the range 0.2 to 2%, peptone and gelatin had density ratios of 0.16 to 0.20, but the ratio was near 0.03 for 0.05% solutions.

According to Nakashima[11], the $[\rho]$ of foams of "Teepol" and sodium oleate solutions increases with concentration until the critical micelle concentration is reached and remains nearly constant beyond it. "Teepols" are mixtures obtained by reacting olefins C_8 to C_{18} with sulfuric acid.

The $[\rho]/\rho_2$ of the foam of 0.1% saponin solution was 0.13 when air injection was rapid (under a pressure of 75 mbar) but only 0.03 when the driving pressure was 60 mbar; unfortunately, the rate of air flow was not measured[12], so that a comparison with Table 7.1 (§108) is not possible.

Naturally the density of a foam is affected not only by drainage but also by the evaporation of solvent. Presumably this effect is important for the $[\rho]$ of egg white foams produced by whipping. Thus, Barmore[13] noticed that a longer beating resulted in a more expanded foam; the values ranged from 0.14 to 0.10 g/cm^3. In other tests, $[\rho]$ was 0.15 to 0.19 g/cm^3, depending on the speed and duration of whipping[2], or 0.09 to 0.21 g/cm^3, depending on the instrument employed[1]. Dilution with water did not markedly affect the $[\rho]$ as long as the mixture contained more than 60% of egg white, but $[\rho]$ reached 0.25 (instead of 0.15) when one part of egg white was diluted with four parts of water[2]. The ratio $[\rho]/\rho_2$ in pneumatic foams of egg albumin was[10] 0.15 for both 0.5% and 2%, but only 0.04 for 0.08% solutions.

The effect of pH on the $[\rho]$ of egg white apparently is poorly reproducible. The $[\rho]$ of egg white foams (produced by whipping) was doubled by about 0.7% cottonseed oil[2] or by about 0.15% olive oil[1]. It is difficult to explain these changes. The $[\rho]$ of foams obtained by beating whole egg powder with 1.8 part of water was not affected by whey solids or lactose, but addition of 0.4 part sucrose lowered[14] $[\rho]/\rho_2$ from 0.24 to 0.15. Why? Small additions (below 0.5%) of commercial surfactants raised $[\rho]/\rho_2$ (up to 20%)[14].

When malt residue was shaken with water, the most voluminous (three-phase) foam was observed at the ratio of malt to water equal to 1:33; and the $[\rho]$ of this foam was[15] about 0.14 g/cm^3.

The ratio $[\rho]/\rho_2$ of pneumatic beer foams, carbon dioxide being the gas injected, was 0.20 to 0.23 for four different beer samples; dilution with water lowered this ratio until it became 0.04 to 0.13 for beer plus nine parts of water[10]. As would be expected, when foam was formed by condensation (§23), beer containing a higher amount of carbon dioxide gave a less dense foam than less carbonated beers[16].

The $[\rho]/\rho_2$ of pneumatic foams, produced by injecting moisture-free carbon dioxide into suspensions of sodium alkylbenzene sulfonate in dioctyl phthalate (see §89) was 0.12 to 0.24, depending on the concentration (0.2 to 5%)[17]. During the life of the foam (15 to 40 min) the density varied as reported in §105.

High densities have been observed in foams obtained by shaking solutions of commercial surfactants in liquids as viscous as glycerol[18]. Thus 1% solution of *N*-acyl-*N,N′,N′-tris*(hydroxyethyl) ethylenediamine hydrate in glycerol had $[\rho] = 0.30$ g/cm^3; sodium "aryl sulfonate" in glycerol at identical concentration showed $[\rho] = 0.48$; and 0.5% and 1.0% solutions of poly(ethylene glycol) in glycol had densities of 0.17 and 0.12 g/cm^3.

Ten lubricating oils of very different foam persistences (from 18 to 4900 min) gave, on whipping, systems whose densities varied only between 0.39 and 0.62 g/cm^3 and which, consequently, would be classified as gas emulsions[19]. As emphasized in §51, mechanical agitation of very viscous liquids often is a poor process for incorporating gas into liquid; presumably, more foam-like structures would have been obtained by using air injection through a capillary or a porous sparger.

§110. *Bubble size.*   The average bubble size and the distribution of bubbles according to their dimensions are important properties of every foam and deserve more attention than is usual.

The "diameter" *l* of bubbles is the quantity most commonly measured. As foam bubbles are not spheres, they have, strictly speaking, no diameter. Usually, the largest distance between two opposite walls of a bubble is identified with *l*. Fig. 7.1 is a simplified drawing from a photograph of a foam. The distance *aa* is the "diameter" of the pentagonal bubble

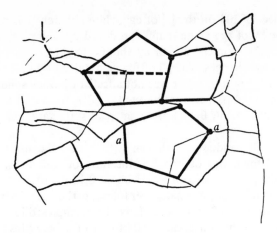

Fig. 7.1. Foam bubbles as seen through a glass wall. The distance *aa* is identified with the "diameter" of a bubble.

near the bottom of the drawing. The *l* generally is determined on photographs taken either from above, that is, of the upper boundary of the foam sample, or through a glass wall of the vessel to which the foam is confined. It should be checked in every system whether the diameters measured in this manner are in fact representative for the whole foam.

This question was settled by Savitskaya[20] for pneumatic foams of 0.125% solution of sodium 6-butylnaphthalene-1-sulfonate. They were placed in a tray whose bottom was cooled with liquid oxygen. The bubble walls solidified before any significant alteration of the dispersity or any marked drainage could take place. The solid foam was then cut along different planes, and bubble dimensions were measured on each section. The frequency distribution (i.e., the relative frequency of the bubbles of a given diameter) proved to be independent of the position and the direction of the section. The most frequent diameter was a little greater for more dilute solutions (120 to 130 μ for 0.06% concentration) than for those less dilute (80 to 90 μ at 2%).

Similar observations have been published by Chang[21]. Air was injected into flowing solutions of (apparently) a heptadecyl benzimidazole, a saponin, impure sodium heptadecyl sulfate, sodium dodecyl sulfate, and a protein hydrolyzate. The foam obtained was forced through a column of saddles to render it more uniform. Then it was frozen by liquid oxygen; the freezing process lasted 15 to 20 sec. The mean diameters in the bulk of the foam differed from those on the surface by not more than 10%. The rate of air injection affected both the expansion factor $1/\varphi$ and the average *l*; for 6% solutions of protein hydrolyzate, *l* was 94, 96, and 120 μ when $1/\varphi$ was, respectively, 11.0, 9.7, and 7.6.

When a stable foam is forced through a tube whose diameter is commensurate with the average bubble diameter, a succession of lamellae, each reaching across the whole tube, forms and moves along the tube. Thus the number of bubbles present in a given volume of foam can unambiguously be measured and the equivalent diameter of the bubbles calculated[22]. Suppose that 1 cm^3 of foam gave rise to $n+1$ lamellae. Then the mean bubble volume was $1/n$ cm^3 and the diameter of the sphere of this volume was $l = (6/\pi n)^{\frac{1}{3}}$.

It is very difficult to produce a monodisperse foam, that is, one in which all bubbles have identical values of l. In practically every foam there is a distribution of bubbles among different bubble volumes. Fig. 7.2 is a histogram based on the data[23] on a foam of an aqueous solution of "Teepol" (see §109) and sodium alginate. The abscissa shows the ranges of bubble radii (in microns), and the ordinate, the number of bubbles in this range. It is seen that the most frequent radius in this foam was near 40 μ, that is, the most common diameter was about 80 μ, very similar to those mentioned earlier in this section.

Fig. 7.3 requires a more detailed explanation. If bubbles are approximated as spheres and if, in unit volume of the foam, there are n_1 bubbles of an approximate diameter l_1, n_2 bubbles of diameter l_2, n_3 bubbles of diameter l_3, and so on, then the total surface area of these bubbles is $S = \pi n_1 l_1^2 + \pi n_2 l_2^2 + \pi n_3 l_3^2 + \dots$ cm^2/cm^3. This S is the "specific surface" of the foam; its dimension is cm^{-1}. If each of the terms in the

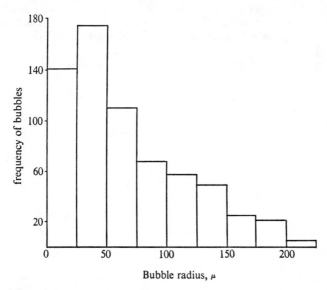

Fig. 7.2. Distribution of bubbles according to their radius. Ordinate: number of bubbles having radii between 0 and 25, 25 and 50, and so on, microns in a foam of "Teepol" (data of reference [23]).

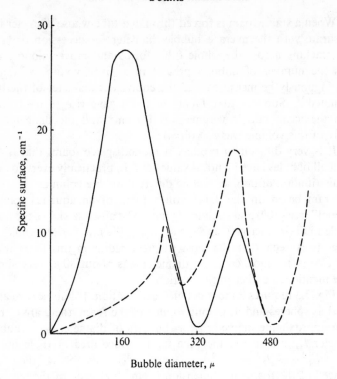

Bubble diameter, μ

Fig. 7.3. Specific surface (cm⁻¹) of a sulfonate foam as a function of the bubble diameter (microns). The continuous curve is for 10-sec-old foams, and the broken line is for foams aged for 3 min (after reference 24).

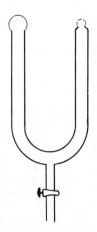

Fig. 7.4. Dependence of capillary pressure on bubble diameter. The large bubble grows at the expense of the small bubbles (after Plateau).

above equation is represented as ΔS (that is, $\Delta S_1 = \pi n_1 l_1^2$, $\Delta S_2 = \pi n_2 l_2^2$, etc.), then $\Delta S/S$ can be plotted as a function of the diameter[24]. This plot is shown in Fig. 7.3 for a pneumatic foam of 0.1% solution of a sulfonate surfactant and trisodium phosphate. The continuous curve is for a 10-sec-old foam. It is seen that the fraction of the total surface contributed by bubbles about 160 μ across is greater than any other comparable fraction. Unexpectedly, there is a weaker maximum near 400 μ. The broken curve is for a 3-min-old foam and is discussed in §113.

§111. *Capillary pressure.*   The earliest work on the capillary pressure in bubbles is by Plateau (see §41). The apparatus used by him is schematically shown in Fig. 7.4. It is a U tube provided with a stopcock. One of the openings of the tube is wide, and the other, narrow. When a soap solution is smeared on both, and additional air is introduced through the stopcock, a big soap bubble forms at the wide, and a small one at the narrow hole. Then the stopcock is closed. The smaller bubble gradually shrinks, and the bigger simultaneously expands until the former almost disappears.

Gravitation affects the thin bubble walls so little that their shape is almost spherical; thus each has only one radius of curvature. Let R_1 and R_2 be these radii for, respectively, the smaller and the larger bubble. Then the capillary pressure in the former is $4\gamma/R_1$, and in the latter, $4\gamma/R_2$; as before, γ is the surface tension of the liquid. The numerical factor is 4 rather than 2 because each lamella has two surfaces and the surface tension of each of these exerts capillary pressure equal to $2\gamma/R$; R is the radius of curvature in general. The pressure difference

$$4\gamma \left(\frac{1}{R_1} - \frac{1}{R_2} \right) \text{ g/cm·sec}^2$$

drives the gas from the small to the big bubble until the former is reduced to a slightly convex (to the outside) lamella whose curvature is equal to that of the remaining bubble. Plateau's famous experiment has been repeated many times (e.g., by Dupré[25]), and both its results and its explanation are certain.

Solids do not possess surface tension analogous to that of liquids and thus do not give rise to a capillary pressure. A bubble was formed from a solution of 2 g "Nekal BX" (an impure di-isobutylnaphthalene sulfonate) and 6 g gelatin in 92 g water[26]. The pressure excess inside the bubble (measured with a gauge) was 50 dynes/cm^2. When the solvent was permitted to evaporate, the gelatinous wall tended to shrink, and the pressure excess rose to 300 dynes/cm^2. After the complete solidification the pressure inside was equal to that outside, as would be expected.

Doubts arise as soon as "three-dimensional foams" (§51) rather than single bubbles are considered. In an ideal foam consisting of regular pentagonal dodecahedrons and Plateau's borders, all bubble walls are plane.

Since there is no capillary pressure difference across a plane boundary, the pressure in all bubbles would be identical, and equal to the pressure outside. In reality, the pressure in different bubbles is different. This shows that the observable foams are not perfect and contain at least some curved lamellae. It would be expected that the pressure differences between different bubbles would be greater the smaller the expansion factor. As long as the thickness of the bubble walls is commensurate with the bubble diameter, the two opposite faces of a lamella can have very different curvatures. When the lamella is very thin, the curvatures of the two faces must be similar. Thus capillary pressure would be more important the nearer the system is to a gas emulsion or the further it is from the ideal dodecahedron foam.

The difference between the pressures inside and outside a foam was proved experimentally by Aleinikov[27], using the apparatus of Fig. 3.5 (§52). A foam column was built up by bubbling. Then stopcocks *9* and *10* were closed so that the space to which the foam was confined communicated with the outside only through manometer *5*. This first registered atmospheric pressure in the measuring tube. Then, as the foam was gradually collapsing, the pressure difference between the inside and the outside increased and reached, for instance, the value of 3 cm of water (i.e., about 3000 dynes/cm^2) for 0.001 M aqueous solution of *o*-cresol[27]. If capillary pressure is 3000 g/cm.sec^2 and if γ is 60 g/sec^2, then (since $P_c = 4\gamma/R$) the radius R of curvature is 0.08 cm. The visible bubbles in a similar foam had similar radii[28]. (See also §117.) Repetition of Aleinikov's measurements and determination of the effect of the expansion factor on the residual overpressure in confined foams would be very welcome.

§112. Diffusion between bubbles. In an ideal foam, bubble walls would be all plane, and the gas pressure everywhere uniform. In real foams, bubbles of different radii of curvature often are in mutual contact. Consequently a pressure difference equal to $4\gamma \left(\dfrac{1}{R_1} - \dfrac{1}{R_2} \right)$ (see §111) exists across the lamella which separates two adjacent bubbles. This pressure difference causes gas diffusion and rearrangement of lamellae, that is, the first process referred to in §43 as being always present in typical foams.

The amount dM of gas which diffuses across the lamella during time interval dt is proportional to the area over which diffusion occurs, to the solubility of the gas in the liquid of the lamella, and to the diffusion coefficient of the gas in the liquid. The solubility and the diffusion coefficient do not vary during the process, so that their product may be expressed by a constant. The area of contact will be approximately proportional to R_1^2, that is, to the surface area of the smaller bubble. Thus

$$\frac{dM}{dt} = 4 k_1 R_1{}^2 \gamma \left(\frac{1}{R_1} - \frac{1}{R_2} \right), \tag{7.8}$$

k_1 being a constant (sec/cm). In many instances, no great error is committed if $1/R_2$ is neglected in comparison with $1/R_1$; this approximation is admissible whenever the large bubble has a diameter at least 10 times as big as the small bubble. Thus

$$\frac{dM}{dt} \approx 4k_1 R_1 \gamma. \tag{7.9}$$

For a perfect gas and a pressure high enough so that capillary pressure may be disregarded,

$$\frac{dM}{dt} = -\frac{3k_2 R_1^2 dR_1}{dt}, \tag{7.10}$$

(see below). In this equation, k_2 is another constant (g/cm^3). From Equations 7.9 and 7.10

$$R_1 \cdot dR_1 = -k_3 \gamma \cdot dt, \tag{7.11}$$

k_3 being an abbreviation for $4k_1/3k_2$. Integration of Equation 7.11 results in

$$R_0^2 - R_1^2 = 2k_3 \gamma t, \tag{7.12}$$

first derived apparently by Dewar[29]; R_0 is the radius of the small bubble at zero time.

According to Equation 7.12, R_1^2 should be a linear function of time. This conclusion was checked by measuring the dimensions of a selected small bubble next to a big one in undisturbed foam. In two instances (not completely described) it was found to be correct[29]. It was confirmed also for foams of an oil-in-water emulsion (chlorobenzene + paraffin oil + water) and of 60% Hevea rubber latex; the expansion factor of the foams was only 13, so that bubble walls were not plane. The average thickness of the walls, calculated from the experimental values of constant k_3, appeared to be a few microns, although direct microscopic inspection gave larger thicknesses[30]; perhaps the approximations needed to derive Equation 7.12 were invalid in this study.

Other reasons are possible, of course, for both this discrepancy and the fact that many other foams studied by Dewar did not comply with Equation 7.12. Dewar surmised that the chemical composition of the wall (a soap solution in aqueous glycerol) changed with time, thus causing an alteration of the surface tension γ; hence the coefficient $2k_3\gamma$ was not independent of time. Gradual change in the bubble wall thickness is another likely cause of deviations from Equation 7.12. The rate of diffusion dM/dt is proportional to the concentration gradient (in the lamella) which, for a uniform lamella, is inversely proportional to its thickness δ. Hence, k_1 of Equation 7.8 and k_3 of Equation 7.12 are smaller the greater δ. If the bubble walls become thicker when the small bubble shrinks (this would be the rule in the absence of draining), then k_3 would decrease in time.

The effect of δ on k_3 presumably is visible in Brown's results[31]. A single bubble, immobile on the top of a convex meniscus, was losing its gas content to the ambient air. When the wall consisted of 0.1% sodium dodecyl sulfate solution, the rate of shrinkage was almost three times as high as when the liquid contained 0.1% of this sulfate plus 0.025% of 1-dodecanol. As reported in §99, the rate of drainage is lowered by this alcohol. Thus, at any given moment, the wall containing dodecanol presumably was thicker than that of sulfate alone, and the gas diffusion through the wall was less rapid.

Another qualitative confirmation of the effect of δ on k_3 was reported by Dewar[32]. The shrinkage of a soap bubble from 10 to 6 cm diameter required 45 hr when the walls were "almost too thick to show color," and only 9 hr when they were an "intense green."

Perhaps the most important reason for the failure of Equation 7.12 is the deviation of the bubble shape from that of a sphere. In Equation 7.9 R_1 means a radius of curvature, and in Equation 7.10 R_1 is the radius of the sphere which has the volume equal to that of the actual bubble. If the bubble is not a perfect sphere, these two R_1 are different from each other, and Equation 7.12 cannot be derived.

In Dewar's experiments[29], the rate $-dR_1/dt$ of contraction was more rapid, the smaller the pressures (p_0 outside the single bubble and $p_0 + (4\gamma/R_1)$ inside it). Thus a bubble, 14 cm across, of 3% ammonium oleate in 30% glycerol shrank in 15 min to 12 cm at 6.6 mbar and to 4 cm at 1.3 mbar. Apparently this observation can be accounted for by the above theory. To see this clearly, the value of k_2 in Equation 7.10 has to be derived. Let n be the number of moles of gas in the bubble; if M is expressed in moles rather than in grams, then $-dn$ and dM are identical. If R is the gas constant and T the absolute temperature, then for perfect gases $pv = nRT$; p is the pressure in the small bubble and v is the volume of the gas space in it. Hence, $dn = (p \cdot dv + v \cdot dp)/RT$. As long as capillary pressure is small compared with the atmospheric, the term $v \cdot dp$ may be neglected in comparison with $p \cdot dv$, and

$$\frac{dn}{dt} \approx \frac{p}{RT} \cdot \frac{dv}{dt}.$$

Substitute $(\frac{4}{3}) \pi R_1^3$ for v. Equation

$$\frac{dn}{dt} \approx \frac{p}{RT} \cdot 4\pi R_1^2 \frac{dR_1}{dt}$$

results. This means that

$$k_2 \approx \frac{4\pi p}{3RT}. \tag{7.13}$$

Since k_3 is $4k_1/3k_2$, it is inversely proportional to pressure p. Consequently at a given duration t (15 min), the difference $R_0{}^2 - R_1{}^2$ ought to be proportional to $1/p$, or $(R_0{}^2 - R_1{}^2)p$ should be constant. From Dewar's data, this product was $(196 - 16) \times 1.3 = 234$ and $(196 - 144) \times 6.6 = 343$ at 1.3 and 6.6 mbar; perhaps it was as constant as could be expected.

Acceleration of bubble shrinkage by lowering the gas pressure (again to about 1.3 mbar) was noticed also by Schwarz[33] for foams of tetrapropylenylbenzene sulfonate $C_{12}H_{25}\cdot C_6H_4\cdot SO_3Na$. He attributed this enhanced velocity to condensation of water vapor on the concave side of the septum (between the two bubbles, see §41) and its evaporation on the convex side (i.e., in the bigger bubble).

When two bubbles coalesce to one, the total gas volume v increases (because the average gas pressure diminishes), while the total area A of the bubble walls decreases. Following Tait (1867), Ross[34] showed that

$$3p\cdot\Delta v + 2\gamma\cdot\Delta A = 0; \qquad (7.14)$$

Δv and ΔA are the changes in v and A associated with the coalescence.

§113.   Instead of measuring the gradual shrinkage of a particular bubble, it is possible to determine the function $n/N = f(l)$ at different times; N is the total number of bubbles (in the foam) surveyed by the experimenter, and n is the number of bubbles whose diameter is confined within the range from l cm to $(l+\Delta l)$ cm. Another function encountered in the literature is $\pi n l^2/S = f(l)$, explained in §110. The change of this function with time is evident by comparing the two curves of Fig. 7.3; in a fresh foam, bubbles of about 160 μ made the greatest contribution to the total area, but after aging for 3 min bubbles of about 350 μ contributed most.

An analogous (and much earlier) graph is reproduced in Fig. 7.5 from Clark[35]. A 5% soap dispersion was used as the liquid. The diameter of the bubbles (in centimeters) is plotted along the abscissa. The ordinate is $\Delta S = \pi n l^2$ of §110. The foam age increased from curve *1* (90 sec) to curve *4* (480 sec). It is seen that the surface area of the bubbles about 150 μ across was greater, after aging for 1.5 min, than any other comparable area. In an 8-min-old foam, the surface area was almost uniformly distributed over bubbles of different diameters; this means that nl^2 was almost constant or that the number of bubbles having diameters between l and $l+\Delta l$ cm was approximately inversely proportional to l^2. The shift of frequency is seen, for instance, at $l = 200$ μ. At $t = 1.5$ min the ordinate is about 78, and at $t = 8$ min it is near 9; that is, the number of bubbles 200 μ across decreased almost to one-ninth in 6.5 min, while the number of larger bubbles simultaneously rose.

The area under the curves of Figs. 7.3 and 7.5 is proportional to the total surface area of foam. It decreases in the process of aging. For instance, the specific area S of a foam was[24] after 200 sec only one-half of what it was after 40 sec. The rate of decrease depends on the gas in the bubble, as

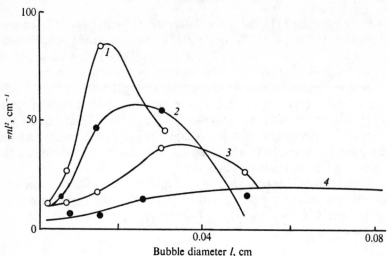

Fig. 7.5. Polydispersity of foams of 5% soap dispersion. Abscissa: bubble diameter, l cm. Ordinate: $\pi n l^2$; n is the number of bubbles having diameters near l in unit volume of foam. The foam age was for curve *1*, 1.5 min; for curve *2*, 2.5 min; curve *3*, 4 min; and curve *4*, 8 min (from reference [35]).

mentioned in §83; it is greater for carbon dioxide than for air or oxygen. This result is in agreement with the theory of §112. The rate of diffusion, that is, constant k_1 of Equation 7.8, should be proportional to the solubility and the diffusion coefficient of the gas in the liquid. The product of solubility (cm^3 of gas in cm^3 of liquid) and diffusion coefficient (10^5 cm^2/sec) in water is for carbon dioxide 0.9×1.8, that is, 1.6; and for oxygen 0.03×2, that is 0.06. The pressure in two contiguous bubbles filled with carbon dioxide should equalize almost three times as rapidly as that in oxygen-filled bubbles.

A particularly simple behavior was observed[30] for foams of "Teepol" + sodium alginate (see §110) and some other foams. The average diameter of the visible bubbles increased almost linearly with time; after 1200 sec it was almost three times as great as immediately after foam formation.

In Fig. 7.6 also an almost linear increase of the bubble diameter with time is seen[36], but there are three separate curves referring to three different levels in the foam (36, 60, and 95 mm above the base of the foam column). The liquid subjected to foaming (by injecting air) was 0.25% suspension of malt residue (see §87). Before the aging, the mean diameter at the 36-mm level was about 0.4 mm, and at 95-mm level, 0.7 mm. After 90 sec the spread was from 1.1 to 1.6 mm. The increase of the average diameter with the height of the level considered has not been satisfactorily explained.

§114. *Compressibility.*   Apparently only one experimental determination of foam compressibility is available in the literature[37]. Potato juice diluted with an equal mass of water and then containing 0.3% protein was converted into foam (of expansion factor near 100), and this was introduced into a tube (closed at one end) to the 100-cm^3 mark. Then the space above the foam was connected with a source of compressed air, and the foam volume was read at different excess pressures, Δp. When this Δp was 125, 240, or 480 mbar, the foam occupied, respectively, 90, 81, and 69 cm^3. An equal volume (100 cm^3) of a perfect gas under 1070-mbar pressure would be compressed to 90, 80, and 69 cm^3 by a pressure excess of 120, 270, and 480 mbar. Thus the foam behaved as a slightly compressed perfect gas. Moreover, the contraction caused by Δp was completely reversible.

This reversibility eliminates one of the possible explanations of the phenomenon. When foam is compressed in a narrow tube, friction along the tube walls (see §117) would lower the volume decrease achieved, that is, simulate a high internal pressure. However, friction would hinder also the recovery of the initial volume.

Three other explanations are available. (1) Because of the capillary pressure P_c, the gas pressure in the above foam really was 1070 mbar instead of the mean atmospheric pressure of 1013 mbar. If P_c was $1070 - 1013 = 57$ mbar, then the radius of curvature of the bubbles was about 25 μ, which is unlikely to be true.

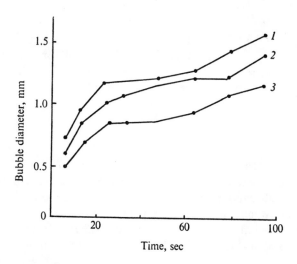

Fig. 7.6. Growth of bubbles in the foam of 0.25% suspension of malt residue. Abscissa: time in seconds. Ordinate: bubble diameter in millimeters. Curves *1*, *2*, and *3* refer to levels 95, 60, and 36 mm above the base of the foam column (after reference [36]).

(2) The oldest theory[38] also relies on the capillary pressure but in a more sophisticated manner (see Bikerman[39]). If the external pressure is P_e, then the internal pressure P_i is $P_e + P_c$. Hence

$$\frac{dP_e}{dv} = \frac{dP_i}{dv} - \frac{dP_c}{dv} ; \tag{7.15}$$

v is the foam volume or, more exactly, the gas volume in the foam. In a perfect gas,

$$\frac{dP_i}{dv} = -\frac{P_i}{v}. \tag{7.16}$$

As P_c is inversely proportional to the bubble radius (in gas emulsions!), i.e., to $v^{\frac{1}{3}}$, the derivative

$$\frac{dP_c}{dv} = -\frac{P_c}{3v}. \tag{7.17}$$

Introduction of Equations 7.16 and 7.17 into Equation 7.15 yields

$$\frac{dP_e}{dv} = -\frac{P_i}{v} + \frac{P_c}{3v},$$

equivalent to

$$-\frac{v \cdot dP_c}{dv} = P_i - \frac{P_c}{3} = P_e + \frac{2}{3}P_c. \tag{7.18}$$

The ratio $-v \cdot dP_e/dv$ is the inverse compressibility, i.e., the compressibility β is $dv/v \cdot dP_e$. It is seen that β depends not only on P_e, but also on P_c, that is, on the surface tension and the radii of curvature. The equation of state for a foam would be, instead of $pv = nRT$,

$$\left(P_e + \frac{\alpha\gamma}{v^{\frac{1}{3}}}\right)v = nRT ; \tag{7.19}$$

n is the number of moles, R is the gas constant, and T absolute temperature; α is a constant depending on the shape of the foam bubble. The bubbles in usual foams are so large that the correction term $(\frac{2}{3})P_c$ may be neglected; and when the lamellae are plane, then $P_c = 0$.

When Equation 7.19 is written in the form $P_e v + \alpha\gamma v^{\frac{2}{3}} = nRT$, it is clear that it is almost identical with the much later equation[34]

$$P_e v + (\tfrac{2}{3})\gamma A = nRT, \tag{7.20}$$

A being the surface area. For a gas emulsion, in which all bubbles are practically spherical, the two formulae become identical when the shape factor α is equal to $2(4\pi/3)^{\frac{1}{3}}(l^3/v)$, l being the edge of the cube enclosing one bubble.

(3) The most plausible explanation[39] is based on the presence of liquid in the foam. The total foam volume v is the sum of the gas volume v_1 and the liquid volume v_2. When the external pressure increases by dP_e, compressibility β is $-dv/v \cdot dP_e$. But the liquid is virtually incompressible, so that $dv = dv_1$ and

$$\beta = -\frac{dv_1}{(v_1+v_2)} dP_e. \tag{7.21}$$

Hence it is smaller than the compressibility $(-dv_1/v_1)dP_e$ calculated without regard to the liquid phase. If, for instance, the expansion ratio is 10, then the compressibility of the foam is $(10/11)\beta_0 = 0.91\beta_0$, β_0 being the compressibility of the gas in the bubbles.

An experimental investigation of the relation between compressibilities and expansion factors would be highly welcome.

§115. Elasticity.   A chain of soap bubbles resists stretching. This resistance must be caused by the surface tension of the bubble walls but no proof for this explanation is known.

The apparatus employed to obtain such chains is shown in Fig. 7.7

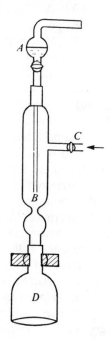

Fig. 7.7. Formation of bubble chains. Soap solution drops from A to orifice B where bubbles are formed by air entering through C. Bubbles expand in D (after reference [32]).

from Dewar[32]; see also Sasaki[40]. The soap solution from reservoir A drops onto the capillary opening B and blocks it. Air injected through C transforms these droplets into large bubbles and pushes these down into the inverted funnel D. When the amounts of liquid and air supplied to B are judiciously chosen, every succeeding bubble remains in contact with its predecessor, and a "unidimensional" foam forms. When an iron particle is attached to the bottom of the lowest bubble and pulled downward by a magnet, the chain extends. Thus a 15-cm-long chain of three bubbles was stretched 1.35 cm by a force of 74 dynes, and a chain of six bubbles, 31 cm long, elongated by 11 cm when a force of 133 dynes was applied. The modulus of elasticity is defined as $E = fl/\Delta l$; f is the tensile stress applied, l is the initial length, and Δl is the length increase caused by f. This equation cannot be used for Dewar's results, since the cross-section of the specimen (i.e., of the films in the chain) is not known. Dewar incorrectly gave the name "modulus of elasticity" to the ratio $Fl/\Delta l$, F being the external force.

Understanding the above values seems possible if it is remembered that surface tension γ of about 30 dynes/cm acts along every face of a soap film. Undeformed bubbles are almost spheres, and their surface area is practically equal to $4\pi r^2$, r being the bubble radius. When the bubble is stretched (in the ratio 16.35:15.0 in the first example), its surface increases because spheres have the smallest possible area for the given volume. The work spent was $74 \times 1.35 = 99.9$ dynes·cm. If it was used up on increasing the surface, then this increase was $99.9:30 \approx 3$ cm^2. Such an increase can readily be achieved by distorting three spheres of 5-cm diameter each.

The 31-cm-long chain broke when the pull exceeded 133 dynes. The perimeter of the horizontal cross-section of the chain presumably was about 13 cm at this moment. As each lamella has two faces, the surface tension force was approximately $13 \times 2 \times 30 = 780$ dynes. The experimental strength was much lower than expected, perhaps because of the poor homogeneity of the solution and the poor symmetry of the system.

A better agreement with expectation apparently was achieved in a recent, very briefly described experiment[41]. A glass sphere, 0.056 cm in diameter, was attached to a dynamometer. When a foam (of unknown composition) was permitted to touch the ball, this performed several oscillations, and the highest force recorded during this motion was about 10 dynes. Presumably this force was exerted by one or a few foam films pulling at the sphere. If the length of the film-glass contact was l cm, then 10 dynes was equal to γl. Setting $\gamma = 50$ dynes/cm, l becomes 0.2 cm; that is, not very different from the greatest circumference of the ball ($0.056\pi = 0.176$ cm).

§116.   The main mechanical constants (bulk modulus $1/\beta$, shear modulus, Poisson ratio, and modulus of elasticity E) of a perfectly isotropic and elastic solid are related to each other by the classic equations of the theory of elasticity. Unfortunately, on no foam have all four constants been measured. A foam's resistance to mechanical deformation was observed many times, but chiefly in a qualitative manner.

Thus the "stability" of whipped whey-protein foams was estimated[42] from the rate of descent of a glass tube inserted in a vertical position into the foam; this rate was lowered, for instance, by adding calcium salts to the whey. The "strength" of malt residue foams was evaluated[43] by placing a vertical glass rod on the bottom of a cylinder filled with foam, releasing the rod, and recording the time needed for the rod to touch the wall. It is not known why this time was longer for 0.6% suspensions of malt residue than for any other concentration. The ability of a foam to preserve sharp edges, to remain in the shape of a tall cylinder, and similar purely empirical properties have been described[44] for foams obtained by beating aqueous sugar solutions containing also lecithin, agar, and so on.

The difference in mechanical behavior between soap and saponin bubbles was noticed many years ago. When a soap bubble is blown at an orifice, it can be made bigger or smaller by injecting additional air or by sucking out some air; and in all these completely reversible changes the surface remains smooth; the bubble walls behave as liquid films. When air is withdrawn from a saponin bubble, its surface becomes, and remains for a time, wrinkled.

A more quantitative indication of the difference between different films was obtained[45] by immersing a metal spring in a soap solution, withdrawing it, and inducing vibrations in the spring now covered with a nearly cylindrical film. When the soap was sodium myristate, vibrations stopped after a few cycles. Sodium laurate films damped the motion so little that 20 to 30 oscillations could be observed.

The saponin-like behavior (or plasticity) was studied in a special apparatus[46]. Horizontal surfactant films were deposited in the annular space between two concentric platinum rings. The inner ring was attached to a dynamometer. When the external ring was pushed down (e.g., 0.1 cm in several seconds), the dynamometer recorded a force. This could be translated into the tension γ' operating in the extended film which had the profile of a catenary (§5). The γ' was independent of the degree of stretching the lamella as long as the concentration of (pure) sodium dodecyl sulfate was above the critical micelle concentration (0.23% by weight); and addition of 1-dodecanol to these relatively concentrated solutions also had no effect on γ'. However, in solutions of 0.05% sodium dodecyl sulfate plus 0.005% dodecanol some irreversibility was seen. Thus when the film area was raised from 3.2 to 9.0 cm^2, γ' increased from 38 to 50 g/sec^2; during retraction, γ' was at least by 5 g/sec^2 lower than during stretching (at an equal area), and was as low as 20 g/sec^2 when the area returned to about 4 cm^2. Evidently, the mixed lamella behaved like a plastic sheet: it resisted initial stretching but had little tendency to contract after the removal of the external force. Surface viscosity of these mixtures is referred to in §99.

§117.   Hydrostatic pressure P in a foam is a property which greatly depends on the attachment of lamellae to solid walls; also the properties discussed in §§118 to 120 depend on this attachment.

It would be expected that the hydrostatic pressure of a foam column containing M g of liquid and having a cross-section A cm^2 would be equal to gM/A g/cm·sec^2; g is acceleration (cm/sec^2) due to gravity. In reality, P is markedly smaller than gM/A and also depends on the value of A in a tube; the smaller A, the smaller the experimental P. No doubt, the foam is more or less suspended on the walls of the tube. If n lamellae touch the wall along unit length and 2γ is the tension acting in each (γ is surface tension and each film has two faces), the force which prevents the descent of the foam must be proportional to $2n\gamma$. $2\pi r$, r being the radius of the tube. If the average angle between the lamella and the vertical wall is θ, then the force is $4\pi n\gamma r \cdot \cos\theta$. The weight of the foam is proportional to the cross-section of the tube, i.e., πr^2. The ratio of the capillary force upholding the foam to the weight pulling it down is proportional to $1/r$; this explains why hydrostatic pressure (that is, the fraction of the force per unit area not supported by the lamellae) increases with r.

For a given tube diameter, P is[12] greater, the taller the foam column and the greater its density $[\rho]$. However, in foams of 0.01 to 0.1% saponin solutions, P was not proportional to $[\rho]$. For instance, for 0.05% solution, P at a definite level was 3.2 cm of water for $[\rho] = 0.002$, and 4.8 cm of water for $[\rho] = 0.010$ g/cm^3; thus a five-fold increase of $[\rho]$ raises P only by the factor 1.5.

The dependence of P on the level examined, without any comparison with $[\rho]$, was studied by Bruckner (see Schiebl[47]). In the "saturation" of beet sugar juice, carbon dioxide is bubbled through the juice contained in a tall vessel. In the liquid and above it (i.e., in the foam) bells filled with air and connected to water manometers were installed. During the steady state of foaming, the pressure indicated by these manometers was greater, the lower the position of the bell; for instance, P was equal to that of 80 cm of water at the level of the initial liquid surface, and 120 cm of water at 80 cm below this level.

When the foam column of a saponin solution was slowly rising during gradual injection of nitrogen, also P at the base of the column was increasing. Unexpectedly, P continued to increase (for 0.005% and 0.01% solutions) also when the foam height reached its steady-state value[48]. The cause of this behavior is not known.

The foam itself may be used as a pressure gauge. Short vertical tubes were filled with a foam, and that tube diameter, $2r_0$, was determined at which the foam still remained in the tube[49]. If H is the height of the foam column, then the weight of the column is $gH \cdot \pi r^2 [\rho]$ and the capillary force holding the foam up is $4\pi n H\gamma . r \cos \theta$. When these two forces are equal, then $r = r_0$, and

$$r_0 = \frac{4n\gamma \cos \theta}{g[\rho]} \qquad (7.22)$$

However this equation was neither tested nor stated.

§118. *Friction and viscosity.* The middle part, 100 cm long, of a glass tube (open at both ends) was filled with the foam (of a diluted potato juice) of the density equal to 0.01 g/cm^3. Then air pressure was applied to one of the ends until the opposite boundary of the foam column started to advance. The minimum pressure p required was[37] equivalent to a water head of 93, 49, 29.5, and 26 cm when the internal diameter of the tube was, respectively, 0.4, 0.8, 1.2, and 1.8 cm. Thus the minimum pressure was roughly inversely proportional to the tube diameter $2r$, so that the force $\pi r^2 p$ resisting motion was proportional to r and, consequently, to the perimeter of the cross-section of the tube.

A very similar law of friction is valid for pellets or indices of a liquid pushed along a tube poorly wetted by the liquid[50]. The resistance to motion is proportional to the tube perimeter $2\pi r$, and the coefficient of proportionality is of the order of magnitude of the surface tension γ of the liquid. In the instance of potato juice foam, this coefficient was much greater than γ. For instance, for the narrowest tube it was equal to $93 \times 981 \times 0.4/4 = 9100$ g/sec^2, that is, about 130 times as great as the γ of the juice. The most plausible explanation of this discrepancy is given in §117: in a liquid pellet only two air-liquid surfaces are present and must be distorted to move the pellet ahead, while in a foam each lamella touching the wall possesses two surfaces. Hence, if n of Equation 7.22 was about 1.3, that is if the foam bubbles were about 100/130 cm across, then the behavior of Siehr's foams was in accord with that of single pellets. Unfortunately neither the bubble dimensions nor the dependence of friction on the length of the foam column have been investigated. According to a recent calculation[51], the pressure drop in a moving foam and its velocity depend above all on the expansion factor.

As expected, the force needed to push a foam column from a wide tube into a narrow tube was greater the smaller the diameter of the second tube[37].

In a slightly more complicated arrangement[52], foam is pressed through an annular space between an immobile glass tube and a coaxial brass rod suspended on a spring. The motion of the foam causes extension of the spring. The traction on the rod was calculated from this extension. It increased with driving pressure, the velocity of the foam, and also with expansion factor. From the theory of §117, the pull on the rod would be greater, the greater the number n, but this aspect has not been investigated.

The foam was obtained by bubbling air through a 6% solution of a hydrolyzed protein stabilized with iron salts.

§119.   Formally, the force necessary to push a foam column of length H relatively to a tube wall or a rod can be expressed as $2\pi rH\tau$, τ being the yield stress (g/cm·sec^2) in shear. If the theory of §117 is valid, then

$$\tau = 2n\gamma \cos \theta. \qquad (7.23)$$

From Wenzel's experiments mentioned in §117, on foams of commercial ammonium dodecyl sulfate (concentration of crude material: 1.5%, of active material: 0.45%), a τ of 2 to 11 dynes/cm^2 was calculated; as expected, it was independent of the tube radius r.

Equation 7.23 can be used to derive a relation between τ and capillary pressure P_c in foam bubbles. As there are n lamellae per centimer of the tube length, the bubble dimension in this direction is $1/(n-1)$ cm. For spherical bubbles, this is equal (more or less) to the bubble diameter $2R$, so that $R = 1/2(n-1)$ cm. Capillary pressure in spherical bubbles is $P_c = 2\gamma/R = 4(n-1)\gamma$. Comparison with Equation 7.23 shows that

$$\tau = \frac{0.5nP_c \cos \theta}{n-1} \qquad (7.24)$$

or, for large n, $\tau \approx 0.5\, P_c \cos \theta$. By a less obvious reasoning, Deryagin[38] obtained the formula

$$\tau = 0.4P_c. \qquad (7.25)$$

According to Equation 7.25, the ratio of τ to P_c is a universal constant, while from Equation 7.24 it depends on bubble dimensions and the angle θ. Apparently only one study of the relation between τ and P_c has been published[53], and its results are subject to doubt. Foams produced by whipping 0.002% saponin solutions served for the tests. Their P_c was determined by measuring the increase in pressure occurring during foam collapse, as reported in §111. It varied between 2900 and 10,100 dynes/cm^2 for different foams. If γ was 70 g/sec^2, then the bubble diameter varied between 0.19 and 0.055 cm if P_c was $4\gamma/R$, and between 0.1 and 0.03 cm if P_c was $2\gamma/R$. These values are small but still reasonable; no measurement of the bubble dimensions was made. To find τ, a horizontal circular disk was suspended (on a metal ribbon) in the surface of the foam. When the ribbon was twisted, the disk turned in its plane over an angle α, and the shear stress in the foam was calculated from an empirical relation between α and the torque on the ribbon. Unfortunately, in the original calculation the deformation of the lamellae under the disk was disregarded, and only that of the lamellae between the vertical walls of the disk and the vessel walls was taken into consideration. Thus the published values of τ presumably are higher than the true ones. They ranged from 810 to

4450 dynes/cm². The ratio τ/P_c was said to be between 0.23 and 0.70, that is, similar to the expected ratio 0.40.

The theories of Equations 7.24 and 7.25 disregard surface viscosity and, consequently, are not likely to be correct for those saponin foams in which the foam "crusts" are almost solid. They broke down[53] for 0.1% saponin foams in which τ appeared to be 22,000, P_c was near 2700, and the dimensionless ratio τ/P_c was approximately 8.

Instead of a horizontal round disk, a vertical rectangular plate can be used. When such a plate is suspended (on a torsion wire) in a foam kept in a beaker, and the beaker is rotated at a constant rate (for instance, eight revolutions per minute), the torque first rises to a maximum and then decreases to a steady value[54]. Assuming that the shear occurred along the cylindrical surface formed by the vertical edge of the rotating rectangle, it is possible to calculate the maximum and the steady-state stress (τ_1 and τ_2). In the foam of 1% hydrolyzed keratin solution, τ_1 was 350 and τ_2 was 230 dynes/cm²; and when the concentration was 10%, τ_1 and τ_2 were, respectively, 2250 and 1100 dynes/cm². There was almost no difference between τ_1 and τ_2 for soap foams, and both had the greatest value (320 dynes/cm²) in 5% solutions. The difference between the liquid soap lamellae and nearly solid protein films is striking. In general agreement with the theory of §117, the shear stresses were greater when the average bubble dimensions were smaller, that is, the number (n) of lamellae per unit length was greater. This n was altered by changing the frequency of vibrations of the perforated plate which was whipping up the foam.

The connection between the maximum stress τ_1 and surface viscosity (or, more generally, surface consistency) was proved[54] by special measurements. A horizontal disk, suspended on a torsion wire, was rotated in the surface of solution (free of foam), as mentioned in §10. The resistance to rotation per unit length of the perimeter was φ dynes/cm (e.g., 3.11 for 5% hydrolyzed keratin solution). The product of this φ and the specific surface S, defined in §110, has the dimension (dynes/cm²) of stress. For three different keratin foams prepared at different vibration frequencies φS was 410, 1180, and 1970, while τ_1 for these foams was, respectively, 307, 937, and 1180 dynes/cm². Thus the ratio $\varphi S/\tau_1$ was almost constant. No doubt, the τ_1 is related to the above τ, but no quantitative comparison is possible at present.

§120. Several series of experiments have been performed in coaxial cylinder viscometers. The annular space between the two vertical cylinders was filled with foam, the external cylinder was rotated, and the torque on the inner cylinder, suspended on a torsion wire, was measured.

In the earliest work[55], all foams were produced by the pneumatic method. The outside cylinder was not continuously rotated but turned once by an angle of φ radians. This caused a turn of the inner cylinder by $\varphi - \varepsilon$ radians and a torque P dynes/cm² on it. The curves of P versus ε, that is, the stress-strain curves of the foam, were of three types. For foams of sodium 6-butylnaphthalene-1-sulfonate, sodium dioctyl sulfosuccinate, and sodium pyridine-hendecane sulfonate, P increased with ε to a maximum (e.g., 10 dynes/cm^2 at $\varepsilon = 0.2$), then decreased, and finally became independent of ε (e.g., 5 dynes/cm^2 between $\varepsilon = 0.9$ and 1.6). During the first stage, P was a linear function of ε and manifested no relaxation; when angle φ remained constant, also ε and P did not change, so that the foam behaved as a truly elastic (Hookean) body. At great ε, P was independent of ε, presumably because slippage of foam at the inner cylinder took place. In the foams of the second type, obtained from saponin solutions more concentrated than 0.1 %, P increased with ε to a plateau; apparently no slippage occurred. More dilute saponin solutions, making up the third type, gave rise to foams in which P increased with ε to a peak and then abruptly dropped when foam started breaking around the internal cylinder. When the expansion factor $\rho/[\rho]$ (§107) was altered by changing the concentration (of butylnaphthalene sulfonate), P varied similarly to $\rho/[\rho]$. The meaning of this observation is not clear, since altering the concentration of a solution affects all properties of its foam.

In a later study[49], the inner cylinder was provided with vanes (or ridges) to prevent slipping, and the external cylinder was continuously rotated at a rate of u cm/sec. The torque τ_0 on the inner cylinder was proportional to u raised to a power, but no correlation with the other properties of the foam could be found for the constants α and β in the equation $\tau_0 = \alpha u^\beta$. Foams (of 1.5 % ammonium dodecyl sulfate) tested in this apparatus were studied also in a cone-and-plate viscometer. To the plate, 32 radial baffles or vanes were attached to eliminate slipping. The plate was rotated 0.004 to 29 times per minute, and the highest shear rate (or velocity gradient) du/dx was 19 sec^{-1}. The cone was suspended on a calibrated torsion wire. The shear stress τ at the cone was expressed as

$$\tau = \tau_y + k\left(\frac{du}{dx}\right)^m, \qquad (7.26)$$

τ_y being the yield stress discussed in §119 as τ, and k and m being two empirical constants. The τ_y was greater when the average bubble diameter l was smaller; the mathematical relation for a typical foam was $\tau_y = 53 l^{-0.65}$, if l was measured in millimeters and τ_y in dynes/cm^2. When the bubbles are smaller, there are more lamellae per unit length; thus the observations are in accord with the theory of §117. The l varied from 0.8 to 8 cm. The coefficient m had values between 0.1 and 0.7; thus the term $k(du/dx)^m$ was different from the shear stress caused by the ordinary viscosity.

The apparent viscosity, determined by means of coaxial cylinders fitted with fins, decreased in the ratio 23:5 when the rate of rotation

increased from 5 to 290 revolutions per minute[56]. It was almost independent of the concentration of the foaming agent

$$C_6H_{13}O \cdot OC \cdot CH_2CH(SO_3Na) \cdot CO \cdot O \cdot C_6H_{13}.$$

It was not determined in this or in the above-mentioned study what is the
difference between the torques on a smooth and a corrugated surface. It is
mentioned in a patent[57] that rough surfaces are better wetted by foams
than smooth solids are.

§121.   Tensile and shear strength of a solid belong to its most
important characteristics and would be expected to be reviewed in a chapter
devoted to mechanical properties. An attempt to correlate the breaking force
of a bubble chain with the surface tension of the lamellae was made, with little
success, in §115. When the lamellae are quasi-solid, as is true for many
saponin and protein foams, a minimum force presumably is needed to cause
rupture, but no direct experiments on these systems are known to the
author.

Usually, no external force is required to break a foam. As pointed out
in Chapter 1, explanations are in order for the fact that so many liquid
foams are long-lived, not for the general rule that they finally collapse.

A striking example of the tenacity, or rather resilience, of foam films
was discovered by Dupré[58]. He dropped lead granules and cork balls
through lamellae of soap solutions in aqueous glycerol and found that no
rupture of the films occurred as long as the impact was not too great.
For instance, corks of 1.0 to 1.2 cm across caused collapse when their fall
height exceeded 25 cm, but passed through without doing any harm when
the fall height was smaller. Rhumbler[59] let iron filings drop through a
foam column and saw no alteration in the latter. Dewar[60] employed
single films as targets, and liquid air, mercury, lead shot, and drops of
water or soap solutions as missiles, again with identical results. He noticed,
however, that very thin films were less resilient than succulent lamellae.

In the above experiments, no difference was detected between different
lamellae and different missiles. Later it was observed[61] that mixed films of
sodium oleate and saponin manifested no resilience; presumably they were
too solid-like. It was claimed also[62] that solids readily wetted by a foam
were less capable of rupturing the latter than nonwettable solids. Thus a
foam obtained by injecting air into molasses was more easily broken by a
glass stirrer coated with paraffin wax than by the one not coated.

The decrease of resilience on aging (which, of course, was accompanied
by thinning) was studied in a quantitative manner[63]. A shower of mercury
droplets, at the rate of 0.3 g/sec, fell from the height of 2 cm onto films of
0.43% solution of disodium α-tocopheryl phosphate monohydrate. The
average persistence of these films (see §47) was 420 and 35 sec when the
diameters of the lamellae were 1.0 and 3.5 cm respectively. The mercury
shower was turned on when the films were t' sec old and continued until

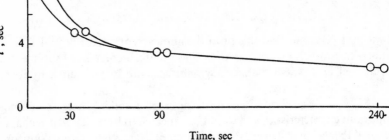

Time, sec

Fig. 7.8. Decrease of resilience of foam films on aging. Abscissa: time in seconds. Ordinate: duration of mercury shower until rupture. Upper curve: age at rupture. Lower curve: age at start (after reference [63]).

the film burst; the duration of the shower is denoted by t'', so that the lamella was $t' + t''$ sec old in the moment of rupture. Fig 7.8 shows the decline of t'' (plotted along the ordinate) when t' (the left-hand curve) and $t' + t''$ (the right-hand curve) increase; the lamellae were 1 cm across. Lamellae of 3.5 cm resisted the bombardment for 4.5 to 5.3 sec when they were 5 sec old at the start of the shower, but t'' was only 2.1 to 2.5 sec when t' was 20 sec. An analogous decrease of resilience on aging took place also for a neutral and an acid sodium laurate solution, a "benzalkonium chloride" solution, and beer.

The explanation advanced for this behavior (see also §18) is that the Marangoni effect is predicated on the existence of bulk solution (possessing the normal surface tension) between the two "crusts" of the lamella. When the lamella, on aging, becomes so thin that the liquid between the two surface layers ceases to have the properties of the bulk, then the Marangoni effect cannot be operative and no healing of the pierced film is possible. Thus the above t'' is the time required for the liquid to drain out so that the film thickness sinks below that needed for the Marangoni effect.

It follows from the above observations that the proneness of foams to mechanical destruction must depend on the age (and consequently also on the density) of the foam. This essential factor is too often disregarded when instruments for mechanical foam rupture are tested. Some devices recommended for this rupture are reviewed in §155.

§122. Heat transfer. The heat conductivity k of air is about 1/25th that of water and about 1/100th that of several minerals. Consequently, air would be an excellent thermal insulator. In reality, heat transfer through large volumes of air is much more rapid than the small value of k would suggest, because in these volumes convection occurs, so that heat is delivered to an (initially cool) point together with warm air. When air is present

in small bubbles, convection is largely suppressed. Hence, foams are poor conductors of heat. This property is utilized on a large scale in solid foams.

The rules established for the k of solid, should be approximately valid also for liquid, foams. A difference may be caused by the mobility of the liquid in the bubble walls and the Plateau borders, but this effect apparently has not been studied yet. A study, in which the system employed is referred to as foam but in reality was, presumably, a gas emulsion, may be mentioned here; see also §193. Steam was condensing on a wall of a copper tube, whose other wall was cooled by running water, flowing air, or water flowing in one, and air injected in the opposite direction (thus forming a gas emulsion)[64]. When the Reynolds number (i.e., also the rate of motion) of the water alone or of air alone increased, the heat transfer was accelerated. However, the cooling effect of the gas emulsion was not only small compared with that of water alone, but also almost independent of the rate of air injection. For other data on the heat transfer through turbulent gas emulsions see Pozin's book[65].

References

1. Bailey, M. I. *Ind. Eng. Chem.* 27:973 (1935).
2. Henry, W. C., and A. D. Barbour. *Ind. Eng. Chem.* 25:1054 (1933).
3. MacMillan, W. P. *J. Imp. Coll. Chem. Eng. Soc.* 13:64 (1960/61).
4. Koizumi, I. *Nippon Kagaku Zasshi* 83:853 (1962). *Chem. Abstr.* 58:13175 (1963).
5. Morris R. M., and A. Morris. *Chem. Ind.* 1965:1902.
6. Constable, F. H., and S. Mergen. *Rev. Fac. Sci. Univ. Istanbul* A12:190 (1947). *Chem Abstr.* 41:6454 (1947).
7. Vilenskii, V. M., and V. Ya. Khentov. *Zh. Prikl. Khim.* 39:1103 (1966).
8. Gleim, V. G., and V. Ya. Khentov. *Zh. Prikl. Khim.* 41:2475 (1968).
9. Gleim, V. G., and V. Ya. Khentov. *Zh. Prikl. Khim.* 42:2608 (1969).
10. Gray, P. P., and I. Stone. *Wallerstein Lab. Comm.* 3:159 (1940).
11. Nakashima, K. *Rept. Osaka Munic. Inst. Ind. Res.* 4, No. 2:49 (1952). *Chem. Abstr.* 47:10952 (1953).
12. Siehr, A. *Kolloid-Z.* 77:27 (1936).
13. Barmore, M. A. *Colo. Agr. Expt. Sta., Tech. Bull.* 9 (1934).
14. Hay, R. L., M. Reid, and J. A. Pearce. *Can. J. Res.* 25F:160 (1947).
15. Mokrushin, S. G., V. I. Borisikhina, and K. G. Potaskuev. *Zh. Prikl. Khim.* 28:107 (1955).
16. Brenner, M. W., R. E. McCully, and S. Laufer. *Am. Brewer* 84, No. 2:39, 49 (1951).
17. Tarakanov, O. G., and E. G. Eremina. *Kolloidn. Zh.* 27:274 (1965).
18. King, E. G. *J. Phys. Chem.* 48:141 (1944).
19. Ross, S. *J. Phys. Chem.* 50:391 (1946).
20. Savitskaya, E. M. *Kolloidn. Zh.* 13:309 (1951).
21. Chang, R. C., H. M. Schoen, and C. S. Grove. *Ind. Eng. Chem.* 48:2035 (1956).

22. Rinkes, H. *Chemie-Ing.-Tech.* 39:301 (1967).
23. de Vries, A. J. *Foam Stability*. Delft: Rubber-Stichting. (1957), p. 29.
24. Balakirev, A. A., and V. K. Tikhomirov. *Zh. Prikl. Khim.* 42:2354 (1969).
25. Dupré, A. *Ann. Chim. Phys.* [4], 7:406 (1866), especially pp. 411–417.
26. Manegold, E. *Schaum*. Heidelberg: Strassenbau, Chemie und Technik. (1953), p. 63.
27. Aleinikov, N. A. *Tsvetn. Metal.* 6:1546 (1931). *Kolloid-Beihefte* 36:82 (1932).
28. Aleinikov, N. A. *Zh. Prikl. Khim.* 22:812 (1949).
29. Dewar, J. *Proc. Roy. Inst. Gt. Brit.* 22:179 (1917).
30. de Vries, A. J. *Foam Stability*. Delft: Rubber-Stichting. (1957), p. 20.
31. Brown, A. G., W. C. Thuman, and J. W. McBain. *J. Colloid Sci.* 8:508 (1953).
32. Dewar, J. *Proc. Roy. Inst. Gt. Brit.* 22:359 (1918).
33. Schwarz, H. W. *Rec. Trav. Chim.* 84:771 (1965).
34. Ross, S. *Ind. Eng. Chem.* 61, No. 10:48 (1969).
35. Clark, N. O., and M. Blackman. *Trans. Faraday Soc.* 44:1 (1948).
36. Mit'kevich, G. P. *Zh. Prikl. Khim.* 21:816 (1948).
37. Siehr, A. *Kolloid-Z.* 85:70 (1938).
38. Deryagin, B. *Kolloid-Z.* 64:1 (1933).
39. Bikerman, J. J. *Foams*. New York: Reinhold. (1953), p. 124.
40. Sasaki, T., and S. Okazaki. *Kolloid-Z.* 159:11 (1958).
41. Balakirev, A. A., and V. K. Tikhomirov. *Zh. Prikl. Khim.* 42:216 (1969).
42. Peter, P. N., and R. W. Bell. *Ind. Eng. Chem.* 22:1124 (1930).
43. Mit'kevich, G. P. *Zh. Prikl. Khim.* 22:846 (1949).
44. Schmalfuss, H. *Kolloid-Z.* 116:161 (1950).
45. Burcik, E. J., and R. C. Neuman. *J. Colloid Sci.* 15:383 (1960).
46. Grabenstetter, R. J., and J. M. Corkill. *J. Colloid Sci.* 18:401 (1963).
47. Schiebl, K. *Chem.-Ztg.* 55:169 (1931).
48. Lauwers, A., and R. Ruyssen. *Conf. Orig. 3ᵉ Congr. Intern. Détergence, Cologne, 1960,* 3:146 (1961).
49. Wenzel, H. G., R. J. Brungraber, and T. E. Stetson. *J. Mater.* 5:396 (1970).
50. Bikerman, J. J. *J. Colloid Sci.* 5:349 (1950).
51. Kudryavtsev, B. K., L. I. Seleznev, and G. V. Tsiklauri. *Teplofiz. Vys. Temp.* 9:141 (1971). *Chem. Abstr.* 74, No. 143739 (1971).
52. Grove, C. S., et al. *Ind. Eng. Chem.* 43:1120 (1951).
53. Deryagin, B., and E. Obukhov. *Zh. Fiz. Khim.* 7: 297 (1936). *Kolloid-Z.* 68: 243 (1934).
54. Clark, N. O. *A Study of Mechanically Produced Foam for Combating Petrol Fires*. London: DSIR (1947).
55. Rozenfeld, L. M., and E. M. Savitskaya. *Kolloidn. Zh.* 13:454 (1951).
56. Marsden, S. S., and S. A. Khan. *Soc. Petrol. Eng. J.* 6:17 (1966).
57. Kodak-Pathé, French 942 087. *Chem. Abstr.* 44:10391 (1950).
58. Dupré, A. *Theorie mécanique de la chaleur*. Paris: Gauthier-Villars (1869), p. 352.
59. Rhumbler, L. *Ergeb. Physiol.* 14:526 (1914).
60. Dewar, J. *Proc. Roy. Inst. Gt. Bri.* 24:197 (1925).
61. Shkodin, A. M., and G. P. Tikhomirova. *Kolloidn. Zh.* 13:134 (1951).

62. Shkodin, A. M. *Kolloidn. Zh.* 14:213 (1952).
63. Bikerman, J. J. *J. Phys. Chem.* 56:164 (1952).
64. Bretsznajder, S., W. Kawecki, and W. Dziedzic. *Bull. Acad. Polon. Sci., Sér. Sci. Chim.* 10:155 (1962).
65. Pozin, M. E., I. P. Mukhlenov, and E. Ya. Tarat. *Foam Gas Purifiers, Heat Exchangers, and Absorbers.* [Russian] Leningrad: Goskhimizdat. (1959).

CHAPTER 8

OPTICAL PROPERTIES OF FOAM

§123. The brilliant colors of soap films have been admired for centuries; especially J. B. S. Chardin (1699–1779) frequently painted people blowing soap bubbles. These colors are caused by the interference of light reflected by the two faces of a lamella, and the elementary theory of this interference is presented in this and the next sections. The physical theory, however, is insufficient to account for the physiological colors, that is, for the impressions obtained by a human eye absorbing the reflected radiation; the physiological considerations are summarized in §125. Almost the only use made of the optical properties of films is determination of film thickness, and the principles of this method are discussed in §126. The light transmission through foams is reviewed in §127.

The main equations of the theory of interference can be deduced by means of Fig. 8.1. This represents a film, δ cm thick, on which a narrow beam of monochromatic light falls along line AB. A part of this radiation is reflected at the phase boundary and continues along line BC. The other part is refracted along line BO, reflected from the liquid-air boundary at O, refracted again at D, and advances in the gas phase along DE. There is a phase difference between the parallel beams BC and DE. It does not vary above the line CD (perpendicular to BC and DE) and can be calculated by comparing the lengths of the paths BC and BOD. As $BC = BD \sin \alpha$ and $BD = 2\delta \tan \beta$, then $BC = 2\delta \tan \beta \cdot \sin \alpha$. But $\sin \alpha = n \sin \beta$, if n is the refractive index of the film. Hence, $BC = 2n\delta \sin^2\beta/\cos \beta$. The length $BO+OD$ is $2\delta/\cos \beta$. However, the *optical* path is n times longer, that is, $2n\delta/\cos \beta$. Consequently the optical path difference is $\Delta = (2n\delta/\cos \beta) - (2n\delta \sin^2\beta/\cos \beta)$, i.e.,

$$\Delta = 2n\delta \cos \beta. \tag{8.1}$$

It is simpler to visualize Δ as a function of the angle of incidence α. Since $\sin \beta = \sin \alpha/n$, then $\cos^2\beta = 1 - (\sin^2\alpha/n^2)$, so that

$$\Delta = 2\delta(n^2 - \sin^2\alpha)^{0.5}. \tag{8.2}$$

The difference Δ is expressed in centimeters. It must be expressed in

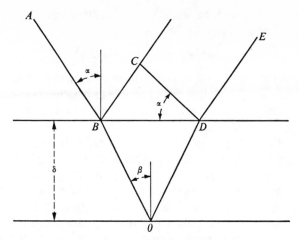

Fig. 8.1. Reflection of light from a foam lamella.

wavelengths λ_0 of the light used (in a vacuum or a rare gas) to understand diffraction. Thus

$$\frac{\Delta}{\lambda_0} = \frac{2n\delta \cos \beta}{\lambda_0} = \frac{2\delta}{\lambda_0}(n^2 - \sin^2\alpha)^{0.5}. \tag{8.3}$$

The actual phase difference between the pencils BC and DE is by $0.5\lambda_0$ greater because the reflection at B occurs from an optically denser, and that at 0, from an optically less dense, medium. Thus, when Δ/λ_0 is equal to 0.5, 1.5, 2.5, etc., the phase difference is λ_0, $2\lambda_0$, $3\lambda_0$, and so on, the two beams are in phase, and the intensity of the reflected light has its greatest value. When Δ/λ_0 is 0, 1, 2, . . . , then the phase difference is $0.5\lambda_0$, $1.5\lambda_0$, $2.5\lambda_0$, . . . , the two beams cancel each other, and the intensity of the reflected light is zero.

Consider a particular level of a vertical foam film. Suppose that, at the first observation, the expression $(2\delta/\lambda_0)(n^2 - \sin^2\alpha)^{0.5}$ for it happens to be 10.5. The spot will appear bright. When, because of drainage, the thickness decreases so that the above number becomes equal to 10.0 the spot will be dark. Such succession of brightness and darkness will continue until the film becomes too thin to reflect light; see §124. An analogous succession can be seen at one time if the film is scanned from top to bottom or from bottom to top. Every level, at which $(2\delta/\lambda_0)(n^2 - \sin^2\alpha)^{0.5}$ is a whole number, appears as a dark band.

Bright and dark areas are visible also when, at a constant thickness δ, the angle of incidence α is gradually varied. Suppose for the sake of simplicity that the ratio $2\delta/\lambda_0$ is a whole number. Then the lamella is dark when $(n^2 - \sin^2\alpha)^{0.5}$ is equal to 1, or to $(\lambda_0/2\delta)$, or to (λ_0/δ), and so on. Hence, the difference between the values of $(n^2 - \sin^2\alpha)^{0.5}$, corresponding to two subsequent dark bands is $\lambda_0/2\delta$. Call the corresponding angles of incidence α_1 and α_2. Then the difference

$$\sin^2\alpha_1 - \sin^2\alpha_2 = (\lambda_0/2\delta)^2. \qquad (8.4)$$

If the film is thick and $\lambda_0/2\delta$ is a small number, then the difference $\sin^2\alpha_1 - \sin^2\alpha_2$ is very small and may be impossible to measure (and to see). Consequently a thick lamella does not show interference effects from whatever direction it is viewed.

Fig. 8.1 is unduly simplified because it treats the incident ray as a line without any thickness, or at least with a thickness negligible compared with that (δ) of the film. In reality, the width of the illuminating beam practically always is much greater than the δ of a foam lamella. Consequently the bundle of rays BC (which finally reaches the eye of the observer) originates not only from AB but also from numerous rays parallel to AB which undergo multiple reflections at the two phase boundaries. The theory of this effect can be found in many books on optics; see, for instance, Preston[1] and Vašiček[2].

The main result for the intensity I of the reflected light is

$$\frac{I}{I_0} = \frac{4\varepsilon^2 \sin^2\Delta_0}{(1-\varepsilon^2)^2 + 4\varepsilon^2 \sin^2\Delta_0} ; \qquad (8.5)$$

I_0 is the intensity of the incident light, ε is the ratio of the amplitudes (not of the intensities!) of the light once reflected to that before the reflection; for small angles of incidence (less than, say, 30°) this $\varepsilon \approx (n-1)/(n+1)$, n again being the refractive index of the film, while the refractive index of the medium outside it (that is, of the air) is equal to unity. Finally,

$$\Delta_0 = \frac{2\pi n\delta}{\lambda_0} \cos\beta. \qquad (8.6)$$

When the wavelength λ_0 and the angle of refraction β are kept constant and only the film thickness δ varies, then $\sin\Delta_0$ (and consequently I/I_0) is zero when $\delta = 0$, or when $\delta = \lambda_0/2n\cos\beta$, or $\delta = \lambda_0/n\cos\beta$, and so on, in agreement with the above comments on Equation 8.3. It may be worth repeating that $I = 0$ when the lamella is extremely thin (i.e., $\delta = 0$); this is the simple reason for the black films (§21) being black.

§124.   When the incident light is white rather than monochromatic, the above reasoning applies separately to every wavelength. This means that the reflected light always is a mixture of different wavelengths. For instance, when $2\delta = 1.350~\mu$, then the ratio $2\delta/\lambda_0$ is 2 for $\lambda_0 = 0.675~\mu$ (a red color) and 3 for $\lambda_0 = 0.450~\mu$ (a blue color); and, as long as the expression $(n^2 - \sin^2\alpha)^{0.5}$ is similar for the two wavelengths, both radiations will be simultaneously enhanced and simultaneously depressed. When, for instance, both are weakened, then the favored wavelengths will be near $0.56~\mu$ and the reflected light will have a yellow hue.

In principle it is possible to divide the visible spectrum into, say, 1000 segments and to calculate the intensity of the reflected light for each of these (assuming that n is known for every λ_0); this would give the spectral composition of the light obtained from the given film; and repetition of this operation for many values of δ would give the dependence of the film color on film thickness. A simplified calculation of this kind was performed by Rayleigh[3] for three "primary" colors (of λ_0 equal to 6302, 5281, and 4569 Å); the results are inexact, however, because the refractive index was treated as a constant independent of the wavelength λ_0. They are reproduced on p. 140 of reference [4] and are still very instructive. The range of the values $2n\delta \cos \beta$ employed in the calculation was 0.27 to 2.5 μ.

Rayleigh's table and analogous graphs compiled later give an indication of the intensity of reflected radiation but still do not supply sufficient information on the hue of this radiation, that is, on its effect on the eye of the beholder. This is so because the sensitivity of human eye is very unequal at different wavelengths; it has a maximum near 0.55 μ (green) and descends from it in a bell-shaped curve to zero near 0.76 and 0.38. The curve of sensitivity versus wavelength depends moreover on the intensity of light. When this intensity is very small, the maximum of sensitivity shifts toward shorter λ_0, as far down as 0.52 μ.

§125.   Usually no attempt is made to characterize the visible color in a quantitative manner. Since Newton, the interference colors of thin layers are described qualitatively as white, red, and so on. The list compiled by Reinold and Rucker[5] for a nearly normal incidence (i.e., $\alpha = 0$) is presented here (Table 8.1). The colors are divided in eight orders, each ending in red; this periodicity exists, of course, because Equation 8.5 contains a squared sine and, consequently, function I/I_0 acquires a maximum value every time this sine is equal to 1.0. The numbers, shown in all except the first order, are the values of the product $n\delta$ (in microns) corresponding to the center of the band having the color specified. Thus, a film 0.218 μ thick of a liquid whose refractive index for purple rays is 1.40 will show the purple color of the second order. The value of $n\delta$ monotonously increases from the black of the first to the red of the eighth order.

The change of color on altering film thicknesses is particularly sudden between two successive orders, especially between the first and the second. The product $n\delta$ corresponding to this "first sensitive violet" or "first sensitive purple" usually is given as 0.280 μ; near this thickness a small increase in δ changes the hue from red to purple.

Lamellae thicker than a few microns do not show any interference colors because their $n\delta$ is equal to a whole number of too many wavelengths (see §124). Suppose that $n\delta$ is equal to 21.000 μ. Then it is equal to $30\lambda_0$ if λ_0 is 0.700 μ, to $31\lambda_0$ if λ_0 is 0.677 μ, to $32\lambda_0$ if $\lambda_0 = 0.656 \mu$, and so on; these equalities are approximate only because n depends on λ_0. In this manner, an aqueous lamella ($n = 1.33$) of about 16 μ would reflect

Table 8.1. Colors of Thin Films

First order	Second order		Third order		Fourth order		Fifth order		Sixth order		Seventh order		Eighth order	
Black	Purple	0.305	Purple	0.559	Green	0.893	Green	1.188	Green	1.479	Green	1.787	Green starts at	2.004
White	Blue	0.353	Blue	0.603	Yellow-green	0.964	Red	1.335	Red	1.627	Red	1.936		
Yellow	Green	0.409	Green	0.656	Red	1.052							Red starts at	2.115
Red	Yellow	0.454	Yellow	0.710										
	Orange	0.491	Red	0.765										
	Red	0.522	Bluish-red	0.815										

rays of approximately 0.700, 0.677, 0.656... microns equally strongly, so that the reflected light would be practically as white as the incident light.

§126. Optical determination of *film thickness* δ is discussed, for instance, by Vašiček[2] and Heavens[6].

As long as δ is smaller than $\lambda/4$, λ being the wavelength of the light employed in the liquid (i.e., not in the air or a vacuum) and equal to λ_0/n, the thickness can unambiguously be calculated from the ratio I/I_0 of Equation 8.5, §122. This follows from Equation 8.6. Let $\cos \beta$ in it be about 1.0. Then $\Delta_0 = 2\pi\delta/\lambda$. As long as Δ_0 is less than $\pi/2$, the value of $\sin \Delta_0$ depends on Δ_0 and δ in an unique way. When $\Delta_0 = 0.5\pi$, then $\delta = 0.25\lambda$. When δ is greater than 0.25λ, a given value of $\sin \Delta_0$ (and of I/I_0) corresponds to two or more different values of δ. The condition $\delta < 0.25\lambda$ is valid for black films (see §21), and their thickness commonly is determined from Equation 8.5; for a recent example see Lyklema[7].

When δ is greater than $\lambda/4$, some counting is needed. Consider a vertical foam lamella whose top layer is already black. The nearest dark horizontal band is situated at the level where the lamella is by $\lambda/2 \cos \beta$ thicker than the black film; at the next-lower dark band δ again is by $\lambda/2 \cos \beta$ greater, and so on. When one spot of the draining lamella is observed, it is seen that it becomes alternatively light and dark (in a monochromatic light). If the number of dark periods before the area is invaded by the black film is x, then the initial thickness was equal to the thickness of the black film plus $x\lambda/2 \cos \beta$.

An analogous counting is used when the incident light is white. If the draining film appeared red x times before it became permanently black, the order of the initial red was x, and the approximate film thickness can be judged from Table 8.1 in §125.

Unfortunately, all the reasoning above, including Equations 8.2 and 8.5, is valid for homogeneous lamellae only; the layer δ cm thick should have the refractive index n everywhere and be along both surfaces in contact with a medium of refractive index $n = 1$. Foam lamellae are not homogeneous. They are sandwiches, and the refractive index n_1 of each "crust" is different from that, n_0, of the internal liquid. Equations for the interference of light in stratified films are known[2,6], but it is difficult to apply them, in a straightforward manner, to foam lamellae because neither the thickness of the "crust" nor its n_1 can be ascertained with precision. On the basis of plausible assumptions it is believed[7,8] that the real thickness of many surfactant films is by about 10 to 20 Å smaller than that calculated from Equation 8.5 for water lamellae. Obviously this correction is irrelevant for all films showing colors. However, when a black film appears to be, say, 100 Å thick, its real thickness may be near 90 Å, and from this length perhaps 30 Å are taken up by the two "crusts", so that the aqueous core may be only 60 Å thick.

Ellipticity of the reflected light is often used to estimate the thickness of thin films such as oxides on metals or oriented water molecules on the surface of liquid water (see, for instance, Bikerman[9]) but is not popular with the students of foam.

In some instances it is easier to measure the optical thickness $\Sigma n\delta$ of several lamellae than that, $n\delta$, of a single film. This was done by Reinold and Rücker[10], who arranged many black soap films one behind the other in a tube and estimated the optical path difference (between the ray reflected from a mirror and that reflected by a series of lamellae) from the color of the reflected beam. The sum $\Sigma n\delta$ can be determined also by the displacement of fringes in a Michelson interferometer[11]. If Δ is this displacement, N the number of black films in the system, λ_0 the wavelength in air, and n the mean refractive index of the film material, then $\delta = \Delta\lambda_0/2N(n-1)$. Naturally, the results also of these methods are inexact because of the three-layer structure of foam films.

§127.　　The optics of foams, that is, of conglomerations of lamellae, has been little investigated. In addition to reflection and refraction considered in §123 and later sections, attention should be paid to light absorption. This absorption, as a rule, is negligible in single films, but it may be essential when dealing with large volumes of foam. If the light beam traverses l cm of foam, if there are x lamellae per centimeter, and if δ cm is their average thickness, then the true absorption in the foams is equal to that by the liquid layer $x l\delta$ cm thick. This absorption can be measured or calculated and used for correcting data on light scattering by foams.

A ray entering a foam is refracted and reflected in every bubble wall. As, in general, the angle between the incident ray and a lamella varies from point to point, the direction of the rays emerging from a cluster of bubbles is nearly random. This means that a large sphere of foam, of surface area A, if illuminated by a pencil of light of intensity I_0 and cross-section A_0, will appear luminous over the whole surface, and the intensity I of the emission will be almost uniform over the surface and equal to $I = I_0 A_0/A$, after the correction for absorption has been made.

However, when the foam layer is not very thick, the fraction of the incident light scattered by reflection and refraction is not a general function of A_0/A but depends on the bubble diameter, or the number x of lamellae per unit length, or the specific surface of the foam. If I_0 is the intensity of white light falling on a foam layer of a small and constant thickness and I is the intensity of the transmitted ray, it is found by experiment[12] that $I_0/I = ax+b$, a and b being empirical constants. It is clear that constant b should be equal to 1.0, as there is no loss in intensity when $x = 0$, but this was nearly true only for a 5% solution of a potassium cocoanut oil soap, while three solutions of hydrolyzed keratin gave b values of approximately -16. In the foam of 5% solution of hydrolyzed keratin, the ratio I/I_0 was 0.02 for $x = 80$; thus 80 bubbles reduced the intensity of light to 1/50th of the initial value.

Similar measurements have been performed by Ross[13,14], and the results used for estimating the rate of decay of various foams. For instance, 200 cm^3 of solution was beaten for 1 min and the foam illuminated. The ratio I_0/I was determined at different times t, as shown in Fig. 8.2. The gradual decrease of I_0/I means that the foam was becoming more transparent. Curve C is for 0.033% Nacconol ARSF (a mixture of sodium alkylbenzene sulfonates), curve D is for 0.1% solution of this surfactant, and it is seen that the two foams decayed at almost identical rates. Tributyl phosphate, $(C_4H_9)_3PO_4$, and 2-methyl-4-pentanol destroy Nacconol foams when present in concentrations much above 0.02%. However, they accelerate the decay also at this concentration; this is indicated by curves A (0.10% Nacconol+0.02% methylpentanol) and B [0.10% Nacconol +0.02% $(C_4H_9)_3PO_4$]. The authors had the impression that the decrease of the ratio $(I_0-I)/I$ in time for all these foams was curvilinear for short times (less than 40 min) and linear later, and it appeared probable that, during the first stage, drainage was important but gas diffusion between small and big bubbles (§122) was not, while this diffusion was the main process during the second stage. Apparently, bubble bursting was not significant at any time.

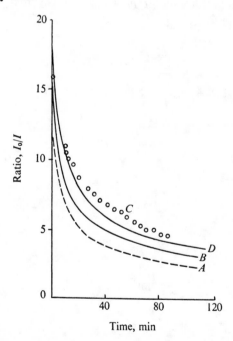

Fig. 8.2. Increase of the transparency of foams with time (minutes). Ordinate: ratio of intensities of incident and transmitted light. A: 0.10% Nacconol +0.02% methylpentanol. B: 0.10% Nacconol +0.02% tributyl phosphate. C: 0.033% Nacconol. D: 0.10% Nacconol (from reference [14]).

References

1. Preston, T. *The Theory of Light*. London: Macmillan. (1912), p. 184.
2. Vašiček, A. *Optics of Thin Films*. Amsterdam: North Holland. (1960), Chap. 3.
3. Rayleigh, J. W. S. *Scientific Papers*. Cambridge: University Press, vol. 2, p. 498.
4. Bikerman, J. J. *Foams*. New York: Reinhold. (1953).
5. Reinold, A. W., and A. W. Rücker. *Trans. Roy. Soc. (London)* 172:447 (1881).
6. Heavens, O. S. *Optical Properties of Thin Solid Films*. New York: Dover. (1965).
7. Lyklema, J., P. C. Scholten, and K. Mysels. *J Phys. Chem.* 69:116 (1965).
8. Clunie, J. S., J. F. Goodman, and C. P. Ogden. *Nature* 209:1192 (1966).
9. Bikerman, J. J. *Physical Surfaces*. New York: Academic Press. (1970), p. 65.
10. Reinold, A. W., and A. W. Rücker. *Trans. Roy. Soc. (London)* 174:645 (1883); 184:505 (1893).
11. Johonnot, E. S. *Phil. Mag.* [5] 47:501 (1899).
12. Clark, N. O., and M. Blackman. *Trans. Faraday Soc.* 44:7 (1948).
13. Ross, S., B. Barth, and J. F. Terenzi. *J. Phys. Chem.* 58:247 (1954).
14. Ross, S., and M. J. Catillas. *J. Phys. Chem.* 59:863 (1955).

CHAPTER 9

ELECTRIC PROPERTIES OF FOAMS

§128. Electric conductance is the only electric property of liquid foams for which experimental data are available. Apparently no one has measured yet their dielectric constants, the dependence of this constant on bubble dimensions, and so forth. First, the electric conductance of single films is reviewed.

In the early extensive measurements by Reinold and Rücker[1-3], direct current and gold needle electrodes were used. Thus the conditions were favorable for marked electrode polarization; it is not known to what extent the published results were distorted by this source of error. A large vertical film was drawn, two gold electrodes (one above the other) were inserted in it, a current of known intensity was sent through the lamella, and the potential difference φ between the needles was measured. Then the needles were immersed to the depth δ_0, at the same mutual distance as in the film, in the solution which gave rise to the lamella, a current identical to the previous was passed through the electrodes, and the potential difference φ_0 between them was determined. The equation employed to calculate the specific conductivity κ ohm^{-1} cm^{-1} of the lamella was

$$\kappa = \frac{\delta_0 \varphi_0 \kappa_0}{\delta \varphi}; \qquad (9.1)$$

in it, κ_0 is the specific conductivity of the bulk liquid and δ is the (optically determined) thickness of the film between the electrodes.

Several objections may be raised against Equation 9.1. It presupposes, for instance, that when δ_0 cm is the depth of immersion of the two electrodes, the path for the ions between them is δ_0 cm thick. In reality, the lines of the electric current spread far below δ_0, so that κ calculated from Equation 8.1 is too low. Moreover, there is only one value of δ in the equation; in other words, the lamella must have a uniform thickness between and around the electrodes. This is an exacting demand, especially in a vertical film. Reinold and Rücker had to humidify the air in which the lamella was suspended. Although they worked with fatty soaps, they noticed no effect of carbon dioxide present in this air (see §129). The calculated κ was independent of current intensity. The ratio $\delta_0 \varphi_0/\delta \varphi$ was independent of temperature between 19 and 26°, so that the temperature coefficient of κ was equal to that of κ_0. When the film was permitted to thin to, for instance, the first sensitive purple (§125), was flooded, and permitted to drain again, the previous value of κ was attained anew; thus the conductivity was reproducible.

The numerical results quoted below are those of the 1893 paper. The ratio $\delta_0\varphi_0/\delta\varphi$ was never smaller than 1 and generally was greater the smaller δ. Thus for a solution of sodium oleate, 1 part to 60 parts of water, this ratio was 4.47, 1.98, and 1.66 at thicknesses of, respectively, 0.097, 0.296, and 0.641 μ. For a film of sodium oleate 1, potassium nitrate 1.8, and water 60 parts, the ratio varied irregularly between 1.0 and 1.27. It was nearly equal to 1, when three volumes of this solution were mixed with two volumes of glycerol.

The results can formally be accounted for by the concept of surface conductance[3]. If the lamella consists of two "crusts," each x cm thick, and a core of thickness $\delta - 2x$ cm, then the specific conductivity κ of the lamella, taken as a whole, is

$$\kappa = \frac{1}{\delta}[2x\kappa_s + (\delta - 2x)\kappa_1];\qquad(9.2)$$

κ_s is the specific conductivity of the "crust," and κ_1, of the core. For the ratio κ/κ_0, the expression

$$\frac{\kappa}{\kappa_0} = \frac{\kappa_1}{\kappa_0} + \frac{2x(\kappa_s - \kappa_1)}{\delta\kappa_0}\qquad(9.3)$$

is obtained. If it is correct, then the ratio κ/κ_0 is a linear function of $1/\delta$ and the ratio κ_1/κ_0 can be determined from the value of κ/κ_0, extrapolated to $\delta = \infty$. Unfortunately the experimental data are neither numerous nor precise enough to test Equation 9.3.

§129.   Reinold and Rücker believed that κ_1 was different from κ_0. At present it seems permissible to neglect this difference. Lamellae of a soap solution are pictured as consisting of two unimolecular layers of, e.g., anions between which a layer of solution is sandwiched; this solution contains the cations compensating the charge of the immobile anions but, otherwise, is nearly identical with the bulk liquid from which the lamella has been withdrawn. On this view, the use of quantity κ_s is misleading because the transport of electricity through an immobile layer of fatty acid or an analogous surfactant radicals is very slow. The quantity to be used is *surface conductivity* χ ohm^{-1}, which exists because of the above-mentioned accumulation of cations next to the "crust," and the electro-osmosis caused by this accumulation. If a film, l cm long, w cm wide, and h cm thick (so that $w \gg h$) is placed between two electrodes so that the two end faces $w \times h$ are in contact with the latter, then Ohm's law acquires the form

$$I = \frac{\varphi}{l}(2\chi w + \kappa_0 wh); \tag{9.4}$$

I is the total current in the film and φ is the potential difference between the electrodes. Thus the apparent conductivity κ of the lamella is

$$\kappa = \kappa_0 + \frac{2\chi}{h} \tag{9.5}$$

and the ratio

$$\frac{\kappa}{\kappa_0} = \frac{\delta_0 \varphi_0}{\delta \varphi} = 1 + \frac{2\chi}{\kappa_0 h}. \tag{9.6}$$

In one of the examples of §128, κ/κ_0 was 4.47 for $h < 10^{-5}$ cm (h must be smaller than δ); hence χ/κ_0 was near 1.7×10^{-5} cm. It is clear from Equation 9.6 that, at a slowly changing or constant χ, the ratio κ/κ_0 is smaller, the greater the bulk conductivity κ_0. Since potassium nitrate increases the κ_0 of soap solutions, the lowering of the ratio κ/κ_0 by this salt seems satisfactorily explained.

The value of χ depends on the potential ζ of the diffuse double layer, as is shown[4,5] in Equation 9.7 valid for small concentrations c and small potentials

$$\chi = \left(\frac{2DRTc}{\pi}\right)^{0.5}\left(\frac{v_c}{A-1} \frac{v_a}{A+1} + \frac{DRT}{\pi\eta vF} \cdot \frac{1}{A^2-1}\right); \tag{9.7}$$

D is the dielectric constant of the bulk liquid, R the gas content, T the absolute temperature, v the common valency of the cations and anions, v_c and v_a their normal mobilities, η the viscosity of the liquid, F the faraday, and

$$A = \frac{e^{\epsilon\zeta/2kT}+1}{e^{\epsilon\zeta/2kT}-1}; \tag{9.8}$$

ϵ is the charge of the positron and k is the Boltzmann constant. Consequently an estimate of ζ is possible when χ is known. Such an estimate was performed by Clunie[6].

Vertical films of 4.5×10^{-4} M solution of "sultaine,"

$$C_{16}H_{33}N^+(CH_3)_2CH_2CH_2CH_2 \cdot OSO_2^-,$$

were obtained between horizontal platinum electrodes kept in a mutual distance of about 1 cm, and an alternating current was conducted along the lamellae. These were about 750 Å thick, and their χ was near 2×10^{-10} ohm^{-1} as long as carbon dioxide was excluded. When sodium bromide was added to the solution, the film thickness was smaller (see §15), but χ reached the value of 3×10^{-9} ohm^{-1}; this increase in χ agrees with

Equation 9.7, which predicts that χ (at a constant A) would be proportional to the square root of the ionic concentration c. However, the calculated A did not remain constant; when c of sodium bromide increased from about 0.005 to 0.1 M, ζ seemed to decrease from 0.12 to 0.05 volt.

The electric resistance of vertical films of 0.05% solutions of sodium dodecyl sulfate plus variable amounts of sodium chloride was measured by Platikanov[7], but unfortunately the film thickness (which presumably was about 40 Å) has not been ascertained. The temperature coefficient[8] of conductance of these films was greater below 30° than between 35 and 50°.

The ζ potential of Equation 9.8 is the cause of electrokinetic effects which, consequently, would be expected to occur in foam lamellae. One of these effects, namely electro-osmosis (already mentioned in this section), that is, flow of liquid caused by an electric field, was observed by Reinold and Rücker[9], but in qualitative manner only. Downward electric current accelerated the drainage of a vertical film of a potassium soap solution, while upward current prevented thinning and sometimes caused oozing of droplets at the wire on which the film was suspended. Thus the soap solution in the core was positive and the two "crusts" were negative, in agreement with the picture indicated at the start of this section.

§130.   The κ of single lamellae was measured also by Hagenbach[10]. The solution contained 1 part of sodium oleate in 40 parts of water and 4 parts of glycerol. Assuming that $\kappa \approx \kappa_0$ in a film as thick as 0.78 μ, the ratio κ/κ_0 appeared to be 1.79, 1.71, 2.58, 2.71, and 2.57, respectively, for film thicknesses δ of 0.300, 0.272, 0.164, 0.140, and 0.121 μ. As this ratio did not increase monotonously when δ decreased, it must be presumed that the precision of the measurements was low. Taking, as a probable value, $\kappa/\kappa_0 = 2.6$ at $\delta = 1.4 \times 10^{-5}$ cm and neglecting the difference between h and δ, it is found from Equation 9.6 (§129) that χ/κ_0 was about 1.1×10^{-5} cm, that is, not very different from the χ/κ_0 calculated in the preceding section.

The above measurements were performed on vertical films. Hagenbach studied also horizontal films spanning a dish. To the lid of the latter, two platinum needles were affixed; they pierced the film when the lid was put on. When the dish, closed with the lid, was rotated about a vertical axis, then centrifugal force acted and the lamella became thinner near the axis and thicker near the periphery. In monochromatic light, bright and dark rings gradually moved outward from the center of the film. When the two needles were in contact with one (bright or dark) ring, the electric resistance between them was determined.

The conductance κ proved to be proportional to the thickness δ of the ring when δ ranged between 0.2 and 1.2 μ. The proportionality between κ and δ, which would imply that surface conductivity χ was negligibly small, may be fortuitous. When the needles pierced a film ring whose average thickness was, say, δ_x cm, the current was not confined to the ring; some lines of current passed through the next ring, thicker than δ_x, and others, through the next inside ring whose thickness was less than δ_x. Thus the real thickness of the pass open to the current may have been different from δ_x.

This objection applies also to Rickenbacher's data[11]. Again, rotating lamellae and two needles were employed. At a constant potential difference between the electrodes, the current intensity was proportional to the average thickness of the ring when this varied between 0.1 and 0.5 μ. However, an indication of a higher ratio κ/κ_0 near the lower end of this range was noticed.

§131. Multicellular foams.   The electric conductance κ of a foam is proportional to the relative volume of liquid in it, that is, inversely proportional to the expansion ratio $\rho_2/[\rho]$ (§107)[12]. However, the coefficient of proportionality is not equal to 1.0. This is seen in Fig. 9.1, in which data[13] on foams produced from five different solutions at several expansion ratios are combined. The curve of κ_0/κ versus $\rho_2/[\rho]$ is almost linear, but the first (dimensionless) ratio is almost everywhere 2 to 2.5 times as great as the second. The conductivities κ_0 of the solutions tested varied between 2×10^{-4} (sodium alginate+saponin) to 5×10^{-3} ohm^{-1} cm^{-1} (ferrous protein hydrolyzate), and the expansion factors ranged from about 2 to 21. Since one curve is obtained for all solutions employed,

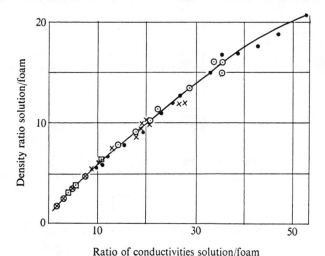

Fig. 9.1. Electric conductivity and density of foams. Abscissa: ratio of conductivities solution to foam. Ordinate: ratio of densities solution to foam. Points refer to sodium isopropylnaphthalene sulfonate + glue (.); same sulfonate + alginate (□); ferrous protein hydrolyzate (×), soap (⊙), and alginate + saponin (⊗) (after reference [13]).

it is clear that dilution of the liquid with air lowers κ in a ratio depending on dilution only, but not on the conductivity of the liquid.

It is possible[14] to account for the experimental fact that $\kappa_0/\kappa > \rho_2/[\rho]$. The electric resistance of a foam is considerable, not only because there is less liquid present, but also because the liquid is distributed in an unfavorable manner. Imagine the foam as an aggregate of cubic bubbles pictured in Fig. 9.2, and let the current flow upward (the arrow!). The horizontal bubble walls (which are perpendicular to the current) will not participate in the transport of electricity. In this manner, two out of the six walls of every cube make no contribution to conductance, so that κ_0/κ would be equal to 1.5 $\rho_2/[\rho]$. Such a relation was observed[15] for a few foams of 2% "Nekal BX" (an impure di-isobutylnaphthalene sulfonate), but it is not in agreement with the more extensive measurements by Clark[13].

Apparently, Fig. 9.3 is a better picture of foam. In it, only three lamellae meet together in a liquid vein (or Plateau border, §42). Again, the current flows in a vertical direction. It is clear that the films perpendicular to the current not only fail to enlarge the path open to ions but, in addition, raise the length of this path, as indicated by the bold black line in Fig. 9.3. The length of this line is again 1.5 times as great as the length of the corresponding path in Fig. 9.2. Hence it may be expected that the ratio κ_0/κ would be $1.5 \times 1.5 \times \rho_2/[\rho]$, i.e., 2.25 $\rho_2/[\rho]$. This conclusion is reasonably well confirmed by Fig. 9.1. In the above explanation, bubble walls rather than Plateau's borders were considered. If the major

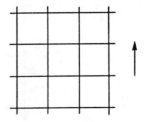

Fig. 9.2.   Scheme of conductance in foams.

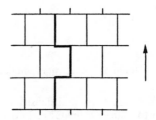

Fig. 9.3.   Scheme of conductance in foams. Because of their cellular structure, the path of the current, see thick line, is longer than expected.

part of the liquid (and, consequently, of the conductance) resides in these borders, the explanation still is valid, as horizontal Plateau borders do not transmit a vertical current but render the path longer. The coefficient 2.25 ought to be independent of bubble diameter; this conclusion also agrees with Clark's results.

When the ratio $\rho_2/[\rho]$ is equal to unity, that is, no foam, only liquid is present, then also $\kappa_0/\kappa = 1$. Hence the above coefficient would be expected to be smaller for gas emulsions than for real foams. The assumption $\kappa_0/\kappa = \rho_2/[\rho]$ made by Vilenskii (see §108) perhaps was not far from truth. Other data on the κ of foams are presented in §109.

Foams have been prepared[16] from 1% solutions of a surfactant (characterized only as a sulfonate) containing also sodium chloride. The ratio κ_0/κ was near 16 when the sodium chloride concentration was 5%, about 5.6 when this was 8%, and about 4.8 for 10% sodium chloride. The lowering of κ_0/κ by a foreign electrolyte is not in agreement with Equation 9.6, and presumably indicates that the expansion ratio of foams was smaller when more sodium chloride was dissolved; unfortunately, the foam density was not determined. The relative temperature coefficient of κ (i.e., $d\kappa/\kappa \cdot dT$) between $-10°$ and $+2°$ was equal to $-d\eta/\eta \cdot dT$ and, consequently, to $d\kappa_0/\kappa_0 \cdot dT$; η is the viscosity of the solution and T the temperature.

§132.   Usually, two soap bubbles can be slightly pressed against each other without coalescence. However, in the presence of an electrified body (such as freshly rubbed stick of sealing wax), they immediately combine to one bubble[17]. Apparently this striking phenomenon has never been studied in a quantitative manner.

References

1. Reinold, A. W., and A. W. Rücker. *Trans. Roy. Soc.* (*London*) 172:447 (1881).
2. Reinold, A. W., and A. W. Rücker. *Trans. Roy. Soc.* (*London*) 174:645 (1883).
3. Reinold, A. W., and A. W. Rücker. *Trans. Roy. Soc.* (*London*) A184:505 (1893).
4. Bikerman, J. J. *Z. Physik. Chem.* A163:378 (1933).
5. Bikerman, J. J. *Physical Surfaces.* New York: Academic Press. (1970), p. 418.
6. Clunie, J. S., et al. *Trans. Faraday Soc.* 63:505 (1967).
7. Platikanov, D., N. Rangelova, and A. Sheludko. *Godishnik Sofiisk. Univ., Khim. Fac.* 60:293 (1965/66).
8. Platikanov, D., and N. Rangelova. *Compt. Rend. Acad. Bulg. Sci.* 21:913 (1968).
9. Reinold, A. W., and A. W. Rücker. *Phil. Mag.* [5] 19:94 (1885).
10. Hagenbach, A. *Arch. Sci. Phys. Nat.* 35:329 (1913).
11. Rickenbacher, W. *Kolloidchem. Beihefte* 8:139 (1916).

12. Miles, G. D., L. Shedlovsky, and J. Ross. *J. Phys. Chem.* 49:93 (1945).
13. Clark, N. O. *Trans. Faraday Soc.* 44:13 (1948).
14. Bikerman, J. J. *Foams.* New York: Reinhold. (1953), p. 148.
15. Manegold, E. *Schaum.* Heidelberg: Strassenbau, Chemie und Technik. (1953), p. 256.
16. Balakirev, A. A., and V. K. Tikhomirov. *Zh. Prikl. Khim.* 41:2762 (1968).
17. Boys, C. V. *Soap Bubbles and the Forces Which Mould Them.* London: Soc. for Prom. Christ. Knowledge, (1890), p. 131.

CHAPTER 10

THEORIES OF FOAM STABILITY

§133. The theories reviewed in Chapter 1 for the stability of single films are equally valid for the foam stability. In spite of the tendency to contract common to all liquid surfaces, foams have a measurable lifetime because of one or several of the following reasons. (a) The viscosity η (or the consistency) of the liquid is so high that every motion, including that necessary for the rupture of bubble walls, is greatly retarded. (b) Surface viscosity η_s is high; it retards drainage and the surface deformations preliminary to bubble bursting. (c) The Marangoni effect resists these deformations. (d) Repulsion of electric double layers helps to maintain the lamellae intact. While the importance of all these factors is generally admitted, in many instances it is not certain which of them are essential for a given foam, and no general quantitative relation still exists between any of them and the experimental foam persistence.

Presumably, quantitative dependence of stability on properties is difficult to ascertain because, as mentioned in §47, bubble bursting is a probability event. Bursting starts from a point at a definite moment of time, and a cause must exist for this point and this moment. As these causes are unknown, it is necessary to turn to probability considerations. In some areas of the bubble wall, the above defence mechanisms (because of the non-uniformity of the lamella) are unusually weak; hence the probability of collapse being initiated in this area is great. Many external agents (dust particles and so on) can act as foam inhibitors; their contact with the lamella is more probable, the greater their concentration in the atmosphere. It may be stated that the probability of bubble rupture is the product of two probabilities, namely that of a particularly weak spot being present in the lamella and that of a foam inhibitor being present in the ambient gas. The former probability is proportional to the area A of the lamella, if this is visibly uniform; this is the reason for the inverse proportionality between persistence τ and A, mentioned in §47. Unfortunately, bubble walls, as a rule, are not uniform. The second probability accounts for (at least a part of) the effect of environment on persistence, as reviewed in §48.

When the liquid is very viscous, the formation of a foam is as slow as its collapse. It may even happen that the method of foam formation selected by the experimenter (for instance, shaking) will give rise to almost no foam, as incorporation of air into the liquid may be largely prevented by the high η of the latter. Better examples of foam stability caused by high viscosity are supplied by liquids whose η is a function of position. When a hot liquid is kept in cool air, the temperature of the surface layer is lower than that of the bulk. Consequently the viscosity near the surface is greater than far from it. A bubble formed in the liquid rises (because of buoyancy) at a reasonable speed as long as the η of the surrounding liquid is small, but the ascent is slowed down in the surface stratum; thus several layers of bubbles (i.e., a multicellular foam) may form above the liquid before any significant coalescence takes place.

In many instances, the gas above the hot liquid has a temperature far below the melting point of the latter. When a bubble tarries in the surface, the heat supply from the liquid to the "dome" of the bubble (see §40) is retarded by the gas in the bubble and the dome may become almost or unquestionably solid. In these systems, the persistence of the bubbles and of the whole foam would be indefinitely long. Foams on metallurgical slags are examples of this mode of formation.

The η of a surface layer may be greater than that of the bulk also when the temperature is uniform. If the heavy liquid (e.g., a molten metal) contains lighter impurities (e.g., metal oxides), these accumulate near the surface because of gravitation. If the melting point or the viscosity of the impurities are higher than those of the bulk liquid, the ascent of bubbles in the surface layer will be slower than in the bulk, and persistent foams may result.

Finally, the bulk viscosity of the liquid may be low during the foam formation, so that this process is not hindered, but may rise in the lamella because of the evaporation of the solvent. The η of the solution sometimes is only a little greater than that of the solvent when the concentration is, say, 0.1%, but a gel is formed when the concentration reaches, say, 2%. Changes of this magnitude may readily occur in a thin film because the relative rate of evaporation in it is so much greater than in the bulk liquid. If the film thickness is δ cm and the bulk liquid is present in a layer h cm thick, the *percentage* of solvent lost by vaporization in unit time is in the lamella h/δ times as great as in the bulk liquid of an identical surface area. As a gel has not only a high viscosity but also a yield stress (see §10), the persistence of a gel-like lamella will be indefinitely long.

§134. The relation between foam stability and *surface viscosity* η_s g/sec has been extensively discussed. Some scientists found almost no correlation. Thus the rate of drainage during the first minute of foam life was almost identical for "olive textile soap" and "white toilet soap," each in 0.1% solution, although the second solution had, and the first did not have, a measurable η_s[1]. This observation contradicts the data collected in §§99 and 100, and presumably should be disregarded; at any rate, the velocity of drainage is a poor measure of stability. Surface viscosity of sodium laurate solutions did not exceed that of water but their foams were, of course, much more persistent[2]. The precision of these measurements was adversely criticized[3], but it is well possible that the foaminess of soaps of a low molecular weight is due mainly to the Marangoni effect and the electrostatic repulsion.

Trapeznikov[4] observed no agreement between the η_s of unimolecular layers on water and the persistence of single bubbles stopped by these layers. For instance, unimolecular films of hexadecanol were 10 to 100 times as viscous as those of oleic acid, but the bubble persistence in the latter system was three to four times as great as in the former. Surface viscosities of ethyl palmitate and hexadecanol were nearly equal, but the lifetime of bubbles under the alcohol exceeded that under the ester by a factor of 5. When the η_s of a film of hexadecanol was raised in the ratio 15:1 by compression, the bubble persistence was lowered from 3.2 to 2.3 sec. Presumably compression gave rise to a nearly solid and brittle film through which the bubbles easily escaped. Brittleness was made responsible also for the fact that the persistence of bubbles under a "monolayer" of stearic acid on water was shorter than on 0.01 N hydrochloric acid; the former film contains stearates of calcium and other metals and is rigid.

On the other hand, in measurements on foams produced by shaking, only those solutions whose η_s was considerable [i.e., aqueous solutions of commercial dyes such as phloxine and crystal violet, of a saponin, an albumin, and a poly(vinyl alcohol)] gave high values for the time of half-collapse[5]; unfortunately, the η_s was measured by a method similar to that of Plateau and shown to be misleading by Marangoni (see §10).

The "average lifetime of gas in the foam," L_g, defined in §60, also was compared with η_s, which this time was measured with a suitable apparatus (of the Couette type)[3]. The foam was produced by an oscillating perforated plate. The solutions contained 0.1 g purified sodium dodecyl sulfate in 100 cm^3 of water, with or without addition of 1-dodecanol whose final concentration (in percents) is indicated in Table 10.1. When this concentration exceeded 0.005%, the surface layer had not only additional viscosity η_s but also a yield stress (see §133), denoted in Table 10.1 by the symbol γ_y (its dimension is g/sec^2, that is, different from that of the τ of

Table 10.1. Foam Stability, Surface Viscosity, and Yield Stress

	Dodecanol concentration, %				
	0	0.001	0.003	0.005	0.008
η_s, g/sec	0	0.002	0.037	0.032	0.032
γ_y g/sec^2	0	0	0	0.054	0.064
L_g min	70	825	1260	1380	1590

§119). It is clear that L_g increases with η_s and γ_y, although no quantitative correlation seems possible.

Similar results have been obtained[6] for saponin foams produced by manual shaking. Two saponin samples were compared; the first had a yield stress of 60 to 90 g/sec² and a much higher η_s than the second whose γ_y was zero. The time needed for the collapse of 80% of the initial foam was, for instance, 480 and 50 hr for 0.05% and 0.5% solutions of the first, and 33 and 9 hr for the equally concentrated solutions of the second saponin. However, the concentration at which γ_y had a maximum value (about 0.2%) was not equal to that of the maximum foam persistence (0.5%).

Apparently, higher consistency of the surface layer enhances foam stability in organic liquids also. Foaminess of many lubricating oils is raised by small additions of water; and the η_s of a 2% "calcium sulfonate" solution in a mineral oil also was greater in moist than in dry nitrogen[7].

Because the dependence of foam stability (in whatever manner defined) on η_s is never quantitative, the suggestion by Plateau that this stability increases with the ratio η_s/γ also has no clear confirmation; γ again is the surface tension of the solution tested.

§135.   The Marangoni effect should operate whenever this γ is considerably smaller than the surface tension γ_0 of the solvent and the diffusion of the surface active solute is slow enough. These two conditions are satisfied for the majority of aqueous solutions of organic substances, from alcohols and acids to synthetic surfactants. Unfortunately, the attempts to express the dependence of foam stability on the magnitude of the Marangoni effect were unsuccessful.

Quincke[8] expected foams to be "more persistent the more the tension of the free surface is reduced by a layer of another liquid (soap water, protein) which covers the free, i.e., adjacent to air, surface." In other words, foam persistence should increase with the difference $\gamma_0 - \gamma$ and have a maximum whenever $\gamma_0 - \gamma$ has one. This expectation is refuted, among others, by the experimental data collected in §§66 and 68. The surface tension of aqueous solutions of aliphatic alcohols and fatty acids mono-tonously decreases when concentration rises; consequently, foam persistence ought to have its greatest value in saturated solutions. In reality, however, saturated solutions do not foam, and maxima of foam stability occur at concentrations at which the quantity $\gamma_0 - \gamma$ does not have any extreme value.

According to another hypothesis[9], "the optimum of foam formation occurs when the boundary layer has the greatest heterogeneity, that is, when the molecules of the solvent and the dissolved frother are present in the boundary layer in equal numbers." It is difficult to understand why a lamella whose "crust" contains equal numbers of water molecules and, say, dodecyl sulfate ions (so that the weight percentage of these ions in the crust is 94%) should be more persistent than a lamella whose crust consists of dodecyl sulfate anions alone; but presumably the hypothesis was proposed for foams of aliphatic alcohols and other poor frothers, not for surfactants. In the systems of alcohol plus water, Gibbs adsorption equation is valid and Bartsch's assumption is equivalent to the statement that Γ of the approximate relation

$$\Gamma = -\frac{c}{kT} \cdot \frac{d\gamma}{dc} \qquad (10.1)$$

has a more or less definite value for each binary system when the foaminess is at its greatest. In this formula, Γ molecules/cm^2 is the excess of solute present in the surface layer, c moles/cm^3 is the bulk concentration, k is the Boltzmann constant, and T the absolute temperature. Using the concept of unimolecular surface layers, it may be stated that the area occupied by one solvent molecule is a cm^2, and by one solute molecule, b cm^2. If the bulk concentration of the foaming agent is small, then $\Gamma = 1/(a+b)$ when the numbers of the two different molecules in the surface are identical. This means that, at the maximum of foaminess, the product $-c(d\gamma/dc)$ is equal to $kT/(a+b)$. No confirmation of this conclusion exists.

A similar suggestion was made by Andrew[10]. The highest foaminess is supposed to occur when the expression

$$\frac{x(1-x)}{1-x+mx}\left(\frac{d\gamma}{dx}\right)^2$$

is a maximum; x is the mole fraction of the frother in the bulk solution, and m is the ratio of the mole volumes of the frother and the solvent. When x is small, this condition can be simplified to "$x(d\gamma/dx)^2$ has a maximum value." Experimental data are not in agreement with this hypothesis either. None of the theories of this section appears applicable to, for instance, solutions of alkylbenzene sulfonates (see Table 4.4, §72). The foam height above the CMC is, wherever measured, greater than below CMC, while $d\gamma/dc$ has a considerable value only below the CMC; at higher concentrations it is almost zero and only for some presumably impure samples has small positive values.

Nakagaki[11] believed that the foam height after shaking and the duration of the subsequent collapse were proportional to the product WT, W being the work required to transfer one molecule of the solute from the surface layer into the bulk liquid. The value of Γ can be measured directly, without recourse to Equation 10.1; the main results of these measurements are collected in §158.

§136.   Maxima of foaminess which cannot be accounted for by the hypotheses of §135 can be explained in a qualitative way by means of surface viscosity or of the Marangoni effect.

When the concentration of the foaming agent increases, the viscosity η_s of the surface layer usually also increases. In some systems the consistency rises so much that this layer becomes brittle, so that the gas confined in the bubble can escape through a fissure in the rigid wall, as indicated in §134. The concentration C_m corresponding to the maximum bubble persistence would then be closely related to that concentration at which the surface film is very viscous but not yet brittle.

The Marangoni effect depends not only on the difference $\gamma_0 - \gamma$ of the equilibrium surface tensions but also on the rate of attaining this equilibrium. An extension of the surface reduces the concentration of the surface active agent in the surface layer (and thus raises γ) *if* the diffusion of this agent from the bulk of the solution into the surface region is not as rapid as the mechanical movement resulting in extension. Thus, if Γ is the excess concentration of the surfactant in the surface, then stretching by a factor of 2 would lower this concentration to 0.5Γ; but if, simultaneously, 0.5Γ molecules of the solute enter the surface layer from the bulk (per square centimeter), then the extension will have no effect on the surface concentration and, consequently, on the surface tension of the liquid. The time needed for the transport of 0.5Γ molecules/cm^2 from the bulk into the surface region is shorter, the smaller the volume of the liquid from which the diffusion must take place. If the solution is relatively concentrated, a layer perhaps 1 μ thick contains so much of the solute that the diffusion path will be of the order of only 1 μ; but if the solution is very dilute, the diffusion path may be many microns long. This means that the equilibrium between the bulk and the new surface will be achieved in a very short time when the bulk concentration is high, and the Marangoni effect will not be operative. The maximum of Marangoni effect (and the maximum of foaminess) should occur then when the frother concentration c is high enough for the difference $\gamma_0 - \gamma$ to be considerable, but not so high that the equilibrium is restored in a small fraction of a second.

A similar reasoning may be formulated in a more mathematical manner. If the volume of liquid, from which diffusion into the expanded surface layer proceeds, contains at least, say, 20 times as much solute as the Gibbs excess, then the diffusion is not retarded by the depletion of the bulk solution. Let δ_0 be the thickness of this volume. The amount of solute in it, per unit area, is $\delta_0 c$ moles/cm^2. It follows from the above condition that $\delta_0 c = 20\Gamma$; here Γ is measured in moles/cm^2, and consequently RT must be substituted for kT in Equation 10.1; R is the gas constant. From this equation,

$$\delta_0 c = -\frac{20c}{RT}\left(\frac{d\gamma}{dc}\right) \qquad \text{or} \qquad \delta_0 = -\frac{20}{RT}\left(\frac{d\gamma}{dc}\right). \qquad (10.2)$$

The derivative $-(d\gamma/dc)$ usually decreases when c increases. For instance, for 1-butanol in water at 25°, it is about 1.56×10^5 g·cm^3/sec^2·mole between $c = 0$ and $c = 0.000135$ mole/cm^3 and only 1.48×10^4 between $c = 0.000810$ and $c = 0.000945$. Hence, δ_0 is 1.3×10^{-4} cm in the more dilute and 1.2×10^{-5} cm in the less dilute solution. The latter thickness may be so small that the equilibrium between the surface and the bulk may be achieved in a time so short that the surface tension forces have no opportunity to act. The three values of C_m for 1-butanol quoted in §66 are approximately equal to 0.000054, 0.000135, and 0.00054 mole/cm^3.

The maxima of foaminess considered so far refer to the curves of foaminess versus concentration. But maxima are seen also on curves of foaminess versus molecular weight; for instance, it is pointed out in §66 that the foaming ability of various pentanols usually exceeds that of all shorter (butanols, etc.) and all longer (hexanols, etc.) aliphatic alcohols. Equation 10.2 does not predict these maxima. In general, the longer the aliphatic chain, the smaller is the solubility in water and, consequently, the greater the average values of $-(d\gamma/dc)$ and δ_0. Also the diffusion coefficient is smaller when the molecule is bigger, so that the Marangoni effect ought to be prominent in solutions of octanol or decanol. Evidently other reasons are needed. Two possible causes come to mind. (1) When the solubility is very small (or δ_0 is very great), then the formation of a complete "crust" may take so long that bubbles collapse before the equilibrium γ has been reached. (2) When the chains are long, their mutual attraction is strong, the unimolecular films on the water surface are not dispersed uniformly but form islands so that gas can escape through the voids between these. No experimental confirmation is available for either of these explanations. See also the following section.

§137. Maxima of foaminess and CMC. In some homologous series the maximum of foaming ability is observed at a concentration equal to, or near to, the critical micelle concentration, CMC (see, for instance, §71). This concentration is analogous to that of a saturated solution; the difference is that some substances, when their solubility limit is reached, form colloidal solutions in which no visible phase separation occurs, while other substances at once give rise to visible precipitates. The former group is characterized by the existence of CMC, while to the latter the common crystalline salts belong. Because CMC is another term for solubility limit, the remarks made in §136 should be applicable to the present problem also.

At the concentrations above CMC, the Marangoni effect ought to be negligible because of two reasons. First, the gradient $d\gamma/dc$ theoretically is zero; consequently also the Γ of Equation 10.1, §135, is zero, and not affected by any extension of the surface. Secondly, micelles are present in the surface region as well as in the bulk and would act as sources of truly dissolved surface active agent, should its concentration in the surface region be lowered by stretching. At the concentrations far below CMC, the Marangoni effect again should be weak, as the difference $\gamma_0 - \gamma$ is small.

Hence, a maximum of the Marangoni effect (and of the foaminess depending on it) would be expected to occur at concentrations below, but not far below, the CMC.

This explanation would be valid for pure surfactants whose solutions have a surface tension independent of concentration above the CMC. The majority of the surfactants tested for foaminess are mixtures, and it is difficult to predict the behavior of their solutions.

§138. Other theories.   The importance of surface viscosity and the Marangoni effect for foam stability is accepted probably by all students of foam, but several additional hypotheses are found in the literature.

A theory based on dimensional analysis was developed by Contini[12]. It is assumed that the foam height h_0 cm of (apparently pneumatic) foams is a function of only five variables, namely foam density ($[\rho]$ g/cm^3), internal pressure (p g/cm·sec^2), viscosity (η g/cm·sec), surface tension (γ g/sec^2), and the acceleration due to gravity (g cm/sec^2). It is difficult to imagine $[\rho]$ and g to be represented in any other way but as a product $g[\rho]$ g/cm^{2}sec^2; however, these quantities are treated by Contini as mutually independent. It is concluded that h_0 should be a function of the dimensionless parameters $h_0 g[\rho]/p$ and $([\rho]/p)^{0.5}(\gamma/\eta)$. In reality, the height of a pneumatic foam is proportional to the linear rate of gas injection (see §53), and just this quantity has been disregarded by the author.

Foam drainage is primarily a hydrodynamic phenomenon. The liquid flows in the bubble walls and Plateau borders at a rate ideally independent of film stability. Thus, if the foam is unstable, it collapses when the drainage has not yet proceeded far and the films are still succulent, while in a stable foam lamellae burst only after a protracted drainage when they are very thin and may be too thin to exhibit interference colors. In other words, very stable lamellae are likely to contain black patches or bands (see §19) before breaking down. This logic was altered by Exerova[13,14], who believed that the ability to form black films was necessary for the longevity of a foam.

It is pointed out in §64 that every liquid capable of foaming must consist of at least two ingredients. Brown[3] suggests that two solutes (and a solvent) are needed for high stability of foams. The solute present in large amounts must be relatively soluble in the solvent but give rise to surface films of low viscosity. The other solute must have a lesser solubility but be strongly surface-active and produce very viscous or almost rigid surface layers.

Gleim[15] maintained that the stability, observed by him, of electrolyte foams was caused by the hydration of ions. It was imagined that the two opposite surfaces of a bubble wall could not approach each other near enough for rupturing because of the resistance of the "hydrated adsorption layers." Table 4.1 of §65 demonstrates that no such effect exists. Ravinskii[16] gave another explanation for the stability of electrolyte foams:

they contain colloidal particles and perhaps are three-phase systems (see §85).

An older hypothesis attributed foam stability to the interplay of surface forces and osmotic pressure[17,18]. Surface tension differences cause accumulation of the solute in the "crust" of the lamella, while osmotic pressure tends to render all concentrations equal everywhere. The equilibrium between the two tendencies was believed to be upset in very thin films, with the result that the mutual approach of the surfaces is hindered by the osmotic pressure of the liquid between them. At present it appears much more likely that this approach is restricted by the mutual repulsion of the two diffuse electric double layers (§13).

Because of evaporation of the solvent, foams are likely to be cooler than the ambient air. This temperature difference may be a factor in foam stability[19,20]. This was concluded[19] from the following experiment. A wire triangle was dipped into turpentine oil, withdrawn, and kept in horizontal position; the film thus obtained lasted for a few seconds only. When the triangle was kept partly immersed in the oil so that one of its sides was in the liquid but the opposite corner was in air, then the lamella obtained was very persistent (as long as 17 min). Presumably, the apex of the triangular film was cooler, as a result of evaporation, than the lower strata. The oil was raised in the lamella because surface tension was greater near the top than near the bottom. In this manner, drainage was retarded and film persistence enhanced.

The lowering of temperature because of evaporation may be overshadowed by other effects. The energy spent on foam formation (by shaking, by stirring with gas bubbles, etc.) must remain for awhile in the foam and keep it warmer than the surroundings. This was observed[21] in foams of 0.01% saponin solution. Collapse of foam, at least in some instances, produced no measurable heat effect[22]. It is easy to show that the surface energy of an aqueous foam, imagined as an agglomeration of cubes of 0.1-cm edge and with walls 1 μ thick, is equal, at $\gamma = 40$ g/sec^2, to 2400 ergs/cm^3; this energy would be sufficient to raise the temperature of this mass (equal to 3 mg) by about 0.02°.

§139. Foaminess of mixtures. Mixtures have been repeatedly referred to in this chapter, for instance in Table 10.1, §134, and in the earlier chapters. In this and the three following sections the possible mechanisms are reviewed by which an additional component may cause enhancement or inhibition of the efficiency of a foaming agent. Many data on foam destruction by chemical, mechanical, and other means will be found in Chapter 11.

A second solute may affect any of the four main causes of foam stability listed in §133. It may raise the viscosity η of the bulk liquid so that foam formation still is possible when a large external force is used (for instance, gas injection from a high-pressure bomb) but foam collapse is retarded because the forces causing it (surface tension, gravitation, and so on) are relatively weak. This is a device employed for many years. Gum arabic, ethyl cellulose, and similar hydrophilic materials of high molecular weight raise the persistence of many industrial foams; and marked increase of η is the most likely cause of their influence. Of the five recipes for persistent bubbles (§77), four contain glycerol and the fifth contains tragacanth and glycerol; and the main effect of these ingredients undoubtedly is enhancement of bulk viscosity. The liquid core of the lamellae of "Aerosol OT" (see §75) becomes thixotropic when the solution contains also sucrose or gelatin[23].

A second solute often raises the surface viscosity η_s of the solution. The best known example (sodium dodecyl sulfate plus 1-dodecanol) is mentioned in §134. Presumably, the enhancement of the foam stability of other surfactants by long-chain alcohols and similar compounds (see, for instance, §74) has an identical mechanism. As mentioned in §81, the bubble persistence under unimolecular layers of myristic acid and its homologs has a pronounced maximum at small coverages. When, instead of pure acids, mixtures (1 mole for 1 mole) of myristic or stearic acid, or of 1-hexadecanol, with an aliphatic hydrocarbon are deposited on water surface, the bubble persistence becomes independent of the coverage (e.g., within the region of 22 to 38 square Å per molecule)[24]; unfortunately, no data on the η_s of these films could be found. This is true also for aqueous solutions of decanoic acid and 1-decanol which give rise[25] to a particularly stable foam (after shaking) when the mole ratio of acid to alcohol is 1:3. Probably an increase in η_s accounts for the high stability (referred to in §78) of foams of (tannin + heptanoic acid), as compared with the transient foams of the two constituents separately.

It is convenient to consider two distinct mechanisms of change of η_s caused by the introduction of a second solute. This compound may enter the surface film (with or without displacing the first solute there) or it may alter the composition (and the mechanical properties) of the film without being present in it. These two mechanisms would be operative also when the second solute alters the Marangoni effect rather than the η_s of the initial binary solution.

§140.   A second solute may change the composition of a surface film without penetrating it, for instance, by affecting the solubility or the critical micelle concentration (CMC) of the first solute. The existence of a correlation between CMC and foaming capacity is indicated in §137. It is clear that any addition which alters the CMC of the foaming agent would be expected to influence foaminess[26,27]. The truth of this conclusion was demonstrated[26] on mixtures of sodium 2-dodecylbenzene sulfonate with some organic compounds; unfortunately, the CMC was determined in absence of inorganic salts while foaming solutions contained about 4 parts of sodium sulfate and sodium tripolyphosphate for

each part of the organic sulfonate. The CMC of the pure sulfonate at 55°
was 0.59 g/l. and the foam volume v of an aqueous solution of 0.64 g/l.
of sulfonate (plus inorganic salts) was 18 cm^3. When 1 part of 2-tetra-
decanol was added to 4 parts of the sulfonate, the CMC was raised to
0.60 g/l. and v (in the presence of inorganic salts) was lowered to 12 cm^3.
Neither CMC nor v were markedly affected by caprylamide,

$$C_7H_{15}CO \cdot NH_2.$$

Dodecyl glyceryl ether $C_{12}H_{25}O \cdot CH_2CHOH \cdot CH_2OH$ and N-β-hydroxy-
ethyl-layrylamide $C_{11}H_{23}CO \cdot NHCH_2CH_2OH$ lowered CMC to 0.29 and
0.31 g/l. and raised v to 32 and 50 cm^3. The v was measured 20 min after a
reproducible shaking.

The apparent solubility of proteins usually is lowered by inorganic
salts. This salting-out presumably is the cause of the increased foaminess of
alpha soybean protein solutions in 0.5 M sodium chloride or calcium
chloride, as compared with distilled water[28].

Changes in real solubility probably were responsible for the effect of a
second solute in organic solvents[29,30]. Some mixtures of cyclohexane
and nitrobenzene give rise to stable foams. Also binary mixtures of piperidine and nitrobenzene have a marked foam stability. However, when
piperidine is added to well-foaming cyclohexane-nitrobenzene solutions,
foaminess is depressed. The suggested explanation is that the miscibility
of C_6H_{12} and $C_6H_5NO_2$ is enhanced by $C_5H_{11}N$. Ethanol has a weaker
effect than piperidine but this may be caused by displacement in the
surface film. In the ternary systems of nitrobenzene, p-cymene, and
cyclohexane, the most stable foams were observed at the composition
60:25:15 parts; no correlation with miscibility was demonstrated in this
case. p-Cymene and methanol are miscible in all proportions, but complete
mixing is prevented by small amounts of water (e.g., 1%). In agreement
with expectation, the foaminess of mixtures of anhydrous methanol and
p-cymene was raised by water as long as the ternary system remained
homogeneous.

A more correct form of the Gibbs adsorption equation (see §135) is

$$\Gamma = -\frac{fc}{RT} \cdot \frac{\partial \gamma}{\partial (fc)}. \tag{10.3}$$

Quantity f is the activity coefficient of the solute whose adsorption is given
by Γ. This f depends on the total composition of the solution, that is, on
the concentration of the second, the third . . . solute. In this manner, the
second (third, etc.) addition alters the composition of the surface film.
This is a mathematical formulation of the above reasoning, valid for
equilibrium systems.

Nakagaki[31] pointed out that inorganic salts accelerated the collapse of congo red foams after shaking in a manner similar to their effect on the coagulation of congo red sols. In both systems, HCl had the greatest effect, followed by $BaCl_2$, NaCl, and $NaNO_3$ (the weakest). If the resistance to coagulation and the stability of foam films both depend on the mutual repulsion of electric double layers, the similarity noticed by Nakagaki is readily understood. Also the salt concentrations corresponding to the most rapid foam collapse and the most rapid coagulation were similar; for instance, 0.0028 N HCl and 0.0043 N HCl.

Polyamides are used to reduce foaming in locomotive boilers (see §145). Their efficiency is raised by addition of chestnut extract, for instance, 100 parts for 0.2 part of polyamide in 10^6 parts of water[32]. The effect is explained as follows. Boilers contain a sludge. The sludge adsorbs polyamides and renders them inactive. Chestnut extract displaces polyamides from the solid surface and thus restores their activity.

The easiest effect to understand is a chemical reaction between the second and the first solute. Thus, calcium chloride suppresses the foaming ability of sodium palmitate, simply by forming calcium palmitate, whose solutions foam very poorly; and hydrochloric acid gives free palmitic acid from sodium palmitate and causes an analogous decrease in foaminess. Many examples of this direct chemical effect are known.

§141. When the second solute becomes a constituent of the surface film, it may join the first solute there or displace it, and the presence of the second solute in the surface layer may influence surface viscosity η_s, or the Marangoni effect, or the mutual repulsion of the two double layers. Thus, many mechanisms of foaminess change are possible. In this section the effects in the absence of displacement are considered.

The increase of η_s by incorporation of a second solute is mentioned in §139. The Marangoni effect also would be influenced by a second substance present near the interface. Suppose, for instance, that the solution of B in A is super-saturated and that microscopic (or submicroscopic) droplets of B are floating in the surface region. The concentration of truly dissolved B may be so low that the surface tension of the solution is almost equal to that of the solvent so that the Marangoni effect remains weak. An additional solute (C), which may be present in the bulk of A in negligible concentration but is a substantial ingredient of the surface layer, may enhance the solubility of B there and thus lower the surface tension. Such may be the explanation for the favorable effect of ethanol on the foam stability of aqueous solutions of nonanoic acid and 3-methylbutanol[10]. If C lowers the solubility of B in A, it may cause formation of emulsions near the interface and thus bring the surface tension of the mixture nearer to that of solvent A; and the Marangoni effect would be weakened. This weakening may account for the observation that benzyl alcohol lowers both the solubility of 3-methylbutanol and *m*-cresol in water and the time of collapse of their aqueous foams.

The effect of salts on the electrostatic repulsion of opposite film faces is reviewed in §15.

§142. Displacement of the first by the second solute was postulated, for instance, by Quincke[8,33], who noticed that ether vapors caused a rapid collapse of beer foam. He believed that displacement took place whenever the relation

$$\gamma_{12} < \gamma_1 - \gamma_2 \qquad (10.4)$$

was satisfied; γ_1 is the surface tension of the initial solution (i.e., solvent plus first solute), γ_2 is that of the mixture of solvent and second solute, and γ_{12} is the hypothetical tension of the interface between the film consisting of this mixture and the bulk almost identical with the initial solution. The rationale of inequality 10.4 is as follows. When ether vapor is absorbed by a spot on the beer surface, the surface tension of this spot becomes equal to γ_2, while the surrounding liquid surface still has the tension γ_1. If γ_1 is greater than γ_2, the Marangoni effect tends to operate, that is, the boundary between the contaminated spot and the main surface is pulled outward with the force $\gamma_1 - \gamma_2$ per centimeter. But, when the spot area increases, also the area between the ether-rich film and the bulk beer increases by an identical amount. Consequently, tension γ_1 has to overcome not only tension γ_2 but also that of the interface, that is, γ_{12}. Thus condition 10.4 is derived. It is clear that neither γ_2 nor γ_{12} correspond to an equilibrium, and the very existence of γ_{12} is somewhat doubtful.

If γ_{12} does not exist or is very small, displacement of the first by the second solute occurs when $\gamma_1 > \gamma_2$; in other words, a more surface-active substance displaces one which is less surface-active. Qualitative data which agree with this simple rule have been published by several investigators[34-38].

In some instances, the foaming behavior of a solution containing considerable amounts of one solute (*B*) and (often small) amounts of a second solute (*C*) is so similar to that of a solution containing *C* only, that it is difficult to doubt that *C* has almost fully displaced *B* from the surface region. For instance, the time of collapse of 0.027 M aqueous solution of *m*-cresol was 25 sec, and of 0.153 M solution of 3-methylbutanol, 12 sec; and when the solution contained both 0.027 moles of *m*-cresol and 0.153 moles of 3-methylbutanol per liter, the above time was 11 sec, that is, practically identical with that for the alcohol alone[39]. The effect of pH on the foaminess (determined with a pneumatic method) of an aqueous solution of ovalbumin (0.5%) and gelatin (10%) was practically identical with that of 0.5% albumin solution alone[40].

In §141, emulsions formed by the first solute are mentioned. It was suggested[36,41] that if the second solute *C* gives rise to an emulsion in the surface layer, it is likely to be particularly efficient as a foam inhibitor. When the surface is stretched, the concentration of *C* in the surface layer decreases. If diffusion of C from the bulk to the surface is necessary to restore the initial concentration, this slow process would permit the Marangoni effect to operate for a considerable time. On the contrary, when a reservoir of C molecules is present in the surface region as emulsion droplets, the restoration of the original concentration would be rapid and the Marangoni effect short-lived.

References

1. Preston, W. C., and A. S. Richardson. *J. Phys. Chem.* 33:1142 (1929).
2. Burcik, E. J. *J. Colloid Sci.* 5:421 (1950).
3. Brown, A. G., W. C. Thuman, and J. W. McBain. *J. Colloid Sci.* 8:491 (1953).
4. Trapeznikov, A. A. *Acta Physicochim. URSS* 13:265 (1940). *Zh. Fiz. Khim.* 14:821 (1940).
5. Kimizuka, H., and T. Sasaki. *Bull. Chem. Soc. Japan* 24:230 (1952).
6. Trapeznikov, A. A., K. V. Zotova, and N. V. Shamrova. *Kolloidn. Zh.* 32:437 (1970).
7. Mannheimer, R. J. *A.I.Ch.E.J.* 15:88 (1969).
8. Quincke, G. *Ann. Physik* [3] 35:580 (1888).
9. Bartsch, O. *Kolloid-Z.* 38:177 (1926).
10. Andrew, S. P. S. *Proc. Intern. Symp. Dist., Brighton, England 1960*, 73. *Chem. Abstr.* 56:4558 (1962).
11. Nakagaki, M. *J. Phys. Chem.* 61:1266 (1957).
12. Finzi Contini, B. *Chim. Ind. (Rome)* 36:542 (1954).
13. Exerowa, D., and M. Bulewa. *Abhandl. Deut. Akad. Wiss. Berlin, Kl. Chem. Geol., Biol.* 6b:938 (1966).
14. Exerowa, D., and A. Sheludko. *Chem. Phys. Appl. Surface Active Subst., Proc. Int. Congr., 4th, 1964,* 2:1097 (1967).
15. Gleim, V. G., and V. Ya. Khentov. *Zh. Prikl. Khim.* 41:2475 (1968).
16. Ravinskii, M. B. *Kolloidn. Zh.* 24:58 (1962).
17. Edser, E. *Fourth Report on Colloid Chemistry.* London: Brit. Assoc. Advance Science. (1922), p. 263.
18. Foulk, C. W. *Ind. Eng. Chem.* 21:815 (1929).
19. van der Mensbrugghe, *Mém. Acad. Roy. Sci. Belg.* 43 (presented in 1878, published in 1882).
20. Neville, H. A., and T. H. Hazlehurst. *J. Phys. Chem.* 41:545 (1937).
21. Schütz, F. *Nature* 139:629 (1937).
22. Durney, H. A., and W. A. Patrick. *Abstracts of 119th Meeting.* American Chemical Society (1951), p. 5G.
23. Trapeznikov, A. A. *Chem. Phys. Appl. Surface Active Subst., Proc. Int. Congr., 4th, 1964,* 2:1133 (1967).
24. Pokhil, P. F. *Zh. Fiz. Khim.* 14:554 (1940).
25. Shah, D. O., and C. A. Dysleski. *J. Am. Oil Chem. Soc.* 46:645 (1969).
26. Schick, M. J., and F. M. Fowkes. *J. Phys. Chem.* 61:1062 (1957).
27. Ross, S., and T. H. Bramfitt. *J. Phys. Chem.* 61:1261 (1957).
28. Perri, J. M., and F. Hazel. *J. Phys. Colloid Chem.* 51:661 (1947).
29. Teitelbaum, B. Ya., and S. G. Ganelina. *Kolloidn. Zh.* 14:267 (1952).

30. Teitelbaum, B. Ya. *Kolloidn. Zh.* 14:292 (1952).
31. Nakagaki, M. *Bull. Chem. Soc. Japan* 23:47 (1950).
32. Gunderson, L. O., and C. M. Bodach. *Ry. Engr. Maintenance* 47:447 (1951). *Chem. Abstr.* 45:8682 (1951).
33. Quincke, G. *Pogg. Ann.* 139:1 (1870).
34. Ross, S., and J. W. McBain. *Ind. Eng. Chem.* 36:570 (1944).
35. Ross, S. *J. Phys. Colloid Chem.* 54:429 (1950).
36. Robinson, J. V., and W. W. Woods. *J. Soc. Chem. Ind.* 67:361 (1948).
37. Pattle, R. E. *J. Soc. Chem. Ind.* 69:363 (1950).
38. Doss, K. S. G., and S. K. D. Agarwal. *J. Proc. Oil Technologists' Assoc., India, Kanpur* 11:41 (1955). *Chem. Abstr.* 51:15974 (1957).
39. Lubman, N. M. *Zh. Fiz. Khim.* 3:204 (1932); *Tsvetn. Metal.* 6:854 (1931).
40. Pankhurst, K. G. A. *Trans. Faraday Soc.* 37:496 (1941).
41. Fiske, C. H. *J. Biol. Chem.* 35:411 (1918).

CHAPTER 11

FOAMS IN NATURE AND INDUSTRY

§143. Cuckoo spit.   Several insects protect themselves by froth. The best known instance is the froghopper, belonging to the genus *Philaenus, Cercopidae.* The nymph of this species, after hatching from the egg, secretes a foam which envelops its body and remains in place for several weeks (usually from May to July in the Northern Hemisphere). If the animal is withdrawn from its armor, it rapidly succumbs. Thus the foam seems to protect it not only from predators (ants and so on). Rain or direct sun rays cause no destruction of this "cuckoo spit" or "spittle". The froth contains only about 0.5% of solid residue, the rest being water; and the residue is mainly inorganic. Apparently the liquid precursor of the froth is secreted through the anus and then transformed into foam, either by injecting air through pores nearby or by stirring the liquid. Nothing is known about the mechanical properties of the cover.

§144. Natural waters.   Systematic annual (or more frequent) determinations of the foaming capacity of a particular spot on water surface, extended for many years, would be a valuable measure of the gradual pollution (or purification) of natural waters; unfortunately, no such measurements have ever been recorded.

Shaking of a sample of ocean water produced (in 1943) a foam almost as voluminous as that of 0.8 M acetic acid in identical circumstances[1]. The foam of a Barents Sea water contained proteins[2]. An unexpected agent[3] responsible for the foaming of the Mediterranean near Izmir is referred to in §89. The biological oxygen demand (BOD) of sea foams was 10 to 20 times that of the original sea water[4] and the BOD of foams formed on Lake Baikal was 2 to 30 times as great as that of an equal mass of the lake water[5]. About 60% of the latter BOD was caused by colloidal material, and the foams obtained from samples of coastal water were more stable than those formed by water far from the shore.

Moorland water often has high foaming ability and contains considerable amounts of fatty acids, saponins, and plant proteins[6].

When surfactants resisting biodegradation (they contain the group —$CH_2C(CH_3)_2CH_2$— in a side chain) were widely used by the population, then rivers and lakes near sewage treatment plants frequently were covered with a blanket of stable foam; a description of this phenomenon can be read, for instance, in Raison's paper[7].

246

§145. Priming and carry-over.   Some aqueous (and non-aqueous[8]) solutions and suspensions give rise to bothersome foams on boiling. Also, liquid droplets are transformed with the steam into the condenser, and this carry-over is objectionable in many applications, to which, for instance, the use of locomotive boilers belongs.

The carry-over, that is, the relative amount of the liquid which reaches the condenser without passing through the vapor stage, is usually determined by chemical analysis or from the electric conductivity of the distillate. If the boiling liquid contains c grams of nonvolatile solutes in a cubic centimeter, and the distillate contains xc g/cm^3, then $100x\%$ of the latter left the liquid as drops rather than as steam. It is also possible to determine the weight percentage of droplets in the vapor by measuring the electrostatic capacity of the vapor phase[9]; this percentage, ideally, is identical with $100x\%$ and designated as "the wetness of the vapor." The volume of foam situated on the surface of a boiling liquid at any time can be measured, for instance, in laboratory boilers made from glass or another transparent material.

The effect of foaming ability on carry-over is not quite clear. According to Gleim[9,10], when the rate of evaporation is accelerated, x or the wetness of the vapor increases to a maximum, then decreases, and in some systems becomes nearly independent of this rate. The decrease of x occurs when foam becomes visible above the boiling liquid; it thus appears that carry-over is hindered by foam. The explanation proposed by Gleim is that droplets are produced by bursting of the bubbles in the liquid surface. As an aside, it may be reported that they originate not only in the bubble "dome" referred to in §40, but also in the "fountain" rising from the bottom of the bubble when the capillary pressure $4\gamma/R$ suddenly vanishes. When a foam blanket is present, the droplets of whatever origin are intercepted in it and never reach the gas phase. Unfortunately it is not explained what happens to those droplets which must form when the bubbles at the top of the foam collapse.

As long as no foam blanket exists, the boiling liquid is a gas emulsion. If it does not contain any foaming agent, the bubbles in it readily coalesce, and the large resulting bubbles rapidly rise to the surface and burst. Consequently, the relative volume of the vapor phase in the liquid is small, the overall volume of the liquid (including the bubbles in it) also is small, and the upper surface of the liquid is relatively far from the channel or duct to the condenser. When a foaming agent is present, the bubbles refuse to coalesce and, hence, rise slowly (see Equation 1.3, §2). Thus the relative volume of immersed bubbles is high, and the upper surface of the liquid is situated near the above duct; as a result, many droplets formed at this surface can be transported by the steam current to the distillate[11].

The importance of the path length between the liquid surface and the channel entrance is emphasized by Durov[12]. When this length increases, the value of x monotonously decreases, but the decrease is steeper for sodium chloride solutions containing a colloidal precipitate of basic magnesium carbonate, $Mg_2(OH)_2CO_3$, than in the absence of colloids, so that, when the vapor-liquid interface is high in the boiler the precipitate enhances x, but when the space above the liquid is high, the carry-over is depressed by the colloid. Presumably, the colloidal precipitate stabilizes the bubbles but renders the droplets heavier and thus prevents them from rising high. When the boiling solution contained sodium chloride and a mixture of colloidal ferric hydroxide and basic magnesium carbonate, the foam volume v had a maximum value and x had a shallow minimum when the mixture was equimolecular. Here again, a high v corresponded to a small x.

Two opposite conclusions can be drawn from the above interpretations. If the absence of coalescence is necessary for a high x, then ordinary foaming agents would enhance carry-over; but if foam intercepts droplets, then carry-over would be depressed by these agents. New experiments are needed before it can be stated which mechanism is more important and under what circumstances. It must be pointed out that a surface active solute usually has at least one additional effect. The dimension of the steam bubble in the moment of formation is to a considerable extent determined by the wettability of the boiler wall (see §§38 and 39). If this wettability is poor (that is, the contact angle θ in the liquid is large), the bubbles are big when they break loose and rapidly rise to the surface; thus the mechanism envisaged by Denman[11] would occur preferentially in liquids which show small contact angles with the hot wall. Many foaming agents directly affect the value of θ and, hence, indirectly the carry-over.

Castor oil was for years the preferred inhibitor of carry-over; unfortunately it is too readily hydrolyzed by boiling water, which often has an alkaline reaction. Small amounts (e.g., 0.001 %) of typical foaming agents (sodium alkyl sulfates, sodium dodecylbenzene sulfonate) also may[13] lower the value of x. At present, more common inhibitors of carry-over belong to the classes[11,14]:

$[R \cdot CO \cdot NHCH_2 -]_2$; the best inhibition is achieved when RCO is stearyl.
$[R \cdot CO \cdot NH(CH_2)_n -]_2$; the number n has little influence on inhibition as long as it is confined to the range 2 to 6.
$[R \cdot CO \cdot NH \cdot CH_2CH_2 -]_2NH$; the best results are observed when RCO is stearyl or palmityl.
$[R \cdot CO \cdot NHCH_2CH_2 -]N \cdot OCR$; RCO should be lauryl or palmityl.
$[C_{17}H_{35}CO \cdot NH \cdot OCO \cdot CH_2CH_2 -]_2$.

Ethers from poly(ethylene oxide) and octanol, dodecanol, or hexadecanol.

Poly(oxyethlene) amines[15]

In Ravinskii's opinion[16], foams on boiling aqueous salts solutions are stabilized by suspended particles of dust, rust, etc., and common surfactants hinder foam formation in these systems because they are absorbed by the particles and suppress the ability of the latter to enhance foaminess (see §90).

An automatic control of priming is described by Henneman[16a].

§146. Pickling liquor.   Ferric chloride is one of the few purely inorganic salts which act as frothers in water; the foam on $FeCl_3$ solutions can be inhibited by diethyl or dibutyl phthalate[17]. More important is the foaming of pickling liquor.

When steel is pickled, that is, its oxide coating is removed by acid, the resulting solution may foam and in this manner may obstruct the process. To find the reason for this foaming ability, a synthetic pickling liquor was prepared from 40 g H_2SO_4, 35 g NaCl, x g SiO_2, and y g $FeSO_4 \cdot 7 H_2O$ in 1 l.[18]. When air was injected into this solution and the time t_e until an "eye" appeared (see §61) was determined, it was found that t_e was raised by heating the liquid to $80°$ for 3 hr and cooling again. For instance, when x was 10 and y was 400, the t_e was 9 sec before and 85 sec after heating. When y was 400 but x was 1 or 5, then t_e after heating was 10 or 50 sec; and when x was 10 but y was 100 or 300, then t_e was 5 or 20 sec. It appears therefore that foam stability rises with the concentration of SiO_2 and $FeSO_4$. Presumably, heating accelerates formation of colloidal silica which acts as a foam stabilizer (see §89). A practical conclusion is that foaming of pickling liquors may be expected whenever the silicon content of the steel is high.

§147. Metallurgic slags.   Foam on slags presumably exists because of their high viscosity, which retards the rise and coalescence of gas bubbles, such as carbon monoxide, so abundant in many furnaces. In addition, there may be a temperature gradient from the bulk to the surface, and the higher viscosity of the cooler surface region stabilizes the foam, as described in §133. Apparently, this excessive viscosity can be eliminated by using a luminous flame to heat the surface of the slags in open-hearth furnaces; both Evans[19] and Tuluyevskii[20] observed that this heating lowered the foam height.

A quantitative study of the relation between this foam height and the viscosity of the slag at different levels would have been instructive; unfortunately, the experimental work known to the author was performed, instead, on the relation between foaminess and the chemical composition of the slag. Meilbaum[21] claimed that blast furnace slags foamed (at 1390 to $1420°$) when they contained more than 35% of silica or more than 44% of (silica + alumina) and less than 50% of calcium oxide. According to Fellcht[22], foaming occurs when the slag contains <4% FeO, <40% CaO, and >25% SiO_2. Layton[23] emphasized the importance of sulfides in the slags.

An extensive investigation was performed by Nechkin[24]. Air was injected into a melt containing large relative amounts of silica, smaller amounts of ferrous oxide and alumina, and various concentrations of calcium oxide. The temperature was 1400°. When the rate CaO/SiO_2 was increased in steps, the greatest foam stability was observed when it was equal to 1.50 to 1.55; and when it exceeded 1.70, the foam was unstable. The surface tension (at 1530°) of well-foaming slags was[22] 388 to 459 g/sec^2, and of slags incapable of foaming: 337 to 369 g/sec^2. The meaning of this difference is obscure.

§148. *Petroleum products.* Objectionable foams frequently form in lubricating oils by mechanical incorporation of outside air into the oil or by cavitation; these processes may take place, for instance, during agitation in a gear box or because of a rapid decrease in atmospheric pressure. More work was performed on suppressing these foams than on understanding the cause of their stability. No doubt, some of them are gas emulsions stabilized by the high-volume viscosity of the liquid, but surface effects are important in some other systems.

Air was injected into four machine and turbine oils at a constant rate[25]. The Σ of §53 was 8 sec at 20° and 1.3 sec at 100° for the best foaming and 1.6 at 20° and 0.5 sec at 100° for the least foaming oil. The Σ values were reduced by 0.5% of an ethyl silicone mixture to, respectively, 2.4, 0.8, 0.8, and 0 sec, and by 1.0% of this inhibitor to 1.2, 0.2, 0.3, and 0 sec. Trautman[26] maintains that some poly(siloxanes) depress the foam height (in a pneumatic method) of light mineral oils only when used in concentrations exceeding that of saturation.

A dynamic foam of the SAE 30 oil was inhibited by adding 1.45 part of methanol, or 0.36 ethanol, or 0.21 2-propanol, or 0.11 3-methylbutanol to 100 parts of oil[27]. Among the more complex hydroxy compounds suggested for destroying petroleum foams, triethanolamine[28], may be mentioned.

Many compounds of other types can inhibit foams of lubricating oils. Examples are poly(hexadecyl methacrylate) plus glycerol monoricinoleate[29], potassium oleate[30], nitro-substituted aliphatic alcohols[31], a complex of titanium and 2-ethyl-1,3-hexanediol[32], and metal salts of copolymers of styrene and maleic anhydride[33].

When petroleum is pumped out of a well, the oil entering the annulus in the borehole often is transformed into foam, which according to two patents[34] sometimes rises to a height of 1500 m; this may be the tallest foam column mentioned in the literature. To break this foam, which retards oil recovery, the inventor recommends first to determine whether the bubbles are positively or negatively charged. This can be decided, for instance, by subjecting the foam to electrophoresis. Positive bubbles coalesce when treated with an anionic surfactant; thus a stick made of poly(methyl siloxane) 50, paraffin wax 25, and "sodium mahagony sulfonate" 25 parts may be dropped into the casing of the well. If the bubbles are negative, their coalescence is accelerated when sticks of, for instance, poly(methyl siloxane) 10, paraffin wax 80, and ricinoleic acid imidozaline 10 parts are used. This treatment raised the daily petroleum production of a well by, e.g., 40%.

It is stated[35] that ionizing radiation raises the foaming tendency of turbine oils.

§149. *Paper manufacture.* According to a review[36], foam in pulping operations is caused by lignin, resin, and fatty acids in the wood; also sulfate soaps derived from pitch may act as foaming agents. The incorporation of gas may occur because of mechanical agitation or by condensation of carbon dioxide (and other) molecules formed when different effluent streams collide and mix. In paper-making operations, sizing materials, dyes, fillers, oxidized starch, proteins, and various surfactants introduced for various purposes may act as frothers. The volume of dynamic foams of various pulping wastes varied from sample to sample, but usually was equal to that of solutions of sodium dodecylbenzene sulfonate having 3 to 18 parts of sulfonate per million parts of water.

Systems as complex and as varied as pulp and paper suspensions evidently are little suited for scientific measurements, and not many scientific studies of their foams have been published. To determine the main foaming agents in jackpine sawdust, this was extracted with water, and the extract (whose pH was between 3 and 4) was foam-fractionated[37]. The foam contained an excess of carbohydrates, pitch, and soluble lignins. Pour tests of each of these ingredients indicated that lignin was the most likely cause of foaming and that pitch stabilized lignin foams. Aging in air depressed the foaming ability, and the foam height at 30° was lower than at 15°.

A lignin-like substance proved to be the main frother also in the sulfite spent liquor from beechwood[38], but the foams of all these liquors are three-phase systems (see §89) because they contain a large number of paper fibers. Foam stability manifested a maximum at pH 7; it increased on aging, but the fiber concentration corresponding to the highest stability decreased. Neither the foam volume nor the foam persistence were affected by moderate temperature changes (0° to 25°). Rosin size slightly raised foam volume but sharply lowered the stability.

Sulfate soap apparently was the main foaming agent in a black liquor containing 158 g of solids in 1 l. at pH 11.5. When 2.9 g of the above solids was sulfate soap, the Σ of §53 was about 100 sec. Dilution in the ratios 100:1 to 400:1 raised Σ, but this was lowered by further dilution. When pH was varied, Σ dropped sharply when alkalinity was lowered from 8.5 to 8.1. A moderate rise of temperature enhanced Σ; also the presence of cellulose fibers (up to 3%) raised it[39].

Many compounds and mixtures can be used to combat foams in the pulp and paper industry. They range from vegetable oils[40] to silicone oils[41] and include materials as unusual as the urethane from dodecyl hexaglycol ether and 2,4-toluene di-isocyanate[42]. Colloidal clay also is recommended[42a].

§150. Sugar. Sucrose and other sugars are not foaming agents but extracts of sugar cane and sugar beet, from which sucrose finally is isolated, have marked foaminess which renders the removal of unwanted ingredients more difficult. According to Saray[43], when diluted to a common density, raw sugar diffusion juice has the greatest foam volume (after shaking), followed by molasses, while thin juice and first carbonation juice have the weakest foaminess.

It is not certain what ingredients act as foaming agents in all these liquids. Proteins and peptides are often considered to be the main culprits. They are said to be more important than saponins[44]. Vlasova[45] claimed that, of the components of molasses, caramel and acids gave rise, on shaking, to more voluminous foams than did melanoids and araban. Insoluble powders or fibers, such as calcium carbonate, usually enhanced foam stability. In agreement with this observation, the foam of molasses was found to be enriched in solids. Thus, after volatilization of water, such a foam contained[46] insoluble matter (rust, sand, etc.) 40%; oxalate, humins, saponins, etc. 25%; fat 10%; lime 10%; and unidentified material 15%.

Among the numerous agents recommended for suppressing foams in sugar plants, superheated steam[47] may be mentioned. Otherwise, the chemical foam inhibitors used in sugar industry are not different from those employed elsewhere; for instance, vegetable oils recommended here[48,49] are similar to those employed in the manufacture of paper. Ethers of sucrose and propylene oxide are foam inhibitors, not only in sugar mills but also for many detergents[50]. An unexpected observation[51] is that traces $(10^{-4}\%)$ of manganese sulfate break the foam on hot sucrose solutions.

Foams which form during saturation and evaporation of intermediate sugar solutions can be destroyed by high-frequency vibrations[52]. The use of foaming for the purification of raw sugar juices is referred to in §179.

§151. Beer. The foaminess of beer would appear irrelevant to one interested in the taste of the beverage, but a surprising number of papers and patents was devoted to the origin of beer foams, their enhancement, and their suppression.

Beer foams are persistent. As mentioned in §47, films 1-cm across lasted for 390 sec and thus had a greater stability than those of 3.7% sodium laurate solutions. The average lifetime of gas (§60) in beer foams was 66 to 146 sec, again a respectable duration[53].

When beer is poured from one into another vessel, the gas in the foam bubbles has a double origin; a part of it is the air entrapped by the falling liquid, and the rest is carbon dioxide released by the lowering of pressure. In some instances, the second source is much more important than the first, and the "total head" of beer is simply proportional to the carbon dioxide amount in it[54].

Different opinions are encountered on what are the main foaming agents in beer. Compounds of nitrogen are accumulated in beer foams; for instance, the nitrogen concentration in the initial beer was[55] 0.064%, in its foam 0.082%, and in the residue 0.060%. The nitrogen accumulated in foam was a constituent of proteins or polypeptides, while the amino nitrogen was present in the foam in a smaller concentration than in the beer itself[56]. Later, the proteins in foam were characterized as high-molecular materials precipitable by tannin or heat[57,58]; according to Mischke[59] they belong to the albumoses. Saruno[60] found that, of all beer protein fractions, that of molecular weights between 3000 and 10,000 had the highest foaminess; this result presumably agrees with the statement by Klopper[61] that the albumin fraction affords foams which collapse slowly (after shaking) while the time of collapse of the peptone fraction foam is short.

The above data would indicate that beer foam is stabilized essentially by proteins. However, the nonproteinaceous ingredients of hops seem to be important as well. In the above foam, which contained 0.082% nitrogen, the accumulation of hop resins was more striking than that of protein; the resin concentration was 0.017% in beer, 0.048% in foam, and 0.01% in the residual liquid. The average lifetime of gas was greater, the greater the amount of hops used[53]. Apparently humulon of the hops is not, but iso-humulon is, a powerful frother[62]. In another study, foam stability was not markedly altered when the amount of hops added to wort was raised from 1.6 to 4.8 g/l.; the pretreatment of the hops (aging, etc.) was more important than their quantity[63].

§152.   Of the foreign substances which affect the foaming of beer, the colloidal nuclei of iron, nickel, cobalt, and tin compounds are probably the least expected; they are said to accelerate the liberation of carbon dioxide occurring on the lowering of pressure[64]. This liberation consequently is retarded when the formation of the above nuclei is avoided by complexing metal ions with sodium ethylenediamine tetra-acetate (1 lb for 50 barrels of beer).

The concentration of iron in various forms (but not of total ash) was greater in the foam than in the untouched beer[58], indicating that at least one of the frothers was an iron compound. This was confirmed by the observation[53] that small additions of ferrous sulfate markedly raised foam persistence. Later also other metals of the iron group were found to enhance the foaminess of beer. Thus[65] the foam stability of a pasteurized beer was raised 18, 43, and 63% by the addition of, respectively, 1, 3, and 5 mg (per liter) of bivalent nickel ion. This effect apparently was related to the hop content, as it was absent in the sweet wort, the unhopped beer, and in peptone solutions, and was more pronounced, the greater the amount of hops (see also §78). Cobalt also is a promoter. According to a patent[66], 1 mg of cobalt (in cobalt-II chloride or cobalt-II nitrate) significantly enhances the foam persistence of 1 l. of beer. In practice, when cobalt, iron, or nickel salts are added to beer, simultaneous introduction of a hydroxy acid or an amino acid is recommended[67]. In the United States the maximum concentration of cobalt ion in beer was fixed at 1.2 mg/l. and the salts permitted were sulfate, chloride, and acetate.

Zinc salts also stabilize beer foams, but larger amounts are needed, namely from 5 to 25 mg/l.[68] Better results are achieved when, simultaneously, gum arabic, or propylene glycol alginate, or carboxymethyl cellulose, etc., is introduced; for instance, the addition of 13 parts of $ZnSO_4 \cdot H_2O$ and 25 parts of the alginate to a million parts of beer is suggested[69]. Another patent[70] recommends mixtures of this alginate with manganese sulfate. Several publications refer, in addition to foam stability, also to "foam adhesion," that is, to the tendency of beer foam to cling to the vessel walls; also this tendency is enhanced by the above mixtures.

Propylene glycol alginate, mentioned in the preceding paragraph, is a foam promoter in its own right[71]. Other organic stabilizers include protein hydrolyzates[72], mechanically degraded poly(ethylene oxide)[73], ethyl cellulose, corn sugar, and a synthetic polysaccharide[74], various algin derivatives[75], and dextrin derivatives[75a].

Bentonite raises the stability of beer foams allegedly because it adsorbs protein of high molecular weight[76].

Destruction of beer foam can be achieved by ether, as noticed a century ago by Quincke (see §142). Higher alcohols act similarly to diethyl ether[53]. According to Lienert[77], beer foaming is inhibited by the following percentages of simple organic compounds: ethanol 10, propanol 4, butanol 0.4, pentanol 0.3, hexanol 0.04, octanol 0.005 %, ethyl acetate 0.5, propyl acetate 0.12, butyl acetate 0.15, amyl acetate 0.05, and hexyl acetate 0.01 %. Foaming ability can be lowered also by treating beer with calcium salts; this operation is supposed to remove oxalic acid from the beverage[78]. In an alternative method, beer is filtered through calcium carbonate, or oxalic acid in it is split by an enzyme into carbon dioxide and formic acid[79].

§153. *Detergents.* The main function of detergents is cleaning. In many instances, cleaning can be achieved in the absence of any foam. Nevertheless voluminous and stable foams generally are desired in the use of detergents, for "sales appeal" of "psychological effect." This tendency exists also when excessive foaming is harmful. When a bulky foam forms in a washing machine, it acts as a quasi-solid support for the load (or a part of the load), and the garments thus supported do not come in contact with the cleaning liquid.

Several examples of enhancement and depression of the foaming ability of typical surfactants are referred to in §§70, 71, and so on. Here some additional instances are given.

The effect of saponins and "builders" on the foaminess of soap solutions was studied by Uppal[80] and Augustin[81]. This foaminess was promoted by phospholipid concentrates obtained from sunflower seeds[82] and by some alkyl sulfates[83].

Many materials enhance the foaminess of synthetic surfactants (e.g., alkyl sulfates or alkylarene sulfonates). Mention may be made of long-chain 1, 2-alkanediols $R \cdot CHOH \cdot CH_2OH$[83], diethanolamides of fatty acids C_{10} to C_{18}, monoethanolamide of lauric acid

$$C_{11}H_{23}CO \cdot NH \cdot CH_2CH_2OH \,^{85},$$

monoethanolamides of acid mixtures[86], stearic acid plus calcium acetate[87], and some more complex mixtures[88]. Wetting agents having a low foaming ability are described by Jung[88a].

Foams of anionic surfactants can be destroyed by those of cationic surfactants[89]. The foam of alkylbenzene sulfonates was broken by irradiation with ^{60}Co.[90] Alkylolamides of long fatty acids, mentioned as promoters in the preceding paragraph, inhibit the foaminess of some detergent solutions[91]. Nonylphenols are recommended for breaking foams of non-ionic or cationic surfactants[92]. Okazaki[93] compared the destruction of existing foams with the prevention of foaming in many solutions of sodium docecyl sulfate plus ethanol, butanol, 3-methylbutanol, phenol, acetone, pyridine, or some other compounds. Tridodecyl isocyanurate and its homologs are recommended as foam depressors in detergent solutions[93a].

§154. *Other industries.* In the fermentative synthesis of antibiotics, the choice of foam inhibitors is particularly difficult because the inhibitor must have no effect on the metabolism of the micro-organisms employed. Nevertheless many materials have been found suitable, ranging from mixtures of aliphatic alcohols, C_8 to C_{20}[94], to a suspension of colloidal silica in a siloxane polymer[95].

The foam produced by shaking potato press juices or potato pulp water is suppressed by silicone oils more readily than by octanol[96]. Aeration of municipal sewage usually causes voluminous foaming, but this can be prevented by, for instance, cyclohexylamine oleate[97]. Trifluoropropyl methyl siloxane polymers are foam inhibitors for perchloroethylene solutions[98]. Benzene reduces the foaming of viscose[99]. Foaming of phosphoric acid is suppressed by the reaction products of aliphatic amines and hydrogen iodide[99a].

The above examples, taken more or less at random, show that no specific correlation can be detected between the chemical composition of the foam inhibitor and that of the foaming liquid; a particular antifoamer usually is efficient in many, very different systems, and the foam in a given system generally can be depressed by many, very different inhibitors.

The truth of this conclusion is confirmed by the much more extensive data available in Currie's review[100]. It was decided not to include here the information gathered by Currie; whoever is interested in the early history of foam inhibition would be well advised to consult Currie's paper. Contrary to the arrangement adopted in this chapter, there the antifoamers are listed according to their chemical nature, not according to the industry or the process in which they are useful. It seems that, then and now, silicone oils have the widest range of application; before their advent, octanol probably was the most versatile inhibitor, and, before octanol, castor oil or milk fat frequently were employed.

The two common methods of estimating the efficiency of a foam inhibitor are: (1) foam is produced in a usual manner and small amounts of the test compound (alone or in solution) are sprayed over it; the smaller the amount which causes a nearly complete collapse, the more efficient the antifoamer; (2) small amounts of the test compounds are mixed with the liquid before foaming, and the least amount which prevents measurable foaming on shaking, air injection, and so on, is determined; the smaller it is, the better the inhibitor. Usually the two procedures afford similar results although Okazaki (see §153) noticed some differences.

A different procedure, whose advantages are not obvious to the present author, was employed by Balakirev[101]. A drop of the antifoamer (valeric acid was used) was placed on a glass plate and surrounded with a foam (of a synthetic surfactant plus sodium phosphate). A cavity formed around the drop, and the volume of the cavity was treated as a measure of the inhibitor efficiency.

In a continuous process, the addition of an inhibitor can be automated. For instance, when the foam rises too high, it reaches two electrodes, and the electric current thus switched on opens a valve admitting a small amount of antifoamer to the foaming liquid[102].

§155. *Nonchemical foam destruction.*   Foams can be broken by nonchemical means. In §121, the classical observation of Dupré is mentioned that projectiles (such as cork balls) pierced soap lamellae without destroying them as long as the rate of fall was small, but bursting occurred when the fall was from a height exceeding 25 cm. It is not known whether this bursting is caused by the cork itself or by the air stream accompanying the missile. At any rate, Dupré's experiments supply a clear instance of mechanical foam destruction. Destruction by blowing a jet of air onto the foam surface was patented[103] more than a century after Dupré's work.

Very similar in principle are the methods based on shock waves, compression waves[104], and so on. These vibrations can be generated, for instance, by causing a membrane to burst[105] or by a siren[106]. Ultrasonic foam breakers have been described by, among others, Adams[107], Dorsey[108], and Fuji Co.[109]. Boucher[110] invented an "aerodynamic defoamer."

Other instruments employ solids rather than air to break a foam.

Centrifugal force is popular; it can be utilized in many different arrangements. Perhaps the simplest is a horizontal disk rotating, for instance, at 15 to 100 revolutions per second, onto which the foam falls from an inverted U tube. This system was studied by Rubin[111]. The material of the disk ["Teflon," poly(methyl methacrylate), or chromium-plated brass] had no significant effect on the foam collapse. A possible reason for this behavior might be that rupture was caused by the air carried around with the disk rather than by the rotating solid itself; if so, the method belongs to the preceding paragraph. A disk provided with baffles was used by Zotagina[112], and a "centrifugal foam separator" by Barishpolets[113].

Instead of just one disk, a disk rotating between two stationary plates[114] or an analogous system[115] may be employed. Walling[116] and Lille-Cail[117] worked with paddle wheels. Forcing the foam through a fine orifice or a wire screen caused its collapse in some instances [118-120]; in other examples this operation lowered the bubble diameter without rupturing the lamellae. Presumably, the difference was that between "dry" and "wet" foams and analogous to that between fresh and aged systems described in §121. A spinning perforated can inside an inverted plastic bottle was preferred by Brunner[121].

A different principle was found workable by the present author. The foam is imbibed by a bibulous solid. For instance, a drum is covered with filter paper and placed horizontally above the liquid capable of foaming. When foam reaches the drum, it is absorbed by the paper. The drum is rotated and rubs against one or two rubber-covered metal bars. The pressure of these bars (as in a wringer) squeezes out the liquid occluded, and the exudate drips back into the main body of liquid.

Foam can be collapsed by heat. The froth on a boiling "alkaline pumping liquor" was broken by a hot wire placed above the rising foam. The position and the temperature of the wire were selected so that the rupture occurred at least 1 cm below the wire; this prevented contamination of the wire and scorching the solute. Presumably evaporation of the solvent was the main cause of collapse[122]. Solvent evaporation was utilized also, among others, by Karger[123,124], and patented in France[125] and Germany[126].

The foam of boiling salt solutions can be broken by forcing it through a spark gap[127]. Soap and surfactant films are ruptured by sparks, by α-particles emitted by polonium[128], by α-particles of uranium-235 or plutonium-240, and so on[129], or by irradiation[90] with cobalt-60.

Polonium was used first by Chaminade[130]. The foaming solution contained sodium dodecyl sulfate (5%), ammonium sulfate (6%), glycerol, sodium alginate, and dodecanol. One α-particle reaching the froth destroyed 0.1 bubble. The irradiation resulted in a depression on the foam surface, and the walls of this depression were 3.8 cm distant from the polonium source; 3.8 cm is the length of the trajectory of these particles in air. Ader[131] found that bubbles ruptured first in 3 cm distance from the polonium source; at this distance the ionizing action of the α-particles has a maximum. When a flat slab of froth was placed on a glass plate near a polonium emitter, the slab was gradually cut in two, and the fissure was approximately 3 cm from the source.

References

1. Miyake, Y. *Science (Japan)* 13:358 (1943). *Chem. Abstr.* 45:9935 (1951).
2. Khailov, K. M., et al. *Gidrokhim. Materialy.* 52:82 (1969). *Chem. Abstr.* 73, No. 18391 (1970).
3. Akcetin, D., and F. H. Constable. *Istanbul Univ. Fen. Fac. Mecmuasi Seri. C* 30: 125 (1965). *Chem. Abstr.* 67: No. 67894 (1967).
4. Skopintsev, N. A. *Dokl. Akad. Nauk SSSR* 18:53 (1938). *Chem. Abstr.* 32:8855 (1938).
5. Votintsev, K. K. *Dokl. Akad. Nauk SSSR* 92:425 (1954). *Chem. Abstr.* 50:573 (1956).
6. van Beneden, G. *Compt. Rend. Journées Hydrauliques*, Soc. Hydrotech. France, 7-emes, Paris 1962, 1:216 (1963). *Chem. Abstr.* 59:6127 (1963).
7. Raison, M. *Centre Belge Étude Doc. Eaux* 227:512 (1962).
8. Lowry, R. P., and M. van Winkle. *A.I.Ch.E.J.* 15:665 (1969).
9. Gleim, V. G. *Zh. Prikl. Khim.* 26:1157 (1953).
10. Gleim, V. G. *Zh. Prikl. Khim.* 28:12 (1955).
11. Denman, W. L. *Ind. Eng. Chem.* 46:992 (1954).
12. Durov, S. A., and N. G. Chen. *Zh. Prikl. Khim.* 30:1096 (1957).
13. Villar, G. E. *Quim. Ind.* Uruguay 2, No. 3:5 (1952). *Chem. Abstr.* 47:3081 (1953).
14. Agre, C. L., G. Dinga, and R. Pflaum, *J. Org. Chem.* 20:695 (1955).
15. Dow Chemical Co., Britain 974 449 (1964). *Chem. Abstr.* 62:4228 (1965).
16. Ravinskii, M. B. *Kolloidn. Zh.* 24:58 (1962).
16a. Henneman, T. J. *Lab. Pract.* 20:420 (1971). *Chem. Abstr.* 75: No. 22991 (1971).
17. Nekervis, W. F., and R. A. Canute. U.S. 3 169 929 (1965).
18. Kuznetsov, V. A., and I. G. Chufarova. *Zh. Prikl. Khim.* 29:688 (1956).
19. Evans, R. W. *Blast Furnace Steel Plant* 32:932 (1944).
20. Tuluyevskii, Yu. N. *Stal'* 16:260 (1956). *Chem. Abstr.* 50:8419 (1956).
21. Mailbaum, D.; see F. Keil. *Stahl Eisen* 62:1091 (1942).
22. Fellcht, K. *Metallurgie u. Giessereitech.* 5, No. 3, Met. Sect. 85 (1955). *Chem. Abstr.* 49:10815 (1955).
23. Layton, W. *Australasian Inst. Mining Met. Proc.* 205:57 (1963). *Chem. Abstr.* 56:8393 (1963).
24. Nechkin, Yu. M., V. A. Kudrin, and V. I. Yavoiskii. *Izv. Vysshikh Uchebn. Zavedenii.* Cher. Met. 7, No. 3:53 (1964). *Chem. Abstr.* 60:15492 (1964).
25. Mamedaliev, Yu. G., A. M. Kuliev, and L. S. Mustafayev. *Zh. Prikl. Khim.* 26:854 (1953).

26. Trautman, C. E. *Lubrication Eng.* 2, No. 4:143 (1946).
27. Villar, G. E. *Bol. Fac. Ing. Montevideo* 4:403 (1951). *Chem. Abstr.* 47:5674 (1953).
28. Borzoff, V. N., and J. O. Clayton. U.S. 2 430 857 (1942).
29. Fenske, M. R., and G. H. Cummings. U.S. 2 407 954 (1946).
30. Smith, H. G. U.S. 2 377 654 (1945).
31. Batchelder, A. H., and F. S. Rollins. U.S. 2 402 487 (1946).
32. Young, D. W. U.S. 3 083 168 (1963).
33. Knecht, A. T. U.S. 3 296 131 (1967).
34. Chittum, J. F. U.S. 3 108 634; 3 108 635 (1963).
35. Kettcamp, C. F., R. P. Nejak, and R. T. Kern. *ASLE Trans.* 2:7 (1959).
36. Carpenter, W. L., and S. Gellman. *Tappi* 50, No. 5:83A (1967).
37. Green, J. W., and R. L. Leaf. *Tappi* 35:468 (1952).
38. Kleinert, T., and M. Ruck-Floriantschitsch. *Papier* 8:13 (1954). *Chem. Abstr.* 48:5494 (1954).
39. Surewicz, W., K. Modrzejewski, and W. Mroz. *Przeglad. Papier* 25, No. 1:1 (1969). *Chem. Abstr.* 70, No. 98082 (1969).
40. Belani, E. *Papier-Ztg.* 56:52, 1296 (1931).
41. Simmons, W. B. *Paper Trade J.* 139, No. 45:32 (1955). *Chem. Abstr.* 50:8206 (1956).
42. Farbenfabriken Bayer A.-G., German 1 069 577 (1959). *Chem. Abstr.* 55:12853 (1961).
42a. Mazminov, K. N., and R. A. Teplitskaya. *Bumazhn. Prom.* 1971, No. 11:21. *Chem. Abstr.* 76, No. 61140 (1972).
43. Saray, Y. *Seker* 15, Sept., 19 (1965). *Chem. Abstr.* 65:17185 (1966).
44. Oldfield, J. F. T., and J. V. Dutton. *Intern. Sugar J.* 70:7 (1968). *Chem. Abstr.* 69, No. 78585 (1968).
45. Vlasova, T. I. *Kolloidn. Zh.* 20:421 (1958).
46. Vavrinecz, G. *Cukoripar* 10:128 (1957). *Chem. Abstr.* 52:15938 (1958).
47. Zhukov, K. S. *Sakhar. Prom.* 27, No. 4:32 (1953). *Chem. Abstr.* 48:2400 (1954).
48. Scheye, H. Z. *Deut. Zuckerind.* 46:626, 645 (1921).
49. Vorob'ev, N. *Chim. Ind.* 32:1405 (1934).
50. Farbenfabriken Bayer A.-G., Britain 928 906 (1963). *Chem. Abstr.* 59: 10349 (1963).
51. Suzuki, K. *Kôgyô Kagaku Zasshi* 61:56 (1958). *Chem. Abstr.* 53:17545 (1959).
52. Perlow, R. *Zucker* 3:468 (1950). *Chem. Abstr.* 45:911 (1951).
53. Kolbach, P. *Brauerei, Wiss. Beil.* 6, No. 12:137 (1953). *Chem. Abstr.* 49:9876 (1955).
54. Helm, E., and O. C. Richardt. *J. Inst. Brewing* 42:191 (1936).
55. Gray, P. P., and I. Stone. *Wallerstein Lab. Commun.* 3:159 (1940). See also W. Kleber and H. Geyer. *Brauwelt* 111:1643 (1971). *Chem. Abstr.* 76, No. 57678 (1972).
56. Lüers, H., K. Geys, and A. Baumann. *Z. Brauwesen* 43:185, 193, 201 (1920). *Chem. Zentr.* 1920 IV 481.
57. Krauss, G., and F. Harreis. *Wochschr. Brau.* 57:33 (1940). *Chem. Abstr.* 36:214 (1942).

58. Nissen, B. H., and C. Estes. *Am. Soc. Brewing Chem., Proc.* 1940, 23. *Chem. Abstr.* 35:3029 (1941).

59. Mischke, W. *Kolloid-Z.* 90:77 (1940).

60. Saruno, R., and K. Ishida. *Hakko Kyokaishi* 21:372 (1963). *Chem. Abstr.* 62:15386 (1965).

61. Klopper, W. J. *Intern. Tijdschr. Brouw. Mout.* 13:114 (1954). *Chem. Abstr.* 48:4765 (1954).

62. Klopper, W. J. *Intern. Tijdschr. Brouw. Mout.* 13:131 (1954). *Chem. Abstr.* 48:6072 (1954). *Wallerstein Lab. Commun.* 18, No. 61, 123 (1955). *Chem. Abstr.* 51:17088 (1957).

63. Kolbach, P., and K. D. Esser. *Brauerei* 11:91 (1958). *Chem. Abstr.* 52: 20878 (1958).

64. Gray, P. P. U.S. 2 711 963 (1955).

65. Luykx, J. M. M., and H. van Veldhuizen. *Intern. Tijdschr. Brouw. Mout.* 18:19 (1958). *Chem. Abstr.* 53:4647 (1959).

66. Thorne, R. S. W. U.S. 2 865 755 (1958).

67. Blue Star Chemicals, Belgium 617 257 (1962). *Chem. Abstr.* 58:9602 (1963).

68. Baxter Laboratories Inc., Belgium 611 810 (1962). *Chem. Abstr.* 57:11677 (1962).

69. Stone, I. M. U.S. 3 223 529 (1965).

70. Brenner, M. W., U.S. 3 266 902 (1966).

71. Steiner, A. B. U.S. 2 659 675 (1953).

72. Frieden, A., and H. H. Geller. U.S. 2 806 791 (1957).

73. Segel, E., and E. J. Wye. U.S. 2 942 980 (1960).

74. Nugey, A. L. U.S. 3 052 548 (1962).

75. Nordman, H. E., and W. H. Mohr. *Am. Brewer* 96:22 (1963). *Chem. Abstr.* 60:8596 (1964).

75a. Marotta, N. G., H. Bell, and G. B. Charlick. U.S. 3 573 928 (1971).

76. Schild, E., and H. Weyh. *Brauwissenschaft* 16:247 (1963). *Chem. Abstr.* 59:10739 (1963).

77. Lienert, H. *European Brewery Conv., Proc. 5th Congr., Baden-Baden 1955*, 282. *Chem. Abstr.* 51:18464 (1957).

78. Brenner, M. W. U.S. 2 763 554 (1956).

79. Brenner, M. W. U.S. 2 878 125 (1959).

80. Uppal, I. S., and R. L. Mehta. *J. Sci. Ind. Res. (India)* 11B:153 (1952). *Chem. Abstr.* 47:9036 (1953).

81. Augustin, J. *Seifen-Oele-Fette-Wachse* 79:76 (1953). *Chem. Abstr.* 47:4796 (1953).

82. Semendyaeva, T. K. *Maslob.-Zhir. Prom.* 18, No. 4:15 (1953). *Chem. Abstr.* 47:9036 (1953).

83. N.V. Bataafsche Petroleum Maatschapij, Dutch 73 501 (1953). *Chem. Abstr.* 48:6721 (1954).

84. Kritchevsky, J., and H. L. Sanders. *Congr. Mondial Détergence Prod. Tensio-Actifs, 1er Congress, Paris, 1954*, 1:133. *Chem. Abstr.* 51:10930 (1957).

85. Dutton, K. R., and W. R. Reinisch. *Soap, Perfumery, Cosmetics* 31:44 (1958). *Chem. Abstr.* 52:9632 (1958).

86. Varlamov, V. S., A. N. Mironova, and G. I. Oyateva. *Maslob.-Zhir. Prom.* 27, No. 4:30 (1961). *Chem. Abstr.* 56:3584 (1962).
87. Wedell, H. U.S. 3 008 905 (1961).
88. Blakeway, J. M., and I. D. Burgess. Britain 921 036 (1963). *Chem. Abstr.* 58:14323 (1963).
88a. Jung, K. Ger. Offen. 2 014 427 (1971). *Chem. Abstr.* 76, No. 87561 (1972).
89. Sebba, F. *Nature* 197:1195 (1963).
90. Spragg, H. R. *Water Sewage Works* 110:163 (1963).
91. Volkova, L. D., and A. G. Shablina. *Maslob.-Zhir. Prom.* 30, No. 11:23 (1964). *Chem. Abstr.* 62:5453 (1965).
92. Grifo, R. A. U.S. 3 215 633 (1965).
93. Okazaki, S., and T. Sasaki. *Tenside* 3:115 (1966). *Chem. Abstr.* 64:20002 (1966).
93a. Stimberg, H. J., J. Galinke, and E. Schmadel, Ger. Offen. 1 965 643 (1971). *Chem. Abstr.* 76, No. 47626 (1972).
94. Rumyantsev, V. I., and V. E. Ezhov. *Maslob.-Zhir. Prom.* 31, No. 7:38 (1965). *Chem. Abstr.* 63:16127 (1965).
95. The Distillers Co. Ltd., Britain. 639 507 (1952).
96. Winkler, S. *Stärke* 7:177 (1955). *Chem. Abstr.* 50:17267 (1956).
97. Scott, J., and R. J. Westland. Britain. 793 737 (1958). *Chem. Abstr.* 53: 624 (1959).
98. Currie, C. C. U.S. 3 115 472 (1963).
99. Kalafut, Š., and J. Kolešar. *Chem. Prumsyl* 9:545 (1959). *Chem. Abstr.* 54: 3941 (1960).
99a. Riggs, O. L. U.S. 3 578 402 (1971).
100. Currie, C. C., in J. J. Bikerman. *"Foams"*, New York: Reinhold. (1953), p. 297.
101. Balakirev, A. A., and V. K. Tikhomirov. *Kolloidn. Zh.* 30:315 (1968).
102. Bartolomew, W. H., and D. Kozlov. *Ind. Eng. Chem.* 49:1221 (1957).
103. Société des Forges et Ateliers du Creuzot, Ger. Offen. 2 051 526. *Chem. Abstr.* 75, No. 7822 (1971).
104. Parlow, R. *Zucker* 13:182 (1960). *Chem. Abstr.* 55:6895 (1961).
105. Karpovich, J., and W. F. Clark. U.S. 3 290 256 (1966).
106. Teknika, Inc., Britain. 1 059 141 (1967). *Chem. Abstr.* 66, No. 106296 (1967).
107. Adams, F. R. *Tappi* 41, No. 5:173A (1958).
108. Dorsey, A. E. *J. Biochem. Microbiol. Technol. Eng.* 1:289 (1959). *Chem. Abstr.* 54:6021 (1960).
109. Fuji Photo Film Co., Britain. 1 075 100 (1967). *Chem. Abstr.* 67, No. 92199 (1967).
110. Boucher, R. M. G., and A. L. Weiner. *Brit. Chem. Eng.* 8:808 (1963). *Chem. Abstr.* 60:6503 (1964).
111. Goldberg, M., and E. Rubin. *Ind. Eng. Chem.*, Process Des. Develop. 6: 195 (1967). E. Rubin and M. Goly, *Ind. Eng. Chem.*, *Process Des. Develop.* 9:341 (1970).
112. Zotagina, S. A. *Gidrolizn. i Lesokhim. Prom.* 9, No. 7 (1956). *Chem. Abstr.* 51:6043 (1957).
113. Barishpolets, V. T. *Khim. Prom.* 1962:121. *Chem. Abstr.* 57:5734 (1962).

114. Forrester, G. U.S. 2 908 652 (1959).
115. S. D. Warren Co., Britain. 813 856 (1959). *Chem. Abstr.* 53:20953 (1959).
116. Walling, C., E. E. Ruff, and J. L. Thornton. *J. Phys. Chem.* 56:989 (1952).
117. Lille-Cail, F. French 1 436 590 (1966). *Chem. Abstr.* 65:19732 (1966)·
118. Haas, P. A., and H. F. Johnson. *A.I.Ch.E.J.* 11:319 (1965).
119. Tarutin, V. P., and G. A. Nikolaev. *Bumazhn. Prom.* 1968, No. 1:23. *Chem. Abstr.* 68, No. 88337 (1968).
120. Scheibel, E. G. U.S. 3 231 251 (1966).
121. Brunner, C. A., and R. Lemlich. *Ind. Eng. Chem. Fundamentals* 2:297 (1963).
122. Gastrock, E. A., and J. D. Reid. *Ind. Eng. Chem. Anal. Ed.* 10:440 (1938).
123. Poncha, R. P., and B. L. Karger. *Anal. Chem.* 37:422 (1965).
124. Karger, B. L., R. P. Poncha, and M. M. Miller. *Anal. Chem.* 38:764 (1966).
125. Soc. Équipement de la Blanchisserie et de la Teinturerie Françaises, French. 1 589 098 (1970). *Chem. Abstr.* 74, No. 23954 (1970).
126. Wacker-Chemie G.m.b.H., German 1 080 956 (1960). *Chem. Abstr.* 55: 11963 (1961).
127. Schnurrmann, R. *Ind. Eng. Chem. Anal. Ed.* 11:287 (1939).
128. Kato, R., and T. Kono, *J. Appl. Phys.* 34:708 (1963).
129. Exerova, D., and D. Ivanov. *Compt. Rend. Acad. Bulg. Sci.* 23:547 (1970).
130. Chaminade, R. *Compt. Rend.* 228:480 (1949).
131. Ader. *J. Phys. Radium* 11:198 (1950).

CHAPTER 12

APPLICATIONS. SEPARATION BY FOAM

§156.　　Foams are used for many and very different purposes. It seems nevertheless that a classification of their main uses is possible. Generally speaking, application of foams is based on

1. difference in composition between the bulk liquid and the collapsed foam
2. immobilization of gas and liquid (sometimes of solids also) in foams
3. their low density
4. their large surface
5. their mechanical properties
6. their thermal properties

A seventh group would be formed by application of foams in laboratory measurements and analysis.

Froth flotation is the main application belonging to the first group. The treatment usually consists of the following steps. The ore extracted from the mine is a mixture of valuable minerals and worthless gangue. First it is comminuted. The powder obtained is suspended in an aqueous solution containing frothers, promoters (also known as collectors), and modifiers, whose function is indicated below. Air is injected into the suspension ("the pulp liquid"), which is vigorously stirred to produce smaller bubbles and to prevent settling of the powder. Reasonably stable foams are produced in this operation. They are mechanically separated from the gas emulsion underneath and then broken by any suitable means. The powder recovered from the collapsed foam (and usually denoted as "the concentrates") has a much higher percentage of the desired mineral and a much smaller fraction of the gangue than the raw powder or the "tailings" which remain in the spent liquor. Naturally, the further treatment of the enriched powder (for instance, its dissolution) is much easier than a corresponding treatment of the initial mixture.

The foam mentioned in the preceding paragraph is a three-phase foam, as described in Chapter 5. The separation of the two main components of the ore occurs because of their different wettabilities. As stated in §85, powder particles well wetted by the liquid never reach the gas-liquid boundary and, consequently, do not rise with the bubbles. The usual gangue, which is chiefly silicates or carbonates, is hydrophilic or can readily be made hydrophilic by a modifier; thus gangue particles to a large extent remain in the pulp liquid. The minerals to be recovered often are sulfides, selenides, and so on, and are poorly wetted by water in their pristine state; but if their wettability is too good, a suitable collector will depress it to make a receding contact angle θ_R near, say, $60°$; with such an angle the particles will remain in the gas-liquid interface and rise with the bubble.

The extensive scientific, technical, and patent literature on froth flotation cannot be reviewed in this book. A detailed review of the literature before 1952 can be found in the article by Booth[1].

§157.   In foam separation no macroscopic solid phase is present. The substances to be separated from each other can all be classified as solutes or solvents. In many instances the particles present in the dispersion medium are too small to be froth-flotated in the usual sense, but too big to be considered molecularly dissolved. Their recovery methods may be thought of as being intermediate between typical froth flotation and typical foam fractionation; they are described in §174.

The results of separation by foam may be grouped under three headings.

1. *Accumulation in foam.* The initial solution is binary and has mole fractions or weight fractions μ_0 (solute) and $1 - \mu_0$ (solvent). It is subjected to foaming. The composition of the collapsed foam is given by μ_1 (solute) and $1 - \mu_1$ (solvent). If $\mu_1 > \mu_0$, accumulation in foam occurred.

2. *Foam fractionation.* The initial solution contains several solutes. If only one of these accumulates (or only few accumulate) in the foam, it is (or they are) separated not only from the solvent but also from other dissolved components. This result usually is more valuable than simple accumulation in foam; in many instances, a single solute can be separated from the solvent by distillation and analogous operations which would not be able to separate the wanted from the unwanted solutes.

3. *Shift of equilibrium in foam.* If the solute is a mixture corresponding to an equilibrium such as $A = B + C$ or $A + H_2O = AOH^- + H^+$, the equilibrium in the surface region may be shifted toward the right or the left, depending on whether A is less or more surface active than B and C or than AOH^- and H^+. Because the volume occupied by surface layers in foam is relatively greater than in the initial liquid, the composition of the solute in the foam is different from that in the liquid; the surface active species is present in the former in a higher concentration than in the latter. These phenomena are reviewed in §§182 and 183.

Identical instruments are used for all three subdivisions of separation by foam; they are frequently referred to as foam towers. A simple foam tower is illustrated in Fig. 12.1. *1* is a large flask partly filled with solution

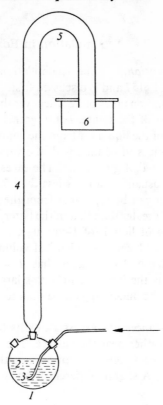

Fig. 12.1. Foam tower for batch fractionation. *1* is the flask containing solution *2*; *3* is the gas diffusor; *4* is the graduated tube; *5* is the connection to the receptacle; *6* is the receptacle.

2; gas (preferably purified and saturated with solvent vapor) is injected through a porous tube *3* (or a spinneret or another sparger); the foam thus formed rises in graduated tube *4* and through the "gooseneck tube" *5* falls into receptacle *6*. The amount and the composition of the collapsed foam are then determined and compared with those of the initial and the residual liquid.

Apparently no instrument manufacturer markets yet a foam tower. Consequently every investigator has to make or to assemble his own apparatus, and the number of different devices described and pictured in the literature is too great for a complete report. The foam towers constructed by the early researchers (Abribat[2], Gray[3], Ostwald[4], Dognon[5], Perri[6]) are not greatly different from the newest variants. At least one foam separation machine has been patented[7].

A book on separation by foam has recently been published[8].

Accumulation in Foam

§158. Adsorption.   The accumulation in foam is the simplest of the three classes of §157 and is discussed first.

This accumulation takes place whenever the surface layer of a solution contains an excess of the solute. Let c (grams or moles) of this solute be present in 1 cm^3 of the liquid far from the vapor phase; then 1 cm^3 (of the same liquid) which is in contact with the vapor phase along the area of A cm^2 contains $c + \Gamma_0 A$ g or moles. The excess $\Gamma_0 A$ is concentrated very near the surface, usually in a layer less than 50 Å thick. Consequently, if the surface region can be separated from the bulk, the concentration in the former will be greater than that of the latter. This separation is achieved in foams. A cube of liquid, of 1-cm edge, right under the vapor-liquid interface, has $A = 1$ cm^2. An identical volume (i.e., 1 cm^3) distributed over foam lamellae of 1 μ thickness has $A = 2 \times 10^4$ cm^2 (on both film faces). Not only is the interface in foams large, but it is much easier to take the foam off the underlying liquid than to cut a thin liquid sheet from the bulk.

In simple instances, the excess Γ_0 exists because surface-active components tend to gather near the surface. The quantitative expression for this tendency is the Gibbs adsorption equation, one form of which is presented in §135. A slightly different form is

$$\Gamma = -\frac{c}{RT} \cdot \frac{\partial \gamma}{\partial c}. \tag{12.1}$$

The symbol Γ is used for the equilibrium excess; Γ_0 need not correspond to any physicochemical equilibrium. Since the gas constant R refers to gram-molecules, Γ is expressed in moles/cm^2.

The values of concentration c and of the variation of surface tension γ with c (that is, $-\partial\gamma/\partial c$) cover a very wide range; thus it might be expected that also the value of Γ would cover several orders of magnitude. In reality, however, the gradient $-\partial\gamma/\partial c$ in many instances is greater, the smaller c, so that the product $-c(\partial\gamma/\partial c)$ varies but little when not only c but also the nature of the solute varies. An example is supplied again, as in §136, by 1-butanol in water. As in this system $-\partial\gamma/\partial c$ is 1.56×10^5 and 1.48×10^4, respectively, when c is 0.000067 and 0.00088 mole/cm^3, the product $-c(\partial\gamma/\partial c)$ is about 11 g/sec^2 in the dilute and about 13 g/sec^2 in the concentrated range; thus it changes by less than 20% when c is multiplied by 13. The examples below confirm that also the dependence of Γ on the solute is less than striking.

As RT at room temperature is about 2.4×10^{10} g·cm^2/sec^2·mole and the product $-c(\partial\gamma/\partial c)$ for butanol is approximately 12 g/sec^2, the Γ

of Equation 12.1 should be near 5×10^{-10} mole/cm². Apparently no experimental confirmation of this result is available, but Γ_0 values of the order of 10^{-10} mole/cm² have been obtained repeatedly for solutes less prone to achieve equilibria than are the small and stable molecules of butyl alcohol.

§159. To calculate Γ_0 from the data on foam accumulation, it is necessary to measure the concentration c of the foaming liquid, the concentration c_f in the foam collected, the volumes of the liquid (V_f) and the gas (V_g) in the foam, and the average diameter l of a bubble. When an instrument of the type of Fig. 12.1 is employed, c decreases during foaming, but the variation may be neglected when the volume of the initial liquid is large and the volume of the collapsed foam (i.e., V_f) is small.

The equation to use is

$$\Gamma_0 = \frac{(c_f - c)V_f l}{6V_g}. \tag{12.2}$$

Its derivation is simple. In the absence of adsorption, V_f cm³ of the liquid would contain cV_f moles of the solute; in reality the amount is $c_f V_f$. The difference $(c_f - c)V_f$ is adsorbed. To obtain the amount adsorbed per square centimeter, this difference must be divided by the total surface area A of the foam before its collapse. Let n be the number of bubbles in this foam. The volume of gas in them was $n(\pi/6)l^3$; it is equal to V_g if the compression of the gas in the bubbles (discussed in §111) is neglected. Since $A = n\pi l^2$, then $A = 6V_g/l$. This gives Equation 12.2. Strictly speaking, the l in the equation $6V_g = n\pi l^3$ is not identical with the l in the equation $A = n\pi l^2$ whenever bubbles of different dimensions are present. Furthermore, both these equations are valid for spheres only, while real bubbles are more like pentagonal dodecahedrons. However, these two approximations are not likely to affect the value of Γ_0 by more than 10%.

Many results for the Γ_0 of typical foaming agents have been published. For instance, several solutions of lithium dodecyl sulfate, sodium dodecyl sulfate, and potassium dodecyl sulfate gave[9] values near 3×10^{-10} moles/cm², that is, very similar to that calculated in §158 for butanol. When the concentration of a sodium dodecylbenzene sulfonate was varied between 0.001 and 0.004 M, the Γ_0 remained[10] about 4×10^{-10}. In a later study[11], this Γ_0, in an identical concentration range, was found to be 3.3×10^{-10} moles/cm². According to Rubin[12], Γ_0 for sodium dodecyl sulfate (at $c > 0.005$ M) and sodium dodecylbenzene sulfonate is, respectively, 4.5×10^{-10} and 3.5×10^{-10}. In view of the well-known difficulty of purifying these compounds, the agreement between different tests is quite satisfactory. Brunner[13] found $\Gamma_0 = 2.5 \times 10^{-10}$ for 0.0005 M−0.003 M solutions of sodium "monobutyl diphenyl monosulfonate," and for sodium palmitoyl methyl taurine $C_{15}H_{31}CO\cdot N(CH_3)CH_2CH_2\cdot SO_3Na$ the value $\Gamma_0 \approx 2.4 \times 10^{-10}$ moles/cm^2 is given[14].

Also senegin foams (see §76) afforded similar values[15]. The Γ_0 was 2.3×10^{-10} for all concentrations c between 0.1 and 1.9 g/l. at pH 5.2. The adsorbed amount was smaller at pH 2.7 but reached 2.8×10^{-10} moles/cm^2 at pH 8.8.

The above order of magnitude of Γ_0 is confirmed by means of radioactive tracers. This method is based on the fact that the radiation emitted by labeled solutes is to a large extent absorbed by solvent layers thicker than, for instance, 0.01 cm (in the case of ^{35}S). Hence the radiation recorded by a counter placed immediately above the surface of the solution originates in a very thin stratum and thus gives information on its composition. When the method is applied to the initial solution and to its collapsed foam, the difference can be ascertained. By this method, the Γ_0 of sodium dodecyl sulfate (containing ^{35}S) was determined[16] as 15×10^{-10} at all concentrations between 0.0010 and 0.0025 M when the foam was relatively "dry" and consisted of dodecahedrons. When the froth was "wet" and the bubbles were almost spherical, the Γ_0 of identical solutions was near 4×10^{-10} moles/cm^2.

It is seen that the experimental values of Γ_0 agree with the expectation (for Γ) in three respects; namely in their order of magnitude, their insensitivity to concentration, and their relative independence of the nature of the foaming agent.

The knowledge of the average bubble diameter l, which appears in Equation 12.2, is needed only because the surface area (A) of the foam must be known. This A can be determined without measuring l. When the bubbles are big enough and the upper part of tube *4* (Fig. 12.1, §157) is made narrow, then it is possible to force the foam to rise in this stretch as single lamellae perpendicular to the tube axis; up to 100 laminae per second can thus be obtained[17]; see also §110. The area of the two faces of each lamella is $0.5\pi D^2$ if D is the internal diameter of the tube. The lamellae are counted. If the sample of collapsed foam was obtained from n films, then the absolute amount of surface excess (or the total adsorbed amount) in it was $0.5 n\pi D^2 \Gamma_0$ moles.

§160. *Accumulation in foam.* Knowledge of Γ_0 is necessary for understanding the distribution of solute between the bulk and the surface, but is insufficient for discussing the distribution between the residual solution and the foam. For this distribution, drainage is the most important process[17a]. As in §10 and several later sections, each lamella is approximated as a sandwich of two surface layers, each x cm thick (see §128) and a liquid core of the thickness $\delta - 2x$ cm. If the "crust" consists of the solute only and the density of the latter is ρ_1 g/cm^3, then $x\rho_1$ g/cm^2 is the mass of the solute per cm^2 of the "crust." The concentration in the "core" usually is nearly identical with that (c g/cm^3) of the foaming liquid.

Hence the total amount of solute present in 1 cm^2 of the lamella is $2x\rho_1 + (\delta - 2x)c = 2x(\rho_1 - c) + \delta c$, and the weight-to-volume concentration c_f in the foam is

$$c_f = \frac{\delta c + 2x(\rho_1 - c)}{\delta} = c + 2\left(\frac{x}{\delta}\right)(\rho_1 - c) \tag{12.3}$$

so that the difference $c_f - c$ of Equation 12.2, §159 is

$$c_f - c = 2\left(\frac{x}{\delta}\right)(\rho_1 - c). \tag{12.4}$$

However, the ratio c_f/c is more useful for judging the efficiency of accumulation in foam than the difference $c_f - c$. This rate is variously denoted as the *lesser accumulation ratio*, enrichment factor, and so on, and in the following is represented by the symbol $\mathscr{R}$. There exists also the greater accumulation ratio equal to c_f/c_r, c_r being the concentration of the residual solution (i.e., the liquid left over after the conclusion of foaming).

From Equation 12.4

$$\mathscr{R} - 1 = \frac{2x}{\delta c}(\rho_1 - c). \tag{12.5}$$

Of the quantities present in the right-hand term of this formula, x and ρ_1 little vary from one to another surfactant. When a definite solution is subjected to foaming, also c changes but slowly. Thus the thickness δ more than any other quantity depends on the experimental conditions; and it is smaller the more time is allotted to drainage (see Chapter 6). If, for instance, Equation 6.1, §95, is valid (and bubble collapse may be disregarded), then $\delta = \delta_0 e^{-kt}$; δ_0 is the average lamella thickness of bubbles emerging from the gas emulsion into the foam. Hence

$$\mathscr{R} - 1 = \frac{2x(\rho_1 - c)}{c\delta_0}e^{kt}; \tag{12.6}$$

it rapidly increases with the drainage time t. Here, t is the time which an average bubble needs to rise from the liquid surface to the bend in the "gooseneck tube" of Fig. 12.1, §157. Let the linear rate of gas injection (that is, the volume rate divided by the horizontal cross-section of tube *4*) be u cm/sec and the length of this tube be H cm; then the available time of drainage

$$t = \frac{H}{u}\ \text{sec} \tag{12.7}$$

or, more exactly, $t = H(l + \delta)^2/ul^2$.

§161. As a result of foaming, the volume (V_l) of the liquid initially present in flask *1* (§157) is divided, ideally, into two volumes. One (V_f) is the volume of the collapsed foam; the solute concentration in it is c_f. The other is the volume (V_r) remaining in the flask; the concentration there is c_r. If no material is lost in the process, then

$$cV_l = c_fV_f + c_rV_r \tag{12.8}$$

Unfortunately, Equation 12.8 frequently is found to be inexact. In many instances, some foam remains attached to the walls of the graduated tube or the "gooseneck"; if these remains are disregarded, the right-hand side of the equation will be found smaller than the left-hand term.

Much more important is evaporation. The specific surface area of a foam is very large. This is a blessing as far as foam accumulation is concerned (see §158), but has the drawback of accelerating the vaporization of solvent (see §133). To reduce the evaporation rate, many investigators humidify the gas used for bubbling before it reaches the diffuser stone (or another sparger). This is a necessary precaution but it is difficult to achieve complete saturation in this manner, and sometimes the experimenter[18] must be satisfied with relative humidity of 90%.

No pre-humidifying of the gas stream can retard the evaporation when receptacle 6 (§157) is open to the ambient air. In the author's laboratory, an approximate correction for this loss of solvent was, for a time, performed in the following way. If the collection of the foam lasted for t_1 sec, then the average time the foam was exposed to the atmosphere was $0.5\,t_1$ sec. The foam was weighed as rapidly as possible, then left alone for $0.5\,t_1$ sec, and weighed again. If the two weights were W_0 and W_1, then the corrected weight would be roughly $2W_0 - W_1$ g. It was easier, however, to weigh the flask with liquid 2 (§157) after collecting each foam sample. If the mass found was M and the initial mass was M_0, then the mass transferred into foam was $M_0 - M$, whatever the (directly determined) mass of the foam.

If not only the solvent but also the solute is volatile, more stringent precautions are indicated. For instance, the receptacle may be continuously cooled to lower the vapor pressure of the two components. In several publications, vaporization was disregarded and, consequently, the accumulation ratios calculated were too large.

Another defect encountered in some papers may be referred to here. The accumulation ratio is not measured at all. The amount c_rV_r (of solute in the residual solution) is determined at suitable intervals, and the experimenter triumphantly announces that, say, 90% of the initial solute was removed in a 30-min foaming. But it takes only few seconds to pour 90% of the liquid from flask 1 into receptacle 6, and the result, as far as the removal of the solute is concerned, would be identical. A foam accumulation experiment is successful not when c_rV_r is small but when c_f is large and V_f is small, as compared with c and V_l.

Suppose that the product c_rV_r is desired to be $0.1\,cV_l$ (meaning 90%

removal). Then from Equation 12.8 $\mathscr{R} = c_f/c = 0.9\ V_l/V_f$. Hence, if $\mathscr{R} = 90$, then $V_f = 0.01\ V_l$; only 1% of the initial liquid is found in the collapsed foam, and this small volume contains 90% of the initial solute. Such a degree of separation would usually be welcomed. On the other hand, if $\mathscr{R}$ is only 3, then $V_f = 0.3\ V_l$. This means that the starting liquid is divided into two fractions, of which one contains fully 30% of the initial solvent and 90% of the initial solute, while the other is made up of 70% of the initial solvent and 10% of the initial solute. This process probably would be found unprofitable.

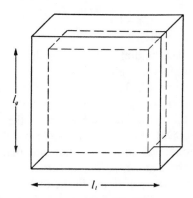

Fig. 12.2. Scheme of a foam bubble; l_g is the "diameter" of the gas phase, and l_l is the edge of the cube containing the gas and half the thickness of each lamella.

§162. A modification of Equation 12.5 is useful because it (1) demonstrates the importance of drainage in another manner, (2) permits a comparison with Equation 12.2, §159, and (3) does not contain the thickness δ, which is rarely (if ever) measured in foam separation experiments.

This δ can be expressed as φ/A_1; φ, as in §1, is the volume of the liquid phase in 1 cm^3 of foam and A_1 is the combined area of all foam lamellae in this volume. The meaning of A_1 which is different from A discussed in §158, is particularly easy to explain when the bubbles are approximated as cubes rather than as dodecahedrons or spheres. In Fig. 12.2, the larger cube (edge l_l) represents the bubble including half the thickness of its walls, and the inside cube (edge l_g) is the gas phase. The outside surface of the larger cube is $6l_l^2$. However, as each face belongs to two bubbles, the contribution of one bubble is only $3l_l^2$. There are $1/l_l^3$ bubbles in 1 cm^3 of foam. Hence $A_1 = 3l_l^2/l_l^3 = 3/l_l\ cm^{-1}$ and $\delta = \varphi l_l/3$. It is easy to find δ also as a function of l_g (which is not much different from the diameter l of §159). The total volume l_l^3 is to the gas volume l_g^3 as 1 to $1-\varphi$.

Consequently, $l_l = l_g/(1-\varphi)^{\frac{1}{3}}$ and $\delta = \varphi l_g/(1-\varphi)^{\frac{1}{3}}$. Instead of Equation 12.5, the following formula is obtained:

$$\mathcal{R}-1 = \frac{6(1-\varphi)^{\frac{1}{3}} \times (\rho_1 - c)}{\varphi l_g c} \qquad (12.9)$$

A similar equation follows from Equation 12.2, §159, when it is remembered that the two volumes V_f and V_g together form the foam falling into receptacle 6 (§157). The ratio V_f/V_g is identical with $\varphi/(1-\varphi)$, so that $\Gamma_0 = (c_f - c)\varphi l/6 \,(1-\varphi)$, which equation is equivalent to

$$\mathcal{R}-1 = \frac{6(1-\varphi)}{\varphi l} \cdot \frac{\Gamma_0}{c}. \qquad (12.10)$$

The quantity Γ_0/c has the dimension of length; in some publications the name of distribution factor is given to it.

Equations 12.9 and 12.10 are not identical because different approximations must be made to derive them. In particular, Equation 12.9 disregards the liquid filling the Plateau borders. If the volume of this liquid is considerable (see §98), it is not advisable to use Equation 12.9. At any rate, both Equations 12.9 and 12.10 indicate that the value of $\mathcal{R}-1$ (or $\mathcal{R}$) is greater the smaller φ. The accumulation ratio is greater, the smaller the relative volume of liquid in the foam, that is, the "drier" this is, that is, the longer the time available for drainage. In short, $\mathcal{R}$ increases with the expansion factor.

The effect of drainage time t visible in Equation 12.6, §160, can be shown also by modifying Equation 12.10. If again Equation 6.1 of §95 is valid, then $\varphi = \varphi_0 e^{-kt}$ as long as there is no coalescence, φ_0 being the relative volume of the liquid in the foam as it emerges from the gas emulsion. Hence

$$\mathcal{R}-1 = \left(\frac{e^{kt}}{\varphi_0} - 1\right)\frac{6\Gamma_0}{cl} \qquad (12.11)$$

The average bubble diameter l is supposed to be invariant during drainage which is, of course, inexact.

§163.   It is indicated in §161 that the rate of decrease of the residual amount $c_r V_r (= G)$ gives no information on the efficiency of the foaming procedure. Nevertheless, this rate was repeatedly measured and also given a mathematical dressing. For instance, Khrustalev[19] based his treatment on two assumptions. First, the concentration c_f in the foam was supposed to be independent of that in the foaming liquid. Consequently, when the amount $G_0 - G$ of solute (G_0 being the amount before foaming) was transferred into foam, the volume of the residual liquid simultaneously decreased from V_0 to V, so that $G_0 - G = c_f(V_0 - V)$. The second assumption, equally arbitrary, was that $dG/dt = -kVG$, t being time and k, a proportionality constant. In other words, the relative rate of decrease of the solute amount in the residual liquid (i.e., $-dG/G \cdot dt$) was said to be proportional to the volume of this liquid at time t. The resulting equation

$$\frac{dG}{dt} = -k\left(V_0 - \frac{G_0 - G}{c_f}\right)G \qquad (12.12)$$

was confirmed by experiments on a non-ionic surfactant. The quantity $V_0 - (G_0/c_f)$ is the liquid volume remaining after the complete removal of the surfactant.

Much simpler is the equation used, for instance, by Grieves[20] and Rubin[21], namely $G = G_0 e^{-kt}$. The value of k naturally depends on the concentration (c), the rate of gas injection, and so on.

§164. The apparatus of §157 is used in the batch process. A small modification of the foam tower is sufficient to render the separation continuous. In Fig. 12.3, 2 again is the liquid, mm is its upper boundary, 3 is a spinneret (or sparger or diffuser) through which the gas enters the

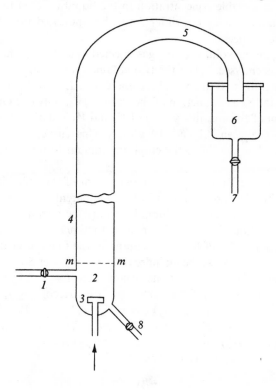

Fig. 12.3. Foam tower for continuous fractionation. *1* is the liquid inlet; *2* is the liquid whose upper level in steady state is at *mm*; *3* is the gas diffuser; *4* is the drainage tube; *5* is the connection to the receptacle; *6* is the receptacle; and *7* is the outlet for collapsed foam.

liquid, *4* is the graduated tube in which the foam rises, and *5* is the duct to the receptacle (*6*). A continuous stream of solution enters the vessel through side tube *1*, and an equal volume leaves the vessel, partly as collapsed foam (through tube *7*) and partly as "bottoms" or "tailings" through tube *8*. After a time (usually 1 to 4 hours in laboratory experiments), steady state is reached. The volumes V_f and V_b (cm^3/sec) flowing through *7* and *8* in unit time remain constant, and so do the concentrations (c_f and c_b) of the solute in the two streams.

Analogously to Equation 12.8,

$$c_0 V_0 = c_f V_f + c_b V_b. \tag{12.13}$$

c_0 and V_0 are the concentration and the volume rate of the supply ("feed") through tube *1*. During the steady foaming, the concentration in the liquid column *2* must vary from point to point; consequently, the accumulation ratio must be calculated using c_0 or c_b. The lesser accumulation ratio $\mathscr{R}$ in a continuous system is c_f/c_0. If the approximation is made[13] that also the variable concentration in the liquid is equal to c_0, then the further treatment does not differ from that indicated in §§159 and 160 for the batch process.

Let V_g be the volume of the gas injected in unit time; thus the dimension of V_g is cm^3/sec. When the two volume rates V_g and V_0 were varied (within the ranges 10 to 35 cm^3/sec and 0.4 to 1.7 cm^3/sec) and the ratio c_b/c_0 was plotted as a function of the ratio V_g/V_0, only one curve resulted, independent of the absolute values of V_g and V_0 and also almost independent of c_0 (between 0.1 and 0.4 g/l.)[18]. This curve, valid for aqueous solutions of hexadecyldimethylethyl ammonium bromide, is reproduced in Fig. 12.4.

The observation that c_b/c_0 is independent of the individual values of V_g and V_0 but depends on the ratio V_g/V_0 apparently can be accounted by drainage, although several assumptions must be introduced on the way. Suppose that Equation 6.3, §95, related to Equation 6.8 there, is valid. Then the volume V_f of liquid remaining in the foam after drainage time t is k_0/t, the constant k_0 being independent of V_g. Let S be the horizontal cross-section of tube *4* and H its height; then $t = HS/V_g$ (see Equation 12.7, §160) or, writing b for k_0/HS, $V_f = bV_g$. Accordingly, $c_f V_f = bc_0 V_g + (6\Gamma_0/lV_g)$ (see §159), and $c_b V_b = c_0 V_0 - c_f V_f = c_0 V_0 - bc_0 V_g - (6\Gamma_0/V_g l)$. The effluent volume $V_b = V_0 - bV_g$. Thus $c_b = c_0 V_0 - bc_0 V_g - (6\Gamma_0/V_0 l)/(V_0 - bV_g)$ and $c_b/c_0 = [V_0 - bV_g - (6\Gamma_0/V_g c_0 l)]/(V_0 - bV_g)$. Denote the ratio V_g/V_0 as x. Then

$$\frac{c_b}{c_0} = \frac{1 - bx - (6\Gamma_0/xc_0 l)}{1 - bx} \tag{12.14}$$

The ratio c_b/c_0 really is a function of V_g/V_0 and independent of the individual values of V_g and V_0.

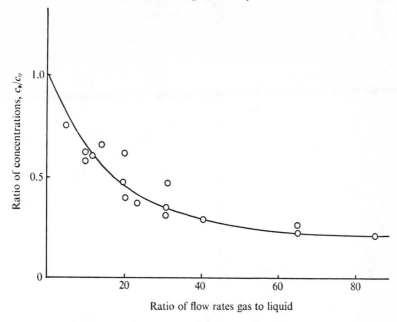

Fig. 12.4. Dependence of the ratio c_b/c_0 on the ratio V_g/V_0 for foams of hexadecyldimethylethyl ammonium bromide (after reference [18]).

The accumulation ratios defined above are not the only expressions for the efficiency of foam accumulation. The "height of transfer unit" is defined[22,23] as

$$[HTV] = \frac{H}{\ln\dfrac{c_0-c_e}{c_b-c_e}} \; ; \qquad (12.15)$$

c_e is the "equilibrium concentration" of liquid 2. As pointed out above, this c_e cannot have a definite value; this renders the values of $[HTV]$ somewhat hazy. Some authors assume that $c_e = 0$.

§165.   Either of the instruments of Figs. 12.1 or 12.3 can be altered to provide an external reflux. In Fig. 12.5, again numbers *1* to *8* indicate, respectively, the liquid supply tube, the liquid reservoir, the gas diffuser, the graduated tube, the duct to the receptacle, the receptacle, the exit tube for collapsed foam, and the tube for recovering the "bottoms." However, a fraction of the collapsed foam is returned to the liquid along tube *9*, which may lead to the foam space or to the liquid reservoir, as indicated by the two sets of dashed lines.

This alteration is wasteful of both gas and time. As only a part of the enriched foam is collected, a greater volume of gas and a longer time are needed to gather the desired part (say, 90%) of the initial solute in

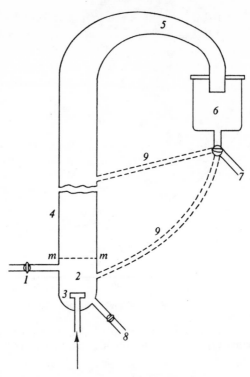

Fig. 12.5. Foam tower for continuous fractionation with external reflux. *1* is the liquid inlet; *2* is the liquid whose upper level in steady state is at *mm*; *3* is the gas diffuser; *4* is the drainage tube; *5* is the connection to the receptacle *6*; *7* is the outlet for collapsed foam; *8* is the outlet for the "bottoms"; and *9*, *9* are two of the possible positions of the reflux tube.

the foam. The reflux raises the steady-state concentration in the liquid reservoir 2, but this result can be achieved simply by lowering the gas flow rate V_g without any change in V_0 or V_b. Unfortunately, the observations recorded[24,25] are insufficient to test the truth of these remarks. It should be remembered also that, the longer the experiment lasts, the greater is the amount of solvent vaporized and the greater the error in the value of $\mathcal{R}$.

Some scientists[10,26] returned the whole of the collapsed foam to the initial liquid, so that the recovery of the solute was nil. The experiments were performed, however, not to recover anything but to have constant concentrations in the liquid and the foam and thus to obtain more exact accumulation ratios $\mathcal{R}$ and distribution factors Γ_0/c. In the usual batch experiments, as mentioned in §159, these concentrations vary in time, and the $\mathcal{R}$ calculated is a time average. When, by leading the collapsed foam back, a steady state is reached, the value of $\mathcal{R}$ has no reason to change with time (if evaporation is completely excluded).

Also internal reflux is mentioned in literature[13,27]. When some bubbles near the top of the foam column burst, the liquid liberated mixes with that in the lower foam regions. This is supposed "to enrich the overflow." In reality, bubble rupture is always objectionable. It causes waste of time and gas; and it counteracts the effect of drainage by rendering the lamellae thicker again. In mathematical language, it increases the value of φ in Equation 12.10, §162, and thus lowers the accumulation ratio $\mathscr{R}$.

Internal reflux was invoked in a report[28] on experiments in which presumably evaporation was more important. A foam of 0.002 M sodium dodecyl sulfate + 0.0001 M 1-naphthylamine hydrochloride was continuously destroyed by heat about 90 cm above the foam-liquid interface. Apparently, evaporation of water was not restricted in any way. After several minutes of this "reflux," solid particles could be seen in the top region of the foam. When the heating was stopped and the foam permitted to reach the receptacle, the surfactant concentration in it was high, as would be expected. A mixture of methyl orange and trimethylhexadecyl ammonium bromide was subjected to a similar treatment, and with similar results[29]. If a hot plate is placed under the receptacle, the concentration of the solute in the recovered foam also is higher than in the absence of heating, but this effect has no relation to the separation in foam.

§166.   It was attempted[27] to improve the efficiency of a continuous foam tower, see Fig. 12.3 (§164), by using the foam as a stripper. The feed enters near the top of the foam column, that is, tube *1* (instead of being near the bottom of tube *4*) is attached a little below duct *5*. Two extreme cases (and any number of intermediate processes) may be considered. If the liquid flows down as one stream (or a few streams) almost without interaction with foam, then the position of inlet *1* is irrelevant; only the bottom liquid is important. This effect is called channeling. If, on the other hand, the liquid which enters tube *4* near its upper end completely mixes with the liquid in the lamellae and the Plateau borders, then the effect on the foam separation would be equivalent to shortening the foam column. Instead of starting at level *mm*, unhindered drainage would start only above the inlet tube. The time allotted to drainage would be smaller and so will be the accumulation ratio $\mathscr{R}$.

The experimental data available are insufficient to check this prognosis. When the liquid inlet tube was placed near the middle of the foam column (instead of the middle of the liquid pool), the flow of "bottoms" V_b slightly increased and their concentration c_b decreased[18]. Hexadecyldimethylethyl ammonium bromide was the foaming agent employed.

In these experiments, the height of the liquid pool had no effect on the surfactant concentration in the residual liquid. It is clear, however, that, when this height is too small, saturation of the bubble surface with the solute will have no time to occur and the surface excess per unit area will be smaller than the theoretical Γ or the experimental Γ_0 discussed in §158. Probably this effect is not bothersome in foam accumulation experiments, because, as long as the bubble surface is far from saturation, foam persistence is small and separation by foaming is not practicable.

When solutions containing 0.1 g/l. of sodium chloride and 0.16 to 0.56 mM/l. of a surfactant described only as monobutyldiphenyl monosodium sulfonate were fractionated in a stripping column, the degree of accumulation of the sulfonate was almost independent of the column height, which varied from 25 to 152 cm[18a].

§167.   As long as one-stage separations, see below, are considered, the following suggestions appear justified by the present state of the art. First, the persistence Σ (§53) of foam bubbles in the selected solution should be ascertained. Secondly, the dimensions of the apparatus and the flow rates of the gas and the liquid should be adjusted so that the duration of drainage is approximately 0.7 to 0.8 Σ sec. During this time, bubble bursting presumably still will be insignificant, the ratio φ (see §162) will have the lowest practically possible value, and the accumulation ratio $\mathcal{R}$ will be as high as feasible. If the time allotted for drainage is much shorter than 0.7 Σ sec, then the possibilities inherent in the method are not fully utilized. At any rate, the $\mathcal{R}$ and related indicators of efficiency should be correlated with the time of drainage, and this time recorded for every experiment. Also the expansion factor of the foam at the outlet point should be determined.

The Σ of commercial surfactants may reach several hours; thus ample time is left for drainage. However, when Σ is only a few seconds, the greatest possible drainage time is so short that the ratio x/δ remains small and very little accumulation is achieved (see Equation 12.5, §160). This conclusion is confirmed by several experiments. For instance, 4.00% aqueous solution of 1-pentanol gave a collapsed foam containing 4.09 to 4.24% alcohol[30]. No accumulation could be detected in the foam of 0.001 to 0.2% 3-methylbutanol[31]. When solutions of 2-methylpropanol, of mole fractions between 0.001 and 0.02, were foamed, the value of $\mathcal{R}$ depended on the relative humidity of the air injected but never markedly deviated from unity[32].

Probably aqueous solutions of lower alcohols are not suitable for any kind of foam accumulation. When Σ is much longer than that of these solutions but much shorter than an hour, the degree of accumulation achieved in a single operation may be insufficient and a repeated treatment may be necessary.

A scheme of multistage instruments is shown in Fig. 12.6; the apparatus is quite similar to that described over 30 years ago[4]. The flow is from left to right. Liquid layer *2* is continuously replenished through inlet *1* and partly transformed into foam by the gas entering through sparger *3*. The dimensions of the instrument should be such that almost

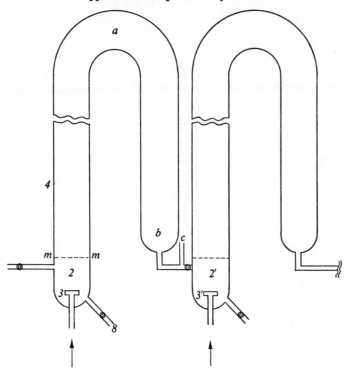

Fig. 12.6. Multiple foam fractionation. The foam collapsed in tube *b* flows into liquid reservoir *2'* and is subjected to foaming again. No bubble collapse should occur left of bend *a*. Air liberated during collapse escapes through side tube *c* (see Fig. 12.3 for the meaning of the other numbers).

no bubble collapse occurs in tube *4*, but the majority of the bubbles burst between points *a* and *b* in the downward duct. The gas liberated in this process escapes through side tube *c*, and the liquid (i.e., the collapsed foam) flows into the next unit in which liquid *2'* is subjected to gas stream from sparger *3'*. Two units are depicted in Fig. 12.6; obviously more than two can be used if necessary. The multistage foam accumulation is not popular at present; for instances of its use see Kishimoto[33] and §171.

§168.   Two of the earliest students[34,35] of accumulation in foam first used a pneumatic method similar to that described in §157, but the effect of evaporation of solvent was so troublesome that both reverted to shaking the liquid in an air-tight vessel. This method gave[34] poorly reproducible and unexpectedly low values for the accumulation ratio $\mathscr{R}$. Thus a 1.29% saponin solution gave a foam which still contained only 1.35% of saponin after three days of drainage. Even after a week-long drainage, $\mathscr{R}$ attained only the value of 3.5.

Some results of a recent study[36] are presented in Fig. 12.7. The initial

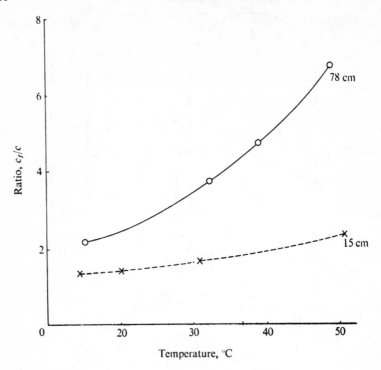

Fig. 12.7. Dependence of the lesser accumulation ratio c_f/c on temperature and the length of drainage tube (78 or 15 cm). Hexadecyldimethylethyl ammonium bromide, 0.0875 g/l. (after reference [36]).

aqueous solution contained 87.5 mg of hexadecyldimethylethyl ammonium bromide per liter, and the foam tower was of the type of Fig. 12.1. The linear rate of gas flow in tube *4* was, apparently, 1.04 cm/sec. The abscissa of Fig. 12.7 means the temperature of the experiment, and the lesser accumulation ratio $\mathscr{R}$ is plotted along the ordinate. The data shown in the lower curve were obtained when the foam was removed (from tube *4*) 15 cm above the foam-liquid interface, and the upper curve was drawn for foams taken out 78 cm above the interface. It is easily calculated that the time allotted for drainage was 15:1.04 = 14 sec for the lower, and 78:1.04 = 75 sec for the upper curve. As expected, the longer the drainage, the higher the ratio $\mathscr{R}$. This relation can be formulated mathematically. It follows from Equation 12.6 (§160) that

$$\ln \frac{\mathscr{R}_2 - 1}{\mathscr{R}_1 - 1} = k(t_2 - t_1); \qquad (12.16)$$

$\mathscr{R}_1$ and $\mathscr{R}_2$ are the lesser accumulation ratios at the drainage times t_1 and t_2. From Fig. 12.7, at 35°, $\mathscr{R}_1$ and $\mathscr{R}_2$ were, respectively, 1.6 and 4.0

for $t_1 = 14$ and $t_2 = 75$ sec. Thus the drainage constant k was 0.026 sec^{-1}; this is a plausible value; compare §§102 and 104.

The importance of drainage time is seen also in the analogous study on an anionic surfactant, namely a sodium alkylbenzene sulfonate[37]. A continuous foam tower, similar to that of Fig. 12.3, §164, was employed. The ratio c_0/c_b increased, for instance, from 2.7 to 5.4 when the duration of drainage rose from 21 to 49 sec. In this arrangement, the distance between inlet tube *1* and effluent tube *8* was important; as would be expected, the ratio c_0/c_b was greater, the greater this distance.

The above values of c_0/c_b and the values of $\mathscr{R}$ seen in Fig. 12.7 (all below 8) presumably are so low because the time allotted to drainage was much shorter than the Σ of the surfactants tested; this Σ probably was of the order of 1000 sec. Higher values of $\mathscr{R}$ are quoted in §172. Also, when a very dilute solution of sodium dodecylbenzene sulfonate was foamed for 3 min, the ratio V_f/V_0 was 0.01 (in a batch experiment) but the foam contained 74% of the initial surfactant[38]; thus the ratio $\mathscr{R}$ was 74. In another batch test[39], the gradual decrease in the concentration c of a sodium alkylbenzene sulfonate in liquid *2* was expressed as $\log(c/c_0) = -kt$; c_0 is the initial concentration, t is time of bubbling (not of drainage), and k is an empirical constant. No data necessary for calculating $\mathscr{R}$ or the drainage time are available. The constant k increased with c_0, which was varied from 0.05 to 0.2 g/l.; and there was only a weak foaming when c_0 was below the critical micelle concentration (0.02 g/l. at pH 6.7). At 5°, k little depended on pH and was near 0.0002 sec^{-1}, but at 60° it was smaller and varied with acidity.

Experiments on a non-ionic surfactant, apparently a derivative of ethylene glycol (see §73) are referred to in §163. It seems that the concentration c_f in foam little depended on the drainage time which was altered by varying the rate of air injection from about 60 to 150 sec; these values are estimated by the present author. When c, i.e., the concentration of the foaming liquid, was 0.08 g/l. (and pH was 2.3), then c_f was about 0.12 g/l., so that c_f/c was only 1.5. When c was 0.02 or 0.19 g/l., the ratio c_f/c was, respectively, 5.0 and 2.6. The increase of c_f with c proves that the hypothesis of c_f independent of the degree of depletion of the initial solution is incorrect. Consequently the values of k of Equation 12.12 recorded in the original publication are not reproduced here.

Foam Fractionation

§169. *Foam fractionation* experiments, in which systems containing more than two components are treated, have been more numerous than those on accumulation in foam. The instruments used are identical in both processes, and also the theories of the operation are almost identical. This applies particularly to the first group of foam fractionation tests, in which the aim is to increase the concentration of a particular ingredient in respect to the solvent, not only in respect to the other solutes. In these instances, the equations reproduced in §§160 through 162 may be used unaltered. In some other applications, the amount of solvent is irrelevant; it is desired to obtain a solution, however dilute, of compound A almost free of compounds B, C, etc., present in the initial liquid. In this group of tests, the greater and the lesser fractionation coefficients $\mathscr{F}$ and f are more useful than the accumulation ratio $\mathscr{R}$. If μ_1 is the mole fraction of A in the initial liquid, μ_1' the corresponding fraction in the foam, and μ_1'' in the residual liquid, then $f = \mu_1'/\mu_1$ and $\mathscr{F} = \mu_1'/\mu_1''$. The solvent is not counted when calculating μ_1, μ_1', and μ_1''. Of course, weight fractions may be used instead of mole fractions.

To the first group belong the numerous attempts to remove dangerous ions from radioactive wastes, the unwanted ions from ground water or streams, and similar processes. The ideal operation would eliminate these ions and nothing else; thus, solvent also counts. None of these ions, however, is surface-active; thus they would not accumulate in the "crust" of a lamella. The easiest way of achieving the necessary surface activity is to add a surfactant which reacts with the undesirable ion and thus gives rise to another surfactant (incorporating the ion). An example of such a reaction is

$$CoCl_2 + 2NaR = CoR_2 + 2NaCl; \qquad (12.17)$$

R is a surfactant anion such as dodecyl sulfate ion $C_{12}H_{25}OSO_3^-$. Instead of one solute, there are now four. The fact that two surface-active species (NaR and CoR_2) are present is not essential; after all the majority of surfactants also are mixtures. On the contrary, the composition of the equilibrium mixture is all important for the success of the operation. If the concentration of CoR_2 is much greater than that of NaR, then the "crust" consists mainly of the cobalt compound and the presence of sodium chloride (and small amounts of cobalt chloride and NaR) in the initial solution will have only a negligible effect on the removal of cobalt. On the other hand, when the amounts of CoR_2 and NaR are comparable, then the question arises which of the two is preferentially anchored in the "crust"; an accumulation of NaR in the surface layer obviously will not help elimination of cobalt with the foam.

The composition of the equilibrium mixture depends on the chemical properties of the compounds involved and cannot be discussed in a book on foams. According to Jorne[40], "the charge and the size of the hydrated ion govern the selectivity in foam fractionation." The dissociation constant of the acid HR is important for the equilibrium such as NaR + HCl = HR + NaCl. The surface activity of HR usually is different from that of NaR; this accounts for the effect of pH on the accumulation ratio or the rate of surfactant removal. Examples of this effect can be found in the following sections.

§170. Radioactive ions frequently are present in the initial liquid in very low concentrations, so that the amount of NaR in Equation 12.17 greatly exceeds that of CoR_2 or the analogous salt. Nevertheless, gratifying accumulation of the unwanted ion was achieved in several series of experiments. The radioactivity of the ions renders their determination possible although their total mass is extremely small.

Gelatin, which is not a surfactant, was utilized as frother in several Russian studies. In some systems the radioactive material may have been present as colloidal particles, so that the results would more properly belong to §174 than to this section. For instance, hydroxides of ^{89}Sr, [134]Cs, [106]Ru, [95]Nb, and [95]Sr were co-precipitated with ferric hydroxide[41]; presumably, the colloidal particles of the latter occluded these ions by the ion-exchange mechanism and carried them into the foam stabilized by gelatin. Ions of [134]Cs were reacted with soluble ferro- or ferri-cyanides and then concentrated in a gelatin foam[42]. At pH 3, ^{90}Sr was accumulated in the foam of 0.01% gelatin solution; at other acidities the extraction was less complete, was hindered by competing cations (from lithium to barium) more the more voluminous the cation, and by anions (Cl^-, NO_3^-, SO_4^{2-}) according to the lyotropic series[43]. Radioactive potassium was enriched in saponin foams[44], as was the ordinary sodium (see §171).

Commercial or purified surfactants are employed most frequently. The non-ionic material mentioned in §§163 and 168 was used[19] to extract [91]Y; the extraction was more efficient, the lower the pH (1.5 to 8.5). Sodium dodecyl sulfate served to remove [45]Ca; when the initial concentration was 2×10^{-6} M, the accumulation ratio $\mathscr{R}$ was[45] from 400 to 2000. Foam extraction of ^{90}Y and ^{90}Sr with alkylarene sulfonates was studied by Pushkarev[46]. The range of acidities in which the removal was possible depended on the purity of the surfactants, and fractionation was depressed by foreign ions; the depression increased in the order LiCl < NaCl < KCl, $CaCl_2$ < $SrCl_2$ < $BaCl_2$, and $NaNO_3$ < NaCl < Na_2SO_4.

Several surfactants were compared[47] as carriers for [89]Sr. N-Dodecylbenzyl diethylenetriamine triacetic acid [presumably

$$C_{12}H_{25} \cdot C_6H_4 \cdot CH_2N(CH_2CO_2H)CH_2CH_2NHCH_2CH_2N(CH_2CO_2H)_2]$$

was particularly efficient when the concentration of foreign ions (e.g., sodium nitrate) was high (e.g., 1 M). Otherwise, "Deriphat 160," which is said to be sodium N-dodecyl-β-iminodipropionate,

$$C_{12}H_{25}N(CH_2CH_2CO \cdot ONa)_2,$$

usually gave better results than dodecyl-β-alanine,

$$C_{12}H_{25}NH \cdot CH_2CH_2CO_2H$$

or "Areskap 100," which is described only as sodium *o*-hydroxyphenyl butylbenzene sulfonate. The effect of drainage on the accumulation ratio $\mathcal{R}$ (for strontium) was shown in the tests with the "Areskap." When the gas flow rate increased from 1.6 to 6.7 cm^3/sec, the density of the collected foam increased from 0.004 to 0.02 g/cm^3 and $\mathcal{R}$ decreased from 22 to 4. According to Equation 12.10, §162, the difference $\mathcal{R} - 1$ should be proportional to the ratio $(1 - \varphi)/\varphi$. In the above experiments this ratio was about 250 and 50, and the corresponding values of $\mathcal{R} - 1$ were 21 and 3. In view of the poor reproducibility of accumulation ratios, the agreement is satisfactory. This is an example of mechanical effects. A probable instance of physicochemical effects is the observation that, when the surfactant concentration increased from 0.16 to 0.7%, the $\mathcal{R}$ was lowered from 11 to 3. Presumably, when the amount of the sodium form of the surfactant greatly exceeds that of the strontium form in the solution, also the "crust" of the foam lamellae contains a greater ratio of sodium to strontium. Probably the Γ_0 of Equation 12.10, that is, the total amount of foaming agent adsorbed per unit area, little depends on the relative concentrations of sodium and strontium compounds, but that part of Γ_0 which belongs to the strontium salt is smaller, the greater the preponderance of the sodium derivative. The $\mathcal{R}$ was independent of pH between 2 and 8 but was low at pH 1; this shows that the salt SrR_2 is more stable than the acid HR, R meaning the sulfonate ion. The existence of an equilibrium of the type of Eq. 12. 17, is rendered probable also by the decrease of $\mathcal{R}$ from 11 to 7 to 2, when strontium concentration increased from 10^{-6} M to 10^{-5} M to 10^{-4} M. The reaction of "Areskap" with the strontium ion was not specific; the $\mathcal{R}$ depended not so much on the concentration of Sr^{++} alone as on the combined concentration of Mg^{++}, Ca^{++}, Sr^{++}, and Ba^{++}.

Radioactive strontium was removed also by means of foaming with sodium dodecylbenzene sulfonate[25]. Continuous foam removal of ^{89}Sr with α-sulfopalmitic acid was described by Davis[48].

Schoen[49] determined the distribution of radium between foam and liquid when the whole of the collapsed foam was being returned to the foam tower (see §165). The foaming agents selected were "Aerosol 22," whose main constituent is said to be

$$NaO_2C \cdot CH_2CH(CO_2Na) \cdot N(C_{18}H_{37}) \cdot OC \cdot CH_2CH(CO_2Na) \cdot SO_3Na$$

and the diethylene triamine derivative

$$C_{12}H_{25} \cdot C_6H_4 \cdot CH_2N(CH_2CO_2H)CH_2CH_2N(CH_2CO_2H) \cdot$$
$$\cdot CH_2CH_2N(CH_2CO_2H)_2.$$

The expansion factor varied between 30 and 60 but the $\mathcal{R}$ for radium usually was near 3, that is, small. Perhaps foreign ions in the solution too strongly competed with radium for the limited space in the lamella surface. The radium concentrations were of the order of 10^{-10} M.

Concentrations of labeled salts between 10^{-6} and 4×10^{-6} M were studied by Moroi[50]. The degree of accumulation was expressed as $\Gamma_1 c_2 / \Gamma_2 c_1$, Γ_2 and Γ_1 being the adsorbed amounts of sodium (from purified sodium dodecyl sulfate) and of the radioactive ion, and c_2 and c_1 their concentrations in the initial solution. The duration of drainage cannot be ascertained from the data published. The above ratio was near 2 for ^{137}Cs in CsCl, between 4 and 6 for $[Co(NH_3)_4(NO_2)_2]Cl$, between 30 and 60 for $[Co(NH_3)_5Cl]Cl_2$, and between 3 and 17 for $[Co(NH_3)_6]Cl_3$. Presumably, the results were affected by precipitate formation when the divalent and trivalent cations reacted with the surfactant anion. If c_1 is the concentration of the cations which escaped precipitation, then $\Gamma_1 c_2 / \Gamma_2 c_1$ was 100 to 400 for $[Co(NH_3)_5Cl]^{++}$ and up to 40,000 for $[Co(NH_3)_6]^{+++}$. The accumulation ratio for the surfactant, which was used in much higher concentrations (all above 0.006 M), was approximately 2.

§171.   Extensive information is available on the removal of non-radioactive ions (whose concentration in the initial solution as a rule is much greater than those referred to in §170) and their accumulation in foam. Apparently the oldest observation of this kind was made by von Zawidzki[51], who foamed saponin solutions containing also electrolytes and found that not only saponin but salts as well were concentrated in the foam; this was confirmed by Chatalan[35]. In 1936, Ermolenko[52] used plant proteins to accumulate iron (from ferric chloride) in froths.

Walling[14] attempted to separate calcium from sodium by foaming. Sodium palmitoyl methyl taurine, $C_{15}H_{31}CO \cdot N(CH_3)CH_2CH_2SO_3Na$, was the frother. When its concentration in the initial liquid was 0.00125 M, and the concentrations of sodium and calcium salts were, respectively, 0.0025 and 0.00025 M, that is, μ_1 was 0.909 and μ_2 was 0.091, see §169, then at the drainage time equal to about 250 sec the corresponding concentrations in the collapsed foam were 0.0089, 0.0042, and 0.0032 M. Thus the accumulation ratio $\mathscr{R}$ for the surfactant was about 7, $f_1 = \mu_1'/\mu_1$ (for sodium) was 0.62, and f_2 for calcium was 4.8. The accumulation ratio for calcium reached almost 13, that is, exceeded the $\mathscr{R}$ for the foaming agent. Magnesium, ammonium, potassium, and ferric ions were transferred preferentially into foam, similarly to calcium, while hydrogen ions preferentially remained in the residual liquid. Also silver could not be extracted; perhaps it was precipitated by impurities in the initial solution. A surfactant similar to, or identical with, the above taurine derivative was employed by Bauer[53] to concentrate scandium.

Magnesium and aluminum accumulated in the foam at the expense of sodium or hydrogen ion also when a sodium alkylbenzene sulfonate was the foaming agent[54]. Zinc and sodium ions were separated by dodecylbenzene sulfonate[54a].

When a solution containing 0.01 g Cu^{++}, 0.5 g sodium dodecyl sulfate, and 0.58 g sodium chloride was foamed, the amount of copper adsorbed was $\Gamma_0 \approx 1.1 \times 10^{-10}$ g-atoms/cm^2 independently of the acidity between pH 3 and 6. When a compound expected to form a complex with cupric ions, namely the ethylene diamine derivative $[(CH_3CHOH\ CH_2)_2N \cdot CH_2—]_2$, was added to the above solution, the value of Γ_0 at pH was not altered but it was depressed both at pH 3 and 9. The ratio of amine to copper had no effect as long as it was varied only between 1 and 7. Presumably, the much larger amount of surfactant determined the value of Γ_0[55]. Sodium dodecyl sulfate served to remove zinc ions at various acidities, ionic strengths, and surfactant concentrations[56], but the accumulation ratios have not been calculated. An analogous removal of ferric iron was achieved by Rubin[57].

The distribution of nickel, introduced as $Ni(NO_3)_2$, between the foam of "Aerosol 22" (see §170) and the liquid (to which the collapsed foam was continually returned) depended on the time of drainage, but when this increased from about 25 to 1250 sec, the $\mathscr{R}$ increased only from 5 to 9, an unexpected result. An increase of the weight ratio of surfactant to nickel from 25 to 100 lowered $\mathscr{R}$ from 7 to 1.8. The surfactant apparently was efficient only near or above its critical micelle concentration (0.066%)[58].

Kepák[59] removed $RuCl_4$ and $Ru(NO)(NO_3)_3$ with gelatin or dodecylamine. In this work, the amount of water lost with the foam was prescribed to be less than 5%. To achieve the high accumulation ratio necessitated by this requirement, 0.004 to 0.016% of foaming agent was needed, although the ruthenium concentration was only 10^{-6} M. The effect of acidity was not simple, because in alkaline solutions ruthenium was present as colloidal hydroxides. A multistage separation by foam (see §167), also was tested; up to five stages were built and additional foaming agent was introduced at each. In this manner the degree of removal was raised from, e.g., 40 to 70%.

§172.   Numerous experiments on the accumulation of calcium, iron, and manganese in foam have been performed in the author's laboratory. They have not been officially published, although some reports on their results got abstracted in *Chemical Abstracts*[60]. A brief review is given in this section.

The batch method of Fig. 12.1, §157, was employed. As expected, longer duration of drainage raised the accumulation ratio $\mathscr{R}$. This is shown in Table 12.1 valid for 0.1% solutions of commercial sodium dodecyl sulfate containing also 3.6 to 72 mg calcium (in calcium chloride) per liter. Both the range of the $\mathscr{R}$ values and the average $\mathscr{R}$ are listed.

When ferrous sulfate was the salt and "Alipal CO-433," whose active ingredient is said to be $C_{10}H_{21} \cdot C_6H_4 \cdot O(CH_2CH_2O)_4 \cdot SO_2 \cdot ONa$, was the frother, the $\mathscr{R}$ for iron was, for instance, 8 when the duration t of drainage was about 1000 sec and 30 when t was 2000 sec. When the

Table 12.1. Accumulation Ratio $\mathscr{R}$ for Calcium and the Duration t of Drainage

t sec	Range of $\mathscr{R}$	Average $\mathscr{R}$
370	3–7	6
730	4–11	7
1230	9–12	11
3450	10–33	18

solution contained $CaCl_2$, $FeSO_4$, $MnSO_4$, and "Alipal EO-526," which seems to be mainly $C_9H_{19} \cdot C_6H_4 \cdot O(CH_2CH_2O)_6 \cdot SO_2 \cdot ONa$, the $\mathscr{R}$ for the three metals was 3, 5, and 2 at $t = 500$ sec, and 13, 13, and 6 at $t = 1000$ sec. An analogous effect is seen in Fig. 12.8 for the mean accumulation ratio of "Alipal CO-433" itself; at a given pH, the $\mathscr{R}$ after 2000 sec of drainage is two to three times as high as after 1000 sec.

As pointed out in §162, this $\mathscr{R}$ is closely related to the expansion

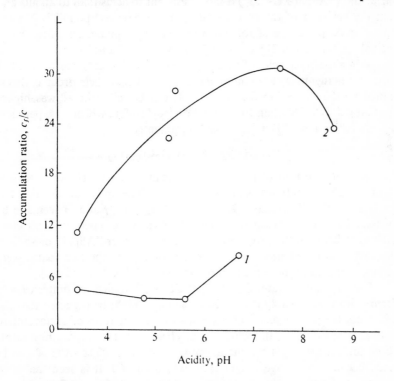

Fig. 12.8. Dependence of the lesser accumulation ratio c_f/c on the pH of the solution of "Alipal CO-433" and the duration of drainage (1000 sec for curve *1*, 2000 sec for curve *2*). Unpublished.

factor $1/\varphi$. When the duration of drainage of an "Alipal CO-433" foam, containing also a small amount of ferrous sulfate, was raised from 1000 to 2000 sec, the $1/\varphi$ rose from 200 to 1580. Simultaneously the $\mathcal{R}$ for the surfactant rose from 3.5 to 15.5 so that the product $(\mathcal{R}-1)\varphi$ changed only from 0.0125 to 0.0091; this change is not particularly significant in view of the large error of measurement (see the ranges in Table 12.1). Another example corroborating Equation 12.10 is supplied by sodium dodecyl sulfate (in the presence of ferrous sulfate). At a drainage time t of 1100 sec, $1/\varphi$ was 1730 and $\mathcal{R}$ of the surfactant was 12. When t was 2200 sec, then $1/\varphi$ was 15,000 and $\mathcal{R}$ was 102. For the product $(\mathcal{R}-1)\varphi$ these two experiments give the very similar values of 0.0064 and 0.0067. The quantity $(1-\varphi)$ in Equation 12.10 needs no consideration at these large expansion factors.

The effect of pH on $\mathcal{R}$ illustrated in Fig. 12.8 cannot be accounted for because the exact composition of the foaming agent is not known. However, the different effect of acidity on iron and manganese extractions clearly indicates that the compound which ferrous ion makes with the surfactant ("Alipal EO-526") is more resistant to acids than to alkalis and that the behavior of the corresponding manganese compound is just the opposite. At pH 2, the $\mathcal{R}$ for iron was 21 and for manganese, 6; at pH 7 the values were 2 and 21; and at pH 9.1 they were 2 and 24.

When the efficiencies of three commercial surfactants (for removal of iron and manganese) are compared, it appears that their order is determined by the expansion factor of the foam. Usually, the $\mathcal{R}$ was higher for "Alipal EO-526" than for sodium dodecyl sulfate which was, perhaps, a little better than "Deriphat 170" [apparently

$$(C_{12}H_{25})_2N \cdot CH_2CH_2CO_2H].$$

At constant duration of drainage, the mean values of $1/\varphi$ were for the "Alipal" about 3.0 times as great as for the sulfate and about 3.5 times as great as for the "Deriphat." The relation between $1/\varphi$ and $\mathcal{R}$ seems to be sustained by the tests on the removal of calcium. There was no clear difference between the $\mathcal{R}$ values for calcium, whether "Alipal" or sodium dodecyl sulfate was used, and the values of $1/\varphi$ for the two foams were almost identical at the three acidities tested (pH 2.5, 3.5, and 4.5).

The amount of water-free "Alipal CO-433" needed to complex 1 g of ferrous iron presumably is near 20 g; hence it might be expected that the $\mathcal{R}$ values for iron would increase with the weight ratio of surfactant to iron at least as long as this ratio remains below 20. This expectation seems to be confirmed: at pH 5.3 and a drainage time of 2000 sec, the $\mathcal{R}$ was 11 and 30 when the above weight ratio was 6 and 60. It is seen that a $\mathcal{R}$ exceeding 10 can be achieved also when the amount of the surfactant is too small to bind all metal ions present. This was true also for "Alipal EO-526" and calcium.

In a few instances, foaming was continued until the foam ceased to reach the "gooseneck" of Fig. 12.1 (§157). For "Alipal EO-526" this occurred, e.g., when its residual concentration was 0.030 g/l. Then the remaining solution was transferred to a beaker and subjected to additional foaming, and the foam layer above the liquid was sucked away by an air stream. This layer naturally was very "wet" but the accumulation of surfactant in it was measurable.

§173.   In §§ 170 through 172, removal of cations was reviewed and naturally, as a rule, anionic foaming agents were employed. This section is devoted to foam accumulation of anions by means of cationic surfactants.

The early work by Grieves on gathering of phosphates and dichromates in the foam of hexadecyldimethylethyl ammonium bromide is summarized in reference [61]. Because this surfactant forms a colloidal salt with the dichromate anion, the system obtained with $Cr_2O_7^=$ belongs to §174. Foaming was continued until the foam ceased to reach the receptacle (6 in §157). For the orthophosphate, the concentration c_r of the residue increased with that (c) of the initial solution and slightly rose with the ratio of surfactant to phosphate. The ratio c_r/c was, for instance, 0.13 at $c = 0.027$ g/l. and only 0.7 at $c = 0.15$ g/l.

Several metals can be extracted as complex anions. When a solution of an uranium salt and a cationic surfactant in 8 N hydrochloric acid was foamed, uranium was concentrated in the froth[62]. The accumulation ratio $\mathscr{R}$ increased, for instance, from 15 to 70 when the air injection rate was lowered from 0.6 to 0.25 cm³/sec. Lowering the uranium concentration from 0.006 to 0.001 g/l. enhanced the value of $\mathscr{R}$.

Fig. 12.9 demonstrates the importance of complex formation in mixtures of ferric chloride (2×10^{-6} M), hydrochloric acid, and hexadecyltrimethyl ammonium bromide[63]. The ordinate is the distribution factor Γ_0/c (see §162) for iron, determined at complete external reflux (§165) in steady state; the surfactant concentration was 0.0001 M and the duration of drainage approximately 290 sec. The abscissa represents the concentration (mole/l.) of hydrochloric acid. In 4 M HCl, the complex formation is still negligible and almost nothing is extracted by the cationic foaming agent. The maximum sorption is achieved in 8 M HCl; the decrease of Γ_0/c in more concentrated acid may be due to formation of other anions such as HCl_2^-. As the highest values of Γ_0/c are near 0.005 cm, the Γ_0 itself is $< 10^{-11}$ mole/cm². This surface excess is smaller than any quoted in §159; apparently, the surface layer is not yet saturated with iron compounds at the ferric chloride concentration as low as 2×10^{-6} M. In agreement with this view, the Γ_0/c for 10^{-4} M $FeCl_3$ (still in 8 M HCl) was 0.0005 cm, so that Γ_0 was 5×10^{-11} mole/cm², that is, nearer to the saturation.

Mercuric chloride affords complexes with hydrochloric acid more

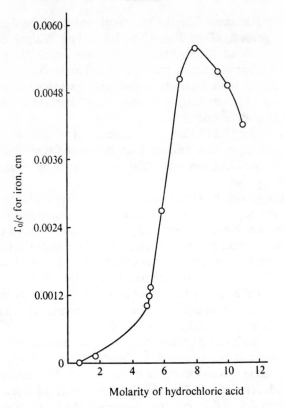

Fig. 12.9. Distribution factor Γ_0/c (in centimeters) for iron in mixtures of ferric chloride, hexadecyltrimethyl ammonium bromide, and hydrochloric acid as a function of the concentration of HCl (mole/l.) (after reference [63]).

easily than ferric chloride does; a maximum of Γ_0/c, analogous to that seen in Fig. 12.9, was observed in 0.3 M HCl. The maximum Γ_0/c for $CoCl_2$ took place in 9 M HCl but was equal to only 0.00002 cm. When both ferric ion and divalent mercury ions were present in hydrochloric acid solutions, mercury was accumulated preferentially to iron from 1 M HCl, and iron was gathered preferentially to mercury from 8 M HCl.

Mercury forms complexes with nitric acid also. When the concentration of mercuric nitrate was 5×10^{-7} M, of surfactant (the bromide of the penultimate paragraph) 0.001 M and of nitric acid 0.1 N, then the Γ_0/c in steady state was 0.0028 cm; and when 1.5 N HNO_3 was the solvent, the Γ_0/c was only 0.0003 cm[26].

Iodine can be removed from aqueous solutions or dispersions with gelatin[63a]; the best results were achieved at pH 1 and, when temperature was varied between 10° and 50°, at 20°.

Dimethylethylhexadecyl ammonium bromide was used for recovery of iodine (from 0.00015 to 0.00030 M sodium iodide), chromate ($HCrO_4^-$ from sodium dichromate), thiosulfate (from $Na_2S_2O_3$), and the complex ions occurring in the mixture of silver nitrate, sodium thiosulfate, and sodium hydroxide. In this work[63b], the residual concentrations c_s of the surfactant and c_a (of the anion) were compared at different times and at different ratios $c_s:c_a$ in the initial solution. In all instances the initial ratio was not far from 1 mole to 1 mole, that is, much smaller than in the tests recorded in §172. The gradual decrease of c_s, i.e., $-dc_s/dt$ (t being time), was similar to that of c_a (i.e., $-dc_a/dt$). The ratio dc_s/dc_a was 1.2 to 1.4 for iodine, near 1.0 for chromate (monovalent anion!), between 2.2 and 2.8 for thiosulfate (divalent anion!), and 1.3 to 2.5 for surfactant and silver. Thus the behavior was such as if the reaction $C_{20}H_{44}NBr + NaI = C_{20}H_{44}NI + NaBr$ (and the analogous reactions) proceeded almost to completion, and the iodide $C_{20}H_{44}NI$ accumulated in foam practically without interference by the inactive salt (sodium bromide).

§174. *Colloids* are reviewed in this section. It is realized that in many experiments described above the surfactant concentration exceeded the critical micelle concentration, so that the liquid contained colloidal particles. The particles treated here are supposed to be too small to be flotated by the wetting mechanism (§156).

In a particularly complex system[64], a non-ionic polymer was used to enhance flocculation of colloidal particles of $R_2Cr_2O_7$, R being the dimethylethylhexadecyl ammonium cation. In this study, instead of bubbling air through, a part of the sol was saturated with air at about 4 bars pressure and then introduced in the foam tower kept at atmospheric pressure; the bubbles formed were much smaller than those usually produced by gas dispersion apparatus. The greater accumulation ratio c_f/c_r exceeded 3000 in the most successful test.

Particles of ferric hydroxide, about 150 Å across, could be removed with either sodium dodecyl sulfate at pH 5.8 or with dimethylethylhexadecyl ammonium bromide at pH 10.8, so that $\log(cV_0/c_rV_r)$ in single batch operation was directly proportional to the duration of foaming[65] (see §163). When foaming was continued as long as feasible, the relative volume of liquid transferred into foam, that is, V_f/V_0 was greater, the greater the concentration of sodium dodecyl sulfate and ferric hydroxide[61]. This presumably means that drainage was slower, the higher the concentration, in agreement with the data of §§101 and 102.

Many classical colloidal substances (sulfides of cadmium, copper, antimony, bismuth, lead, mercury, and silver, and hydroxides of iron and aluminum) have been concentrated in the foam of (preferably 0.02%) gelatin[66]. The success of extraction was affected by foreign substances in the sol. Thus the accumulation of antimony trisulfide in foam was accelerated by the nitrates of calcium, strontium, or barium, and retarded by thorium nitrate and zirconium nitrate[67]. It depended on the pH of the foaming liquid and usually exhibited a shallow maximum between pH 5 and 6. This maximum was noticeable also in the sols of gold, ferric hydroxide, and titanium hydroxide, and presumably was related to the isoelectric point of gelatin. The reason for the weak maximum near pH 3.2 observed in some systems remains unknown[68]. Other unexpected results were published later[69]: the removal of Sb_2S_3 and $Fe(OH)_3$ by gelatin was speeded up by ethanol or glycerol but slowed down by butanol and 3-methylbutanol. The extraction of colloidal $K_2Cu_3[Fe(CN)_6]_2$, $K_4Co_{10}[Fe(CN)_6]_6$, and $KFe[Fe(CN)_6]$ in the presence of potassium ferrocyanide in 1 % gelatin foam was more rapid at higher temperatures[70]; for instance, when the drainage time was about 70 sec, the relative amount removed in 3 min was 35 % at 7°, 65 % at 30°, 75 % at 40°, and 80 % at 58°. The rate of removal of ferrocyanide particles from a solution of potassium ferrocyanide (0.1 g/l.) and gelatin (0.2 g/l.) was rapid for $Mn_2[Fe(CN)_6]$, less rapid for the copper salt, and slow for the ferric salt[71].

Colloidal silica[72] and colloidal stannic acid[73] also can be accumulated in foam.

When colloidal sulfides were prepared (PbS, CdS, and Sb_2S_3) not in water but in 0.02 % gelatin solution, the amount Γ_0 moles/cm^2 adsorbed in steady state depended on the age of this solution and on its acidity[74]. For instance, the adsorption of lead sulfide reached 10^{-7} g/cm^2 (or 4×10^{-10} mole/cm^2) at pH 6 to 7, but was only 7×10^{-8} g/cm^2 at pH 4 and 9. When the gelatin solution was one to seven days old (the aging was conducted at 10 to 13°), the Γ_0 usually was 0.5 to 0.7 of the Γ_0 measured when the aging lasted for four days.

A model for the foam separation of colloids was proposed by Grieves[75].

§175.   Those *organic compounds* which are not foaming agents themselves can be gathered in the foam of another substance, as described in §§170 through 174 for inorganic compounds. Phenol in alkaline solutions exists to a considerable extent as phenolate anion and thus can be extracted with a cationic surfactant. In one study[76], dimethylethylhexadecyl ammonium bromide was the frother. The ratio of the amount of phenol in foam (i.e., c_fV_f) to its total amount in the initial solution (i.e., cV_0) was greatest when the pH of this solution was between 11 and 12; foaming was continued as long as the Σ(§53) of the liquid exceeded 200 sec. The ratio of the concentration c_r of the residue to the initial concentration c was a function of $c_s/c^{0.5}$, not of c_s/c (c_s being the initial surfactant concentration) when c_s varied from 0.05 to 0.6 g/l. and c from 0.011 to 0.672 g/l. In other words, the quantity $c_s/c^{0.5}$ remained almost constant when the ratio c_r/c was constant. Suppose that $c_r/c = 0.01$ is aspired to. This degree of removal of phenol could be achieved when c was 5 mg/l. and c_s was 200 mg/l. Consequently, it would be attained also when c is 20 mg/l. and c_s is 400 mg/l. It is seen that the second 200-mg

Table 12.2. Accumulation Ratios and Expansion Factors

Duration of drainage (arbitrary units)	53	29	21
Expansion factor, $1/\varphi$	4470	811	209
Accumulation ratio, $\mathscr{R}$	105	28	12
$(\mathscr{R}-1)\,\varphi$	0.023	0.033	0.053

portion of surfactant is better utilized than the first 200 mg. Sodium sulfate was found to hinder the accumulation of phenol in foam.

In a simultaneous investigation[77], phenolate ion was extracted with nitrogen as gas and trimethylhexadecyl ammonium bromide as frother. Because the necessary raw data have been recorded, it is possible to construct Table 12.2. According to the theory of §162, the product $(\mathscr{R}-1)\varphi$ ought to be almost independent of the duration of drainage and the expansion factor. The trend of the values of $(\mathscr{R}-1)\varphi$ evident in the table could not be explained by the present author; at any rate this product varies much less than do $\mathscr{R}$ and φ separately. Unexpectedly $\mathscr{R}$ was not affected by variation in the molecular ratio of phenol to bromide from 1.0 to 0.1 and decreased only when this ratio was lowered to 0.05. When the proportion was 1:1 but the concentrations themselves were raised from 0.00014 to 0.010 M, the $\mathscr{R}$ decreased, for instance, from 80 to 2, and the volume of the collapsed foam rose from 1 to 3 cm³; this presumably is another example of the inverse relation between concentration and rate of drainage. A depth of 2 cm of the liquid column was sufficient to obtain the highest $\mathscr{R}$ values.

In the above experiments, long-chain cations were used to collect organic anions. As expected, non-ionic surfactants and surfactants whose active ion has the same sign as the ion to be removed are less suitable. Thus when solutions of 1-naphthoic acid plus surfactant were foamed at pH 9.5, the $\mathscr{R}$ was 4.1 for "Ethoquad 18/25" (described as a poly-ethoxystearyl methyl ammonium chloride), 3.4 for "Tergitol NPX" [a nonylphenyl poly(ethylene glycol) ether], and 1.3 for commercial sodium dodecyl sulfate[78]. At pH 2.5, the accumulation ratios were 3.0, 3.6, and 2.9, that is, almost independent of the surfactant type. On the contrary, this type proved important for removal of 1-naphthylamine at pH 2.5, the three values of $\mathscr{R}$ being 1.3, 4.4, and 10.5. It is seen that the naphthylamine cation was extracted best by a surface active anion. The salt formed by the two probably was precipitated when the foam was concentrated by vaporization on heating (see §165).

Simple electrostatic attraction cannot account for the enrichment of 1-dodecanol and similar non-ionic compounds in the foam of synthetic surfactants. In an early study[79], the substances removed from the foaming agent by foaming were not identified. Commercial dodecylsulfuric acid, which showed a minimum of surface tension γ (about 34 g/sec^2) in 0.007 M concentration, was foamed. The collapsed foam was diluted to 0.007 M and then showed a γ of 29 g/sec^2, while the residual liquid at the identical concentration had $\gamma = 41$ g/sec^2 and its surface tension manifested no minimum. The behavior of sodium dodecyl sulfate was similar. At the 0.007 molarity, the solution of the commercial salt had $\gamma = 24$ g/sec^2; the collapsed foam, 22 g/sec^2; and the residual liquid, 39 g/sec^2. On the curve of γ versus concentration of the residual solution no minimum was visible.

Sporck[80] subjected, among others, a solution containing 0.011 mole of purified sodium dodecyl sulfate and 0.00022 mole of 1-dodecanol per liter to a stream of nitrogen bubbles of an average surface area 0.64 cm^2. The solution was supersaturated in respect to the alcohol, as its solubility in water is only about 0.00001 M. Although the molecular ratio of sulfate to alcohol was 50 in the liquid, in the adsorbed layer it was only 2. One molecule of dodecanol was present, on the average, for each 80 square Å of the bubble surface. Also Kishimoto[33] and several other investigators employed foaming to purify commercial sodium dodecyl sulfate.

§176.   Separation of *dyes* presumably is the most spectacular experiment on foam fractionation. Apparently its earliest example[4] was described in 1940. Patent Blue (a triphenylmethane dye) is in aqueous solution a better foaming agent than Scarlet Red (an azo dye, also known as New Coccine, Ponceau 4R, etc.). When nitrogen gas was bubbled for 10 min through 100 cm^3 of solution containing 0.01 g of each Blue and Red, 3.5 cm^3 of collapsed foam and 96.5 cm^3 of residue were obtained. The concentrations in the former were 0.22% (Blue) and 0.02% (Red), and in the latter, 0.003% (Blue) and 0.01% (Red). The overall amount of Blue comes out to be too great, perhaps because of evaporation. If this is disregarded, the lesser coefficient of fractionation (in weight units) of Patent Blue was 1.8, and the greater coefficient, 4.0. The foam was blue and the residual liquid was red.

In the above test, the initial concentrations were equal. Separation did not succeed when the surface-active ingredient was present in excess (and the foaming was too protracted). When the starting solution (100 cm^3) contained 0.92 g of Patent Blue and 0.04 g of Scarlet Red, the collapsed foam (96 cm^3 collected in 60 min) contained 9.4 mg/cm^3 Blue and 0.4 mg/cm^3 Red, while the residual liquid had 7.7 mg/cm^3 Blue and 0.4 mg/cm^3 Red. The lesser fractionation coefficient for Blue was only a little above 1.0. On the contrary, when nitrogen was bubbled through a solution of 0.008% Blue and 1.05% Red, that is, the amount of the surface-active compound was relatively small, then the concentrations in the collapsed foam were 0.025% (Blue) and 1.08% (Red), so that the lesser coefficient of fractionation was 3.0 for the Blue.

In the summary[81] of Dubrisay's work it is stated that a violet mixture of Night Blue (a diphenyl-1-naphthyl dye) and Basic Fuchsin (a triphenylmethane dye) can be separated into a blue foam and a red residual liquid.

§177. Dubrisay[82] performed also separations of some soap mixtures. When a solution containing 0.1 g each of sodium oleate and sodium laurate in 100 cm^3 was foamed at room temperature with nitrogen, the acid mixture isolated from the collapsed foam was 60% oleic acid plus 40% lauric acid, while there was only 19% of oleic acid in the residue. The lesser fractionation coefficient (in weight units) was 1.2 for sodium oleate. Sodium laurate depresses the surface tension of water at 45° more, and at 90° less, than does sodium stearate. Hence, it may be expected that the former soap would be accumulated in foam at 45°, and the latter at 90°. This expectation was qualitatively confirmed.

Partial separation of oleic from stearic and palmitic acids was accomplished[83] by frothing their strongly alkaline solutions at 50° (for potassium soaps) or 75° (for sodium soaps). Foam separation of fatty acids from sulfonated fatty acids also was attempted[84].

Purification of synthetic surfactants by foaming is mentioned in §175. No doubt, it was easier to accumulate sodium butylnaphthalene sulfonate in foam preferentially to sodium sulfate and sodium chloride present in the crude material[85]. Another separation of surfactants from inorganic substances was patented by Bieber[86]. Sodium hydroxide is used to extract phenols and naphthenic acids from petroleum oils. If the extract is subjected to foaming, the organic compounds gather in the foam, and the alkaline residue may be pure enough to be reused. If necessary, Patent Blue dye is added as a foaming agent. In one test, when the initial liquid contained 0.11% "acid oil," the concentration of the acid in the collapsed foam was 1.81%, and in the residue, 0.06%. Foam fractionation of secondary sodium alkyl sulfates is described by Evans[87].

As mentioned in §159, the Γ_0 of sodium dodecyl sulfate and sodium dodecylbenzene sulfonate was, respectively, 4.5×10^{-10} and 3.5×10^{-10} mole/cm^2. When the solution contained both sulfate and sulfonate in the molecular ratio 10:1, their mole fractions in the foam were nearly equal; thus the sulfonate had a greater tendency to gather in the foam than the sulfate[12]. This was noticed already by Ito[45]. A mixture of three surfactants can sometimes be foam fractionated; this was attempted[88] for sodium dodecyl sulfate plus sodium benzene sulfonate and sodium dodecylbenzene sulfonate.

Non-ionic surfactants of the general formula

$$R \cdot C_6H_4 \cdot O(CH_2CH_2O)_nH$$

(see §73), in which R is an alkyl radical, are complex mixtures. Such a surfactant in which R was iso-octyl and the average value of n was nine, was subjected to foaming, with the result that the mean n was different in different foam fractions[89]. Removal of alkylarene sulfonates from aqueous solutions in a continuous process is described by Zwierzykowski[89a].

§*178.*　*Polymers* can in some instances be fractionated by foaming. The mean acetyl content of a poly(vinyl alcohol) was 0.5%. The acetyl content of the first foam fraction (which made up 18% of the total) was 0.6%, and of the residue (40% of the total), only 0.35%. When the acetyl content of all chains was almost constant but the molecular weights (MW) were different, the early foam fraction had MW = 44,000, and the residue, MW = 21,000[90].

Imai[91] did not achieve any fractionation of poly(vinyl alcohol) when the usual pneumatic method was employed but was successful when, e.g., 1% aqueous solutions of these polymers were shaken and then left alone for several hours. The intensity of the color reaction with iodine was greater in the foam than in the initial liquid. For instance, when the foam contained 5% of the starting amount of the polymer, this intensity was four times as high as that of the raw material. The authors believe that the coloration is given by syndiotactic in preference to isotactic polymers. This view seems to be bolstered by the observation that the iodine reaction is more intense in polymers prepared at 0° than in those made at 60 to 80°; the stereoregularity generally is greater, the lower the polymerization temperature. The MW of the first foam fraction usually was identical with that of the starting polymer (about 70,000).

The chains obtained in the co-polymerization of methacrylic acid and methyl methacrylate have varying ratios of carboxyl to methyl. When such a copolymer was fractionated by foaming, the first fraction had a higher ratio, $CO_2H:CO_2CH_3$, than the crude material[92].

Fractionation is feasible also in nonaqueous solutions. Nitrogen saturated with benzene was bubbled through 0.64% solution of commercial poly(dimethylsiloxane) in benzene. Three foam fractions and a residue, each of approximately identical mass, were collected. The intrinsic viscosity of these samples was 0.98, 0.70, 0.61, and 0.53, while that of the initial solution was 0.81. Apparently some separation according to MW took place[93].

§*179.*　Foam fractionation of *natural liquids*, waste water, mixtures of biological interest, and so on, is reviewed in this and the two following sections.

The composition of sea foam undoubtedly is highly variable. The chemical and the biological oxygen demand of the samples studied by Skopintsev[94] were 3 to 20 times as great as those of the sea water under the foam (see also §144). An attempt to extract uranium from sea water by foaming was described by Yamabe[95].

Sodium chloride solutions, subjected to electric current to manufacture sodium hydroxide and chlorine, often give rise to troublesome foaming. To avoid it, air containing 10% chlorine is bubbled through the crude brine in a continuous process. About 1% of the liquid is transferred into froth, and the rest gives rise to unstable foams only[96]. It is supposed that "some form of tannin" present in the brine acts as frother. This process, however, cannot be considered as purification by foaming only, because similar foaming in which no chlorine is used is much less efficient; it seems that chlorine destroys a part of the foaming agent.

The major part of proteins of potato juice can be removed by bubbling carbon dioxide (but not nitrogen) through the liquid[97]. For instance, 350 cm^3 of the juice (protein content 0.06%) could be divided in this manner into 20 cm^3 of collapsed foam (0.6% protein) and 320 cm^3 of residue (0.02% protein). Apparently, 10 cm^3 of water evaporated during the process. Thus the corrected protein concentration of the collapsed foam was $0.6 \times 20/(20+10)\% = 0.4\%$, rather than 0.6%, and the lesser accumulation ratio was about 7. When additional carbon dioxide was injected into 300 cm^3 of the residual liquid until only 275 cm^3 remained, the protein content of the second residue was as low as 0.009%. Siebert[98] suggested utilization of this treatment in starch manufacturing plants.

Removal of frothers from beet root juice would simplify the isolation of sugar (sucrose), as many ingredients which act as foaming agents (see §150) at the same time retard or prevent crystallization of sucrose. Both laboratory experiments[97,99,100] and large-scale attempts[101] on purifying sugar juice by foaming have been recorded, but this promising process was disregarded more recently.

Proteinaceous materials often can be separated in five fractions designated, respectively, as metaproteins, proteins, proteoses, peptones, and amino acids. When partly hydrolyzed soybean protein was foamed, the percentages of the total nitrogen in the foam, occurring in these fractions, were 2.5, 26.3, 23.8, 23.7, and 23.7%. The corresponding numbers in the residue were 0.0, 6.1, 25.6, 34.1, and 34.2%. It is seen that metaproteins and proteins accumulated in the foam; presumably they acted also as frothers[6]. A similar fractionation of apple proteins was performed by Davis[102].

Some results of the foam fractionation of beer are reported in §151.

In principle, the composition of foam may be different at different levels. This kind of separation was described in two patents[103,104]. The column of a well-drained foam is divided into sections, and each section is analyzed. Different polypeptides are found in different sections. In a steady-state foam, saponin is detected above gelatin, and gelatin above sodium oleate. A foam tower with a roughened internal surface is said to be more suitable than a smooth tube.

§180.   Nitrogen compounds are concentrated in the foam of egg white. The foam contained, for instance, 2.0 to 2.3% nitrogen while the exuded liquid had 1.5 to 1.6% N$_2$[105].

The foaming behavior of commercial bile preparations is complex[106]. When 0.4 to 0.5% solutions of a commercial sodium salt were foamed, the bubble persistence gradually decreased, and the rate of nitrogen flow had to be raised to send the foam into the receptacle. This foam contained mainly fatty acids and their soaps. Then the persistence increased again and the second foam was rich in bile acids and their salts. Of the two main bile acids, one (taurocholic) does, and the other (glycocholic) does not contain sulfur. The earlier fractions of the second foam contained 0.82% sulfur, and the latter, 4.21% sulfur; the sulfur content of the starting salt was 1.10%. Evidently, glycocholic acid was transferred into foam preferentially to the taurocholic. Acids were collected in the foam preferentially to their salts; this was indicated by an increase in pH (e.g., from 6.5 to 8.5) which accompanied the foaming. It was necessary to keep the liquid acidified to achieve a good yield of foam.

Several publications exist on the foam accumulation of enzymes[107]. An additional difficulty in this process is that some enzymes are inactivated by the bubbling operation; thus up to 30% of the streptokinase activity was lost in a foaming test, although the temperature was kept at 4° and nitrogen was the gas injected[108]. Still using nitrogen, diastase was concentrated in the foam by a factor of 1.5, and pancreatic lipase by a factor of 1.2; and the foaminess depended on acidity[109]. Also the choline-esterase of horse serum could be purified by frothing; here, however, the activity of the residue was greater than that of the foam[110].

Crystalline pepsin has also a rennin activity, and the question arose whether the two activities belonged to the same compound. It was apparently decided in the negative by foam fractionation. When pepsin was subjected to foaming at pH 1.7 to 2.0 at 30°, the pepsin activity was collected in the foam more than was the rennin activity (e.g., the accumulation ratios were 3.3 and 1.3)[111].

If n_1 and n_2 are the nitrogen percentages (which are a measure of the protein content) of the solids obtained, respectively, from the foam and the initial material, and if u_1 and u_2 are the corresponding jackbean urease activities, then the ratio $u_1 n_2 / u_2 n_1$ was particularly high (e.g., 10) at pH 5.0 when carbon dioxide was the gas used[112]. The accumulation ratio of urease was greater than that of catalase; and when both enzymes were present together, urease was transferred into foam more completely than the latter[113]. The catalase activity in the collapsed foam was concentrated about elevenfold but that of amylase, in identical conditions, showed no accumulation; and when a mixture of both enzymes was frothed at 4°, the accumulation ratio for catalase again was near 11, and for amylase again near 1.0[114].

The gonadotropic hormones present in the urine of pregnant women could be concentrated in foam. Nitrogen was bubbled through 100 cm^3 of filtered and slightly acid urine for 15 to 45 min. The collapsed foam was subjected to second foaming in a smaller apparatus. Collapse of the second foam yielded 1 cm³ of turbid liquid which was as active as 10 to 20 cm³ of the urine. However, a large percentage of the hormones was lost in the process[115]. Accumulation of the bacteriophage activity in foam was achieved by Ruska[116].

Recovery of bacteria from their suspensions belongs perhaps to froth flotation rather than to foam fractionation. For instance, the difference between *Escherichia coli* and *Mycobacterium tuberculosis* in this process was attributed by Dognon[5] to the different wettabilities of their surfaces. Recently, transfer of bacteria into foam was studied, for instance, by Gaudin[117], Grieves[20], and Rubin[118].

§181. Lanolin can be removed from wool scour liquid into foam. Thus in 100 min, 3 tons of this liquid could be processed to give 40 kg of collapsed foam containing 25% of wool grease[119].

Foaming treatment has been applied to numerous waste liquids. For petrochemical wastes see Grieves[120]. Kresta[121] considers the recovery of surfactants from laundry wastes impractical because fats, proteins, etc., present there often inhibit foaming, and because the volume of the collapsed foam is too great and may be as great as 50% of the waste volume. Unfortunately, the original of this pessimistic publication is not available to the present author. Removal of surfactants from waste water of tanneries was described by Kondratovicius[122]; their concentration in the collapsed foam was as high as 0.4 g/l. and over a half of the surfactant initially present could be transferred into froth. More satisfactory results have been achieved later[123]; nearly 80% of the surfactant could be removed, and the volume of the collapsed foam was only 8 l. for 1000 l. of waste liquid treated, so that the lesser accumulation ratio apparently was $(0.8/8)/(1/1000) = 100$. Also dye wastes were subjected to foam treatment to reduce their coloration, their chemical oxygen demand, and their biological oxygen demand[124]. Foam purification of waste waters of textile plants was attempted by Klimenko[124a].

Partial removal of surfactants from municipal and industrial waste waters to avoid ungainly foam on rivers and lakes was successful several times[125,126]. Raison[127] treated up to 240 l. of primary or secondary effluents per hour and could remove up to 90% of the surfactants initially present, although the volume of the collapsed foam was only 1/1000th of the starting volume. Other pilot plants were described by Brunner[128] and Rose[129]. Instead of bubbling air through, Garrett[130] saturated secondary sewage effluent with air at 11 to 28 bars; when pressure was released, bubbles formed by condensation and the resulting foam was mechanically scooped off.

The concentration of alkylbenzene sulfonates in sewage water was lowered from 0.002 to 0.003 g/l. to 0.0003 to 0.0005 g/l. when 1.5 l. of air was spent on 0.001 g of surfactant removed[131]. Reduction of the above concentration to 0.001 g/l. was described by Jenkins[132], but in his experiments foaming eliminated many other ingredients in addition to the surfactant; for instance, also nitrogen-containing compounds (proteins?) became concentrated in the foam although at a lower rate than the sulfonates. A similar comparison of the removal of surfactants and other oxidizable material was published by Rose[23]. Although other substances were present in the collapsed foam in addition to the detergents, the foam still may have been reused as a detergent in industry, but no such attempt is known to the present author.

§182.   The *shift of equilibrium* by surface forces is amenable to theoretical treatment. If the chemical equilibrium is represented by the equation $A+B = C+D$, and the activities of these components at equilibrium are [A], [B], [C], and [D], then in the absence of surface forces the equilibrium constant $[A][B]/[C][D]$ is K. In the surface layer it is K_1. The relation between the two constants is

$$\ln \frac{K_1}{K} = \frac{S}{RT}(\gamma - \gamma_1); \qquad (12.18)$$

S is the surface area per mole, R is the gas constant, T the absolute temperature, γ surface tension of the mixture in which $[A][B]/[C][D] = K$, and γ_1 is the surface tension when this ratio is equal to K_1. Surface energy shifts the equilibrium so as to lower the surface tension of the liquid.

Since the value of RT at room temperature is about 2.5×10^{10} g·cm^2/sec^2 and the difference $\gamma - \gamma_1$ is unlikely to exceed 10, then $\ln(K_1/K)$ usually will be less than $4 \times 10^{-10} S$, if S is given in cm^2/mole. Thus the equilibrium shift is negligible in usual liquid samples in which S is small. On the other hand, in a foam S may be as great as, say, 10^8 cm^2/mole, and this would give $K_1 = 1.04 K$.

The first quantitative experiments on the equilibrium shift have been conducted by Perrin[133]; unfortunately, their description is very brief. Air was bubbled through 0.2% soap solution. The collapsed foam was acid, while the residue was alkaline and had surface tension about twice as great as the collapsed froth. Soaps of rosin and gamboge behaved similarly to fatty acid soaps. The surface coverage of bubbles in sodium oleate foams was calculated as indicated in §158, and the thickness (see §160) of the oleic acid "crust" on each lamella was between 17 and 21 Å, corresponding to about 7.5×10^{-10} mole/cm^2.

§183.   More extensive results were obtained by Laing[134]. The ratio of acid to base in the foam of sodium oleate solutions varied in the same direction as in the initial solution, but was considerably greater in the froth. Thus it was 1:1 when the solution was 0.047 N in respect to the sodium ion and 0.042 N in respect to the oleate ion, so that the above ratio in the liquid was 0.89. When the solution was 0.05 N as regards both Na^+ and $C_{18}H_{34}O_2^-$, the solid in the froth could be formulated as 1 mole sodium oleate plus 0.61 mole oleic acid. At constant external conditions the accumulation of solid in the foam was more pronounced when the starting liquid was acid or alkaline than when it was neutral.

The hydrolysis of sodium oleate in foam was more marked when the initial solution was more dilute[135]. Thus the mole ratio of acid to base was near 2.7 and 1.5 (in foam) for concentrations (in liquid) 0.0002 and 0.01 N. It is interesting that, when the collapsed froth (from a solution more dilute than 0.002 N) was redissolved in the residual liquid, the surface tension of the resulting liquid was higher than that of the initial liquid instead of being equal to it. Presumably, this paradox was caused by the slowness of colloid-chemical processes in soaps. Similar results have been obtained for sodium laurate[136].

The above ratio was little affected by the nature of the base[137]. Aqueous 0.125% solution of lauric acid was mixed with 0.9 equivalent of NaOH, or 1.12 equivalent of triethanolamine, or 1.28 equivalent of KOH. In the foam of the resulting soap solutions this ratio was 1.27, 1.12, and 0.95. The surface tension of the residual liquid was in all these instances about 1.25 times that of the collapsed foam.

Foaming of 0.4% sodium cholate solution yielded crystals of cholic acid in one of the foam fractions[106].

A green 0.02% solution of Bromothymol Blue afforded a yellow foam and a blue residual liquid[138]; perhaps this also was an example of equilibrium shift caused by surface energy.

References

1. Booth, R. B., in J. J. Bikerman. *Foams*. New York: Reinhold. (1953), p. 243.
2. Abribat, M. *Compt. Rend.* 209:244 (1939).
3. Gray, P. P., and I. Stone. *Wallerstein Lab. Commun.* 3:159 (1940).
4. Ostwald, Wo., and W. Mischke. *Kolloid-Z.* 90:17 (1940).
5. Dognon, A. *Rev. Sci.* 79:613 (1941).
6. Perri, J. M., and F. Hazel. *Ind. Eng. Chem.* 38:549 (1946).
7. Lowe, J. C. U.S. 3 295 688 (1967).
8. Lemlich, R. *Adsorptive Bubble Separation Techniques*. New York: Academic Press. (1971).
9. Weil, I. *J. Phys. Chem.* 70:133 (1966).
10. Newson, I. H. *J. Appl. Chem.* (*London*) 16:43 (1966).
11. Wace, P. F., P. J. Adler, and D. L. Banfield. *Chem. Eng. Progr.*, *Symp. Ser.* 65, No. 91:19 (1969).
12. Rubin, E., and J. Jorne. *Ind. Eng. Chem.*, *Fundamentals* 8:474 (1969).
13. Brunner, C. A., and R. Lemlich. *Ind. Eng. Chem.*, *Fundamentals* 2:297 (1963).
14. Walling, C., E. E. Ruff, and J. L. Thornton. *J. Phys. Chem.* 61:486 (1957).
15. Joos, P., E. Moerman, and R. Ruyssen. *Compt. Rend. XXIe Congr. Intern. Sci. pharmaceutiques* (Pisa, 1961).

16. Lauwers, A., P. Joos, and R. Ruyssen. *Conf. orig. IIIe Congr. intern. détergence, Cologne, 1960*, 3:195.
17. Maas, K., *Separ. Sci.* 4:69 (1969).
17a. Bikerman, J. J. *Separ. Sci.* 7:647 (1972).
18. Grieves, R. B., and R. K. Wood. *A.I.Ch.E.J.* 10:456 (1964).
18a. Goldberg, M., and E. Rubin. *Separ. Sci.* 7:51 (1972).
19. Khrustalev, B. N., V. V. Pushkarev, and E. N. Mikheev. *Zh. Prikl. Khim.* 41:2463 (1968).
20. Grieves, R. B., and S.-L. Wang. *Biotechnol. Bioeng.* 8:323 (1966).
21. Rubin, A. J. *J. Am. Water Works Assoc.* 60:832 (1968).
22. Haas, P. A., and H. F. Johnson. *A.I.Ch.E.J.* 11:319 (1965).
23. Rose, J. L., and J. F. Sebald. *Tappi* 51:314 (1968).
24. Lemlich, R., and E. Lavi. *Science* 134:191 (1961).
25. Schonfeld, E., and A. H. Kibbey. *Nucl. Appl.* 3:353 (1967).
26. Miller, M. W., and G. L. Sullivan. *Separ. Sci.* 6:553 (1971).
27. Lemlich, R. *Ind. Eng. Chem.* 60:16 (1968).
28. Poncha, R. P., and B. L. Karger. *Anal. Chem.* 37:422 (1965).
29. Karger, B. L., R. P. Poncha, and M. M. Miller. *Anal. Chem.* 38:764 (1966).
30. Benson, C. C. *J. Phys. Chem.* 7:532 (1903).
31. Ostwald, Wo., and A. Siehr. *Kolloid-Z.* 76:33 (1936).
32. Kevorkian, V., and E. L. Gaden. *A.I.Ch.E.J.* 3:180 (1957).
33. Kishimoto, H. *Kolloid-Z.* 192:66 (1963).
34. von Zawidzki, J. *Z. Physik. Chem.* 42:612 (1903).
35. Chatalan, J. *J. Chim. Phys.* 20:123 (1923).
36. Grieves, R. B., and D. Bhattacharyya. *J. Am. Oil. Chem. Soc.* 42:174 (1965)
37. Grieves, R. B., C. J. Crandall, and R. K. Wood. *Air Water Pollution* 8:501 (1964).
38. Skomoroski, R. M. *J. Chem. Educ.* 40:470 (1963).
39. Gassett, R. B., D. J. Sproul, and P. F. Atkins. *J. Water Pollution Control Feder.* 37:460 (1965).
40. Jorne, J., and E. Rubin. *Separ. Sci.* 4:313 (1969).
41. Pushkarev, V. V. *Radiokhimiya* 3:498 (1961). *Chem. Abstr.* 56:153 (1962).
42. Pushkarev, V. V., L. D. Skrylev, and V. F. Bagretsov. *Radiokhimiya* 1:709 (1959). *Chem. Abstr.* 54:15003 (1960).
43. Pushkarev, V. V., and V. G. Berezyuk. *Radiokhimiya* 10:497 (1968). *Chem. Abstr.* 69, No. 92090 (1968).
44. Zveverzheeva, V. F., *Gigiena i Sanit.* 26, No. 12:42 (1961). *Chem. Abstr.* 56:10513 (1962).
45. Ito, K., and K. Shinoda. *Kogyo Kagaku Zasshi* 65:1226 (1962). *Chem. Abstr.* 58:1240 (1963).
46. Pushkarev, V. V., V. G. Berezyuk, and B. N. Khrustalev. *Radiokhimiya* 9:438 (1967). *Chem. Abstr.* 68, No. 117600 (1968). V. V. Pushkarev and V. G. Berezyuk. *Radiokhimiya* 10:319 (1968). *Chem. Abstr.* 69, No. 61766 (1968).
47. Schnepf, R. W., et al., *Chem. Eng. Progr.* 55, No. 5:42 (1959).
48. Davis, B. M., and F. Sebba. *J. Appl. Chem.* (*London*) 16:293, 297 (1966).
49. Schoen, H. M., E. Rubin, and D. Ghosh. *J. Water Pollution Control Feder.* 34:1026 (1962).

50. Moroi, Y., and R. Matuura. *Mem. Fac. Sci., Kyushu Univ., Ser. C* 7:51 (1970).
51. von Zawidzki, J. Z. *Physik. Chem.* 35:77 (1900).
52. Ermolenko, N. F. *Discussions Faraday Soc.* 18:145 (1954).
53. Bauer, D. J. *U.S. Bur. Mines, Rept. Invest.* No. 5942 (1962). *Chem. Abstr.* 56:13870 (1962).
54. Koizumi, I. *Kogyo Kagaku Zasshi* 65:1343 (1962). *Chem. Abstr.* 58:5281 (1963).
54a. Onda, K., H. Takeuchi, and M. Takahashi. *Kogyo Kagaku Zasshi* 74:1721 (1971). *Chem. Abstr.* 76, No. 47684 (1972).
55. Dick, W. L., and F. D. Talbot. *Ind. Eng. Chem., Fundamentals* 10:309 (1971).
56. Rubin, A. J., and W. L. Lapp. *Separ. Sci.* 6:357 (1971).
57. Rubin, A. J. *J. Am. Water Works Assoc.* 60:832 (1968).
58. Bhandarkar, P. G., and M. G. Rao. *Indian J. Technol.* 7:65 (1969).
59. Kepák, F., and J. Křivá. *Separ. Sci.* 5:385 (1970).
60. Bikerman, J. J. *Chem. Abstr.* 70, No. 69557 (1969); 72, No. 136252 (1970).
61. Grieves, R. B. *Brit. Chem. Eng.* 13:77 (1968).
62. Iacobelli-Turi, C., S. Terenzi, and M. Palmera. *Ind. Eng. Chem. Process Des. Develop.* 6:162 (1967).
63. Karger, B. L., and M. W. Miller. *Anal. Chim. Acta* 48:273 (1969).
63a. Shveikina, R. V. *Tr. Sverdlovsk. Sel'skokhoz. Inst.* 15:345 (1969). *Chem. Abstr.* 75, No. 26040 (1971).
63b. Grieves, R. B., and D. Bhattacharyya. *Separ. Sci.* 7:115 (1972).
64. Grieves, R. B., and G. A. Ettelt. *A.I.Ch.E.J.* 13:1167 (1967).
65. Grieves, R. B., and D. Bhattacharyya. *J. Appl. Chem. (London)* 18:149 (1968).
66. Mokrushin, S. G., and K. G. Potaskuev. *Kolloidn. Zh.* 18:215 (1956).
67. Shveikina, R. V., and S. G. Mokrushin. *Kolloidn. Zh.* 20:233 (1958).
68. Mokrushin, S. G., and R. V. Shveikina. *Zh. Prikl. Khim.* 31:943 (1958).
69. Shveikina, R. V., and S. G. Mokrushin. *Izv. Vysshikh Uchebn. Zavedenii, Khim. i Khim. Tekhnol,* 1958, No. 4:8. *Chem. Abstr.* 53:3839 (1959).
70. Skrylev, L. D., and S. G. Mokrushin. *Kolloidn. Zh.* 23:304 (1961).
71. Skrylev, L. D., and S. G. Mokrushin. *Zh. Prikl. Khim.* 37:211 (1964).
72. Shveikina, R. V., and S. G. Mokrushin. *Izv. Vysshikh Uchebn. Zavedenii, Khim. i Khim. Tekhnol.* 5:897 (1962). *Chem. Abstr.* 58:13611 (1963).
73. Grieves, R. B., and D. Bhattacharyya. *A.I.Ch.E.J.* 11:274 (1965).
74. Krasnova, G. S., L. D. Skrylev, and S. G. Mokrushin. *Zh. Prikl. Khim.* 41:137 (1968).
75. Grieves, R. B., D. Bhattacharyya, and C. J. Crandall. *J. Appl. Chem. (London)* 17:163 (1967).
76. Grieves, R. B., and R. C. Aronica. *Air Water Pollution* 10:31 (1966).
77. Iacobelli-Turi, C., and M. Palmera. *Gazz. Chim. Ital.* 96:1432 (1966).
78. Karger, B. L., and L. B. Rogers. *Anal. Chem.* 33:1165 (1961).
79. Brady, A. P. *J. Phys. Colloid Chem.* 53:56 (1949).
80. Sporck, C. R. *J. Am. Oil Chem. Soc.* 30:190 (1954).
81. Dubrisay, R. *Bull. Soc. Chim. France* 1953:280.
82. Dubrisay, R. *Compt. Rend.* 194:1076 (1932).

83. Otero Aenlle, E., and S. Garcia Fernandez. *Anales fis. y quim.*, Ser. B, 44:191 (1948). *Chem. Abstr.* 42:7549 (1948).
84. Gomez Errera, C., and R. Guzman Garcia. *Grasas aceites* (Seville, Spain) 5:5 (1954). *Chem. Abstr.* 48:10360 (1954).
85. Chemische Werke Hüls G.m.b.H., German 825258 (1954). *Chem. Abstr.* 49:7272 (1955).
86. Bieber, H., I. A. Eldib, and M. A. Weiss. U.S. 2 998 382 (1961).
87. Evans, H. C., and S. R. Epton. *1er Congr. Mondial Détergence Prod. Tensio-Actifs, Paris 1954,* 1:105. *Chem. Abstr.* 51:10928 (1957).
88. Hargis, L. G., and L. B. Rogers. *Separ. Sci.* 4:119 (1969).
89. Prins, A. *Rec. Trav. Chim.* 82:329 (1963).
89a. Zwierzykowski, W., et al., *Przemysl Chem.* 50:40 (1971). *Chem. Abstr.* 75, No. 24936 (1972).
90. Devin, C., and M. Minfray. *Compt. Rend.* 255:116 (1962).
91. Imai, K., and M. Matsumoto. *Bull. Chem. Soc. Japan* 36:455 (1963).
92. Bolewski, K., T. Tomaszkiewicz, and M. Olbracht. *Polimery* 15:15 (1970). *Chem. Abstr.* 73, No. 45943 (1970).
93. Gaines, G. L., and D. G. LeGrand. *J. Polymer Sci.*, Part B6:625 (1968).
94. Skopintsev, N. A. *Dokl. Akad. Nauk SSSR* 18:53 (1938). *Chem. Abstr.* 32:8855 (1938).
95. Yamabe, Y., and N. Takai. *Seisan-Kenkyu* 21:530 (1969). *Chem. Abstr.* 71, No. 128532 (1969).
96. Sisko, W. J., and B. H. Simmons. U.S. 3 413 220 (1968).
97. Ostwald, Wo., and A. Siehr. *Kolloid-Z.* 79:11 (1937).
98. Siebert, K. *Ernährungsforschung* 2:127 (1957). *Chem. Abstr.* 51:18664 (1957)
99. Dole, R. A., and J. W. McBain. U.S. 2 162 379 (1939).
100. Spengler, O., and W. Dörfeldt. *Z. Wirtschaftsgruppe Zuckerind.* 92:279 (1942). *Chem. Abstr.* 37:6923 (1943).
101. Meyer, W. *Centr. Zuckerind.* 48:449 (1940). *Chem. Abstr.* 36:5375 (1942).
102. Davis, S. C., C. R. Fellers, and W. B. Esselen. *Food Technol.* 3:198 (1949).
103. Kodak-Pathé, French 859 735 (1940). *Chem. Abstr.* 42:3225 (1948).
104. Kodak-Pathé, French 942 087 (1949). *Chem. Abstr.* 44:10391 (1950).
105. Barmore, M. A. *Colo. Agr. Expt. Sta.*, Tech. Bull. 9 (1934).
106. Bader, R., and F. Schütz. *Trans. Faraday Soc.* 42:571 (1946).
107. Kauffmann, F. L., and W. M. Urbain. U.S. 2 492 580 (1949).
108. Holmström, B. *Biotechnol. Bioeng.* 10:551 (1968).
109. Ostwald, Wo., and W. Mischke. *Kolloid-Z.* 90:205 (1940).
110. Bader, R., F. Schütz, and M. Stacey. *Nature* 154:183 (1944).
111. Andrews, G., and F. Schütz. *Biochem. J.* 39:LI (1945).
112. London, M., M. Cohen, and P. B. Hudson. *J. Am. Chem. Soc.* 75:1746 (1953).
113. London, M., M. Cohen, and P. B. Hudson. *Biochem. Biophys. Acta.* 13:111 (1954).
114. Charm, S. E., et al., *Anal. Biochem.* 15:498 (1966).
115. Courrier, R., and A. Dognon. *Compt. Rend.* 209:242 (1939).
116. Ruska, H. *Kolloid-Z.* 110:175 (1948).
117. Gaudin, A. M., N. S. Davis, and S. E. Bangs. *Biotechnol. Bioeng.* 4:211 (1964).

118. Rubin, A. J. *Biotechnol. Bioeng.* 10:89 (1968).
119. Marinkov, N., and M. Vodenicharov. *Tekstilna Prom.* (*Sofia*) 8, No. 8:23 (1959). *Chem. Abstr.* 54:6158 (1960).
120. Grieves, R. B., and R. K. Wood. *Chem. Can.* 15, No. 5:13 (1963).
121. Kresta, V., and M. Koubik. *Vyzkum. Ustav Vodohospodarsky, Prace a Studie* 121 (1968). *Chem. Abstr.* 71, No. 24596 (1969).
122. Kondratavicius, V. *Kozh. Obuv. Prom.* 11:15 (1969). *Chem. Abstr.* 70, No. 108994 (1969).
123. Laskov, Yu. M., and V. Kondratavicius. *Izv. Vysshikh Uchebn. Zavedenii, Stroit. i Arkhitekt.* 12. No:4, 141 (1969). *Chem. Abstr.* 71, No. 73834 (1969).
124. Michelsen, D. L. *Text. Chem. Color.* 1:179 (1969).
124a. Klimenko, N. A., and N. P. Panchenko. *Khim. Tekhnol.* (*Kiev*) 1971, No. 5: 54. *Chem. Abstr.* 76, No. 49549 (1972).
125. Klotter, H. E. *Vorträge Originalfassung Intern. Kongr. Oberflächenakt.*, 3, Cologne, 1960, 3:302 (1961).
126. Eldib, I. A. *Am. Chem. Soc.*, *Abstracts of Papers* 143, p. 7E.
127. M. Raison in a book *La pollution des eaux*.
128. Brunner, C. A., and D. G. Stephan. *Ind. Eng. Chem.* 57:40 (1965).
129. Rose, J. L., and J. F. Sebald. *Ind. Water Eng.* 4, No. 11:36 (1967).
130. Garrett, D. E. U.S. 3 448 044 (1969).
131. Rubin, E., and R. Everett. *Ind. Eng. Chem.* 55, No. 10:44 (1963).
132. Jenkins, D. *J. Water Pollution Control Feder.* 38:1737 (1966).
133. Perrin, J. *Ann. Phys.* [9] 10:180 (1918).
134. Laing, M. E. *Proc. Roy Soc.* (*London*) A109:28 (1925).
135. Raison, M., *Compt. Rend.* 232:1660 (1951).
136. Raison, M. *Mém. Serv. Chim. État* (*Paris*) 37, No. 1:65 (1952). *Chem. Abstr.* 47:10874 (1953).
137. Dubrisay, R., and P. Drouot. *Compt. Rend.* 196:1392 (1933).
138. Thiele, H. *Kolloid-Z.* 118:172 (1950).

CHAPTER 13

OTHER APPLICATIONS

§184. Radius of molecular action. Utilization of soap films to solve scientific questions was inaugurated by Plateau[1]; the question was, what is the radius of the "sphere of molecular action" which appears in Laplace's theory of capillarity. If the foam lamella is so thin that the forces emanating from the molecules adjacent to one surface still have a significant value at the opposite surface, then, according to Laplace, the experimental surface tension should be smaller than that, γ, of thicker lamellae. Suppose that a lamella thickness δ_m is found such that the surface tension of thicker films has the usual, and that of thinner lamellae, an abnormally low value of surface tension; then the above radius is equal to $0.5\delta_m$.

Plateau blew soap bubbles at one end (A) of an open vertical U tube half-filled with water. The meniscus in the other branch (B) stood higher than in branch A because the pressure above the meniscus in A was by $4\gamma/R$ greater than the atmospheric (in B); R is the bubble radius. The bubble was permitted to drain, and its thickness from time to time was measured by optical means (see §123). If at any time before bubble collapse the meniscus in A would rise, this would indicate that the γ decreased and the wall thickness reached δ_m. This did not happen. The thinnest bubble wall observed was $0.114\ \mu$ thick, so that the radius of molecular action proved to be shorter than 570 Å.

Similar experiments have been performed by G. L. van der Mensbrugghe (1866, 1867). Reinhold and Rücker[2] blew two soap bubbles at the opposite ends of a glass tube. One of the bubbles was allowed to drain, while the liquid in the other was frequently replenished. It was expected that, as soon as the wall thickness of the first bubble descended below δ_m, there would be a difference between the two capilary pressures, and some air transport would be noticed. No motion of this kind occurred. The thinnest lamellae were only 120 Å thick (optically determined). The sensitivity of the instrument was claimed to be such that 0.5% change in γ would be detected; thus the molecular force at the distance of 60 Å was still negligible.

Drude[3] used curved soap films as a magnifying glass. When the film was old enough to show patches of different colors, including black, he viewed an object through different film areas. The magnification was identical. Consequently, the curvature was independent of thickness. This implies a constant γ because there can be only one value of the capillary pressure $4\gamma/R$ inside the bubble. The thinnest spot employed in this work was 170 Å thick.

Johonnot[4] compared the pull exerted by a thick cylindrical film (connecting a horizontal ring with the bulk liquid) with the pull of two black films (120 and 60 Å thick) and found no difference exceeding 2% of the normal value. These tests are similar to those reviewed in §20.

   §185. *Immobilization in foam.*   Presumably, fighting fires with foams is the most important instance of this use. In this application, foam is mainly a barrier preventing oxygen of the ambient air from reaching the burning materials and retarding the escape of combustible vapor into the atmosphere. An identical result can be achieved, for instance, by using a blanket, but foam is lighter, is made *in situ*, and consists mainly of an inexpensive material, namely water. It was supposed that the cooling effect produced by the evaporation of water from the foam lamellae was an essential factor in the action of foam fire extinguishers, but this view seems to be incorrect. It was also believed that filling the bubbles with carbon dioxide or another nonflammable gas would make the froth much more efficient than one filled with air, but the difference seems to be insignificant. The extensive literature on fire-fighting foams has been reviewed earlier[5], but is left out of this monograph.

   The idea of retarding the evaporation of a liquid by foam was patented at least twice[6,7]; the liquids considered in the patents were petroleum derivatives and fractions. A related idea is to use foam as a stopper to block the escape of natural gas from underground storage reservoirs. The "stopper" is generated *in situ* and thus is easier to apply than any preformed closure[8]. A model of a reservoir was constructed from sandstone, later coated with a plastic material[9]; when a solution of a foaming agent [e.g., $C_8H_{17}\cdot C_6H_4\cdot O(CH_2CH_2O)_nH$] was injected into the stone, the permeability of the later was reduced (in favourable instances to 0.01 of the initial value) and remained low for a time which was determined by the rate of adsorption of the surfactant by the rock. The pressure gradient needed to start gas flow through a similar system was determined for four surfactants at several concentrations and for several sandstones and unconsolidated sands[10].

   According to a patent[11], the flow of an unwanted liquid into a petroleum well can be prevented by interrupting production or drilling, and injecting a foaming agent into the bore; again, the foam formed there acts as a plug. If natural gas streams in an undesirable direction, its path can be blocked by a foam plug and the stream diverted into other paths[12,13].

   Fluxes used in galvanizing may contain frothers (sawdust, tallow, etc.).

The function of the latter seems to be to retard the evaporation of the flux (usually, ammonium chloride) and to hinder the spattering of molten zinc. Pentaerythrol is claimed[14] to give stable foams without unduly contaminating the flux.

When cleaning liquids are applied as foams, the unwanted mixing, overspraying, and contact with human skin are readily avoided; mixtures of sodium dodecylbenzene sulfonate, sulfonated sodium oleate, etc., are employed as foaming agents[14a].

§186.   Fumigants, that is, vapors or gases toxic to fungi, objectionable insects, etc., are readily blown away by the wind, and liquid pesticides often drain down the plant before completing their task. To keep them in place, they are used as the gas or the liquid phase in foams. For instance, plants can be sprayed with foams produced by injecting air into an aqueous solution of calcium cyanide and saponin[15,16]. Methyl bromide, a soil fumigant, is immobilized by foams covering the soil; as their foaming agents, saponin, methyl cellulose, gelatin, or shaving soap may be employed[17]. Rat poison was applied in foam form[18].

Nicotine sulfate, which is an insecticide, can be incorporated in a foam obtained by mixing sodium carbonate or bicarbonate with tartaric acid or aluminum acetate, both components being distributed in a matrix of gum tragacanth and glycerol[19]. Karreth[20] claimed application of insecticidal foams to human skin; foam formation occurs when the skin raises the temperature of the salve. Gelatin is the stabilizer, and the gas is liberated in a reaction between tartaric acid or aluminum sulfate with sodium carbonate or sodium bicarbonate, or between tartaric acid and sodium sulfide, or between hydroxylamine and sodium nitrite.

Substances capable of destroying poison gases achieve a more complete decontamination when their foams are spread over the area attacked. Thus, oxidizing salts such as $NaClO$, $NaClO_2$, or $CaCl(OCl)$ are dissolved in water, together with a soap or a synthetic surfactant, the solution is transformed into foam, and this is used to blanket the area[21].

A jelly containing a drug, sodium carbonate, and gum karaya is mixed with a similar jelly of gum karaya and glycoboric acid. The foam, whose gas phase is carbon dioxide, is injected into a body cavity and continues to act there for a long time[22]. Contraceptive foams have been patented in Europa[23] and America[24].

§187.   Apparently, it is not known what is the main function of shaving lather. Does it act mainly as a lubricant? Maybe its task is to immobilize water so that hair has more time to swell? It is also possible that the rigidity of the foam is utilized; the flexibility of each hair is reduced by surrounding it with a resisting medium. At any rate, at present, large amounts of shaving lather are produced as "aerosol foams". This term is very unfortunate because aerosol in physical chemistry means a fine suspension of solid or liquid particles in a gas and because the same word has already been misused as a trade name for a group of synthetic surfactants. Pressurized package foams would be a clearer designation. Usually, a can is filled with an emulsion, whose dispersion medium is an aqueous solution (which will become the liquid phase of the foam) and disperse phase consists of a liquid ("propellant") whose boiling point at the atmospheric pressure is below the range of room temperatures. Consequently, this liquid and the whole content of the can are under strong compression. When the can is opened, the dispersed liquid volatilizes and pushes the aqueous phase out of the opening as a foam.

In the common pneumatic method of foam formation, the gas pressure and the gas velocity are kept constant so that nearly identical foams are produced one after the other. The internal pressure in the can rapidly decays during the gradual emptying, and the properties of the foam change in the process. Foams obtained immediately after the can is opened have a lower density (that is, a higher expansion factor) than those collected when the can is nearly empty. For the mean density $[\rho]$ and the mean expansion factor $1/\varphi$, approximate equations are easily derived. Let 1 cm^3 of the can content consist of x cm^3 of aqueous solution and $(1-x)$ cm^3 of propellant. If the density and the molecular weight of the latter are ρ and M, and if the vapor of the propellant may be considered to be a nearly perfect gas, then the volume of the vapor (given by the propellant) is approximately $22,000\rho(1-x)/M$ cm^3 and

$$\varphi = \frac{Mx}{22,000\rho(1-x)+Mx} ; \qquad (13.1)$$

similar equations can be found in Sanders' book[25].

1,2-Dichloro-1,1,2,2-tetrafluoroethane is a popular propellant; its M is 171 and its ρ is 1.53. Consequently φ should be 0.13 at $1-x = 0.033$ and 0.017 at $1-x = 0.395$; and similar values really are quoted in the above monograph. The fact that organic vapors rather than air or nitrogen constitute the gas phase in the foam does not alter some of the foam's properties; for instance, the rate of drainage of foams stabilized by sodium dodecyl sulfate was drastically reduced when a comparable amount of 1-dodecanol or 1-tetradecanol was added to the sulfate (see §99).

Also three-phase foams can be produced from pressurized containers. A solid derivative of 1-octadecanol is soluble in some mixtures of water, ethanol, and propellants of the type mentioned in the preceding paragraph. However, when the propellant evaporates, the solid is precipitated and acts as a solid stabilizer. Apparently, nothing is known of the wettability of this solid.

The liquid phase in the foam need not be aqueous. The above monograph[25] contains, for instance, the following recipe: mineral oil (which is the solvent in the future liquid phase of the foam) 78.3%, lanolin 0.5%, hexadecanol 3.0%, octadecanol 3.0%, psysiologically active ingredients 0.2%, and propellant 15.0%; presumably, lanolin and the higher alcohols act as foaming agents in this system.

§188. *Low density of foam.*   The large expansion associated with foam formation has been used for years in pyrotechnics. "Black nonmercury snakes" consist, for instance, of picric acid, which gives a large volume of gas on ignition, and pitch (plasticized with linseed oil), which supplies the material for the foam lamellae[26].

Because of this low density, it is easier to achieve a uniform coating by means of foams than by means of a liquid. For instance, if a solid is covered with a foam layer five bubbles thick and the thickness of each lamella is $1\,\mu$, the thickness of the coating after the decay of the foam is only $5\,\mu$. It would be more difficult to spread a liquid in a uniform layer of $5\text{-}\mu$ thickness. This artifice was employed, for instance, for applying adhesives[27]. An aqueous melamine-formaldehyde solution (100 parts) was agitated with 1 part of saponin, 1 part of ammonium chloride, and 8 parts of water until the volume was approximately doubled. One volume of this foam (or gas emulsion) could be substituted for one volume of the initial solution, thus effecting a saving of 50%. The foam was stable for several days and was not destroyed by heating up to 100°. A special agitator for adhesive foams has been patented later[28].

Uniform webs of fibers are obtained when a fiber suspension is transformed into foam and the foam is filtered through a screen[29,29a].

To achieve a uniform etching, the corrosive solution is transformed into foam, and the detail to be treated is immersed in the latter[30,30a]. Pickling in foam is very similar. A mixture of 4 kg sulfuric acid, 0.9 kg nitric acid, 6 g of surfactant $C_8H_{17}\cdot SO_2\cdot NHCH_2CH_2N(CH_3)(C_2H_5)_2I$, and 230 g of $CClF_2\cdot CClF_{22}$ was heated to 80°. The fluorine compound volatilized and formed, as in §187, the gas phase of the foam. A corroded metal plate was lowered into the froth and there cleaned by the acid mixture. Then the foam was cooled to 16°, returned to the first vessel as a liquid, and reused[31]. The action of a photographic developer was rendered more uniform by transforming the liquid into foam and spreading this over the negative; several frothers were found suitable to stabilize these foams[32]. See also §191.

Wells of natural gas or petroleum often become waterlogged, and the tall column of water or brine prevents further removal of oil or gas. Dunning[33] suggested to render the aqueous liquid lighter by transforming it into foam. When suitable foaming agents are injected in the well, the gas escaping from the rock (or formation) gives rise to a froth. "The lightened column then is lifted from the well by gas pressures too low to lift a column of water. Furthermore the foam is quite rigid and by capturing the gas prevents it from bypassing water in large casings." Usually, nonionic surfactants of the type $R\cdot C_6H_4\cdot O(CH_2CH_2O)_nH$, R being an aliphatric radical, have been employed, and in some instances the gas production rate was doubled. A similar treatment was advocated by Lissant[34].

Some well brines are hot; but a surfactant efficient at a temperature as high as 98° was found[35]. In some wells, addition of alcohols to the foaming agent was helpful[36]. The use of quaternary ammonium salts was found successful later[37]. Some brines are highly concentrated and depress the foaming ability of many common surfactants, but some non-ionic agents are active in these liquids also[38]. If there is not enough natural gas to give rise to a foam, gas in drowned wells can be evolved by injecting sodium nitrite and sulfamic acid (which together react to liberate nitrogen) or a similar pair of compounds[39].

The liquid blocking a well may be an oil or a mixture of water (brine) and oil. Consequently, a versatile foaming agent should be active in both media. Several suitable materials have been tested by Anderson[40]; for instance, when 2.5 parts of β-oleylamino-propionic acid were dissolved in kerosene 490 plus water 10 parts, and the mixture was foamed for 20 min, 340 parts of the liquid was transformed into foam. Several additional surfactants have been patented[41] for use in gas and petroleum wells. Surfactants fortified with a polymer are mentioned by Holman[42].

In the laboratory of the present author, oil slicks on water were transformed into foam by adding an oil-soluble surfactant and releasing a stream of bubbles under the oil patch; the foam could be removed by suction with only a small admixture of water.

§189. *Large interface.* The rate of dissolution of a gas in a liquid or of volatilization of dissolved gas from a liquid usually is proportional to the extent of the gas-liquid interface. This area in a foam may be easily 100 or 1000 times as great as in bulk liquid. Hence, foams ought to be suitable systems for rapid mass transfer from gas to liquid or liquid to gas. This was observed long ago[43] for gaseous nitrogen oxides and aqueous solutions of hydrogen peroxide, potassium iodide, and so on.

Later, an extensive investigation of this problem was performed by Pozin and his collaborators; a part of their findings was summarized in a Russian book[44]. Some of the advances achieved are very impressive, but it is doubtful whether they should be reviewed in a book on foams. The foam-like systems studied by Pozin contained no (deliberately added) foaming agent and were obtained by bubbling a gas through a liquid at a rate between 1 and 5 m/sec, that is, higher than the rate of rise of usual bubbles in usual liquids. The thickness of the initial liquid layer, as a rule, was so small that the liquid was all lifted in the turbulent mixture with the gas injected. Perhaps, "gas emulsions in turbulent régime" would be a better description for these systems than "foams". Of the later papers of this group, only two[45,45a] can be referred to here.

A real foam was used for absorbing sulfuric acid mist evolved in the process of concentrating sulfuric acid[46]. "Green acid", which is the sludge obtained in the customary treatment of petroleum with concentrated sulfuric acid, was the foaming agent. Mists containing 1 to 3 mg/l. of H_2SO_4 had to spend 10 sec in the froth to lose 90% of their acid content. More concentrated mists (e.g., 30 mg/l.) needed a longer sojourn in the foam (e.g., 12 sec or more).

The rate of absorption of diluted sulfur dioxide gas in a foam of an aqueous solution of basic aluminum sulfate was lowered five- or ten-fold when the foam was broken down by ether; apparently the foaminess of the initial liquid was considerable also when no frother was deliberately added[47]. When the initial liquid layer was 2 cm thick, the tallest foam column was 19.5 cm at the linear rate of air injection equal to 4 cm/sec. Thus the Σ (§53) of the liquid was about $19.5/4 \approx 5$ sec. A large-scale foam absorber of sulfur trioxide is described by Žurakowski[48]. It is not clear whether real foams or turbulent gas emulsions were used for absorbing carbon dioxide from air bubbling through ethanolamine or sodium hydroxide solutions[49] or ethanolamine solutions[50], or for absorbing moisture in sulfuric acid[51,52].

Drying and degassing of oils in a vacuum are accelerated by adding a foaming agent to the oil[53] because the froth has an extensive gas-liquid boundary; the high velocity of foam drying was already pointed out above for instance, in §168. The effect of foam height on the evaporation of water containing no (deliberately added) frother was measured by Rodionov[54]. Desorption of carbon dioxide and ammonia from their aqueous solutions was more rapid when 0.2% saponin was dissolved in the water and air was injected; the foam moved in a horizontal channel[55]. Coffee and tea extracts, instead of being spray-dried, can be transferred into foams by injecting carbon dioxide, after which the lamellae are dehydrated[56]. These foams, however, are very near to gas emulsions; their density usually is between 0.1 and 0.7 g/cm^3.

§190.   When a solid aerosol, that is, a suspension of fine solid particles (smoke, dust) in a gas, is bubbled through a liquid, the degree of interception of the particles usually is very low, unless the liquid is capable of foaming. Interception of aerosols in foams is a relatively efficient process. This may be due simply to the fact that a bubble exists in foam longer than in a nonfoaming liquid; consequently, every dust particle has more time to reach the gas-liquid interface and to become attached to the liquid wall by capillary forces. A support for this explanation is supplied by the observation that the removal of dust of ferrous sulfide and barium sulfate in a foam of aqueous sodium oleate is more complete, the taller the foam layer[57].

Dust interception in the turbulent gas emulsions of §189 frequently is satisfactory also in the absence of foaming agents[44] but the percentage of removal (of carbon dust) could be raised by adding 0.05% of a frother[58]. Gases formed in the hydrolysis of wood generally carry pitch mist whose particles are about 0.1 μ across. These particles can be entrapped in a foam in which pitch oil acts as a stabilizer[59].

In coal mining, dust can be intercepted before it becomes air-borne[60]. The coal face is coated with a foam layer, and drilling and cutting are performed through this foam. Many frothers can be used in this operation, for instance, sodium alkylbenzene sulfonate fortified with poly(vinyl alcohol). Rozenberg[61] studied the persistence of several of these foams. A foam scrubber for metallurgical dusts was described by Erler[62].

Pure water spread over a dusty plate removed a smaller percentage of dust than did the equal volume of 1% surfactant solution. When this solution was transformed into foam (of an expansion factor 30) and the foam was placed on the plate, an almost quantitative removal of the dust occurred[63]. In this instance, foam demonstrated a powerful detergent action. In general, however, no definite relation seems to exist between detergency and foaminess; see also §153. Non-ionic surfactants on the whole give less stable foams than those of anionic surfactants but, as detergents, the former often are superior to the latter[64].

§191. Chemical reactions between a gas and a liquid often are particularly rapid when the two compounds are components of a common foam. An example is supplied perhaps by the observation that, when air and iodine vapor were bubbled through 2.5% solution of ammonium dodecyl sulfate, over 8% of iodine was absorbed in 8 min; but when the liquid contained also 0.5% sodium thiosulfate, then 72% of the initial iodine was intercepted[65].

The oxidation of paraffin wax by air in foams has repeatedly been studied. A catalyst, such as manganese naphthenate, is dissolved in the molten wax, the solution is heated to, for instance, 150°, and air is bubbled through it at a volume rate of, for instance, 3 cm^3/sec; the yield of fatty and hydroxy acids is improved by foaming[66]. Oxidation in foam required, to achieve equal yields, smaller amounts of oxygen, and the average molecular weight of the acids obtained was greater than in the absence of foam[67]. To raise the foaminess of molten wax, oxidized paraffin wax or a long-chain alcohol (2 to 5% of the melt) can be added; and the air volume reacting at any time with the liquid may in the foam be up to 140 times as great as in a gas emulsion[68]. The oxidation of paraffin wax foams can be performed so that a high yield of primary alcohols and a low yield of carbonyl and carboxyl compounds and esters are achieved, all in a shorter time than without foaming[69]. Equipment for foam oxidation of paraffin wax was described by Nesmelov[70]. Unfortunately it is not sure whether the systems described in the above papers deserved the designation of foams; apparently their expansion factor frequently was as low as 1.8; the specific surface area of this gas emulsion was only 8 cm^{-1}, so that the mean bubble diameter was about 0.3 cm.[71] Expansion factors between

1.5 and 4.4 were obtained by Kuznetsova[72]. Air was injected at 100°
into toluene containing 0.05% cobalt stearate and large percentages of
benzaldehyde and benzoyl benzoate; and the yield of benzaldehyde
increased with the extension factor.

A combination of a foaming agent with an oxidant apparently was
achieved in 1, 4-naphthoquinone-2-sulfonic acid; aqueous foams of this
compound intercepted and destroyed hydrogen sulfide diluted with air[73].

Liquid foams seemingly do not burn as such; bubbles in a kerosene
foam burst before the flame front reached them[74].

The two following examples could have been described in §188. Foam
was used for bleaching and dyeing textiles. For instance, a foam, whose
gas phase contained chlorine or whose liquid phase contained a dye, was
forced through a fabric[75]. This procedure, it is claimed, resulted in a more
uniform change of the fabric, in enhanced penetration of the active
material into the fibres, in decreased time of treatment, and in less damage
caused by handling. Another method of dyeing employs foam on boiling
liquids[76]. A surfactant is added to the dye solution and the liquid is
heated to boil. The cotton fabric to be dyed is placed into the foam.

Lipolysis of milk is said to be accelerated by foaming; this enzymatic
reaction was suppressed by a silicon foam inhibitor[77].

Foam can help in preparing crystals of desired size. In the dissolution
of phosphate rock by sulfuric acid, a foam layer usually is obtained. If
additional sulfuric acid is spread over the surface of this layer, the hemi-
hydrate crystals, $CaSO_4,0.5\ H_2O$, precipitated are bigger and more easily
filtered than when the acid is injected below the foam or both acid and
rock slurry are spread over it[78]. This method was tested in a pilot plant, and
it was found that the foam layer had to be 35 to 50 cm tall. If the "natural"
foam was too voluminous, it was partly collapsed by a sulfonated oleic
acid[79].

§192.   Spreading of oil over the extensive surface of a foam may
have some relation to the fact that foams sometimes displace petroleum
from porous rock more completely than air or water do[80]. One of the
difficulties encountered in the practical application of this observation
was that the foam acted as a plug (see §188), thus "making subsequent fluid
injection impracticable." It was found necessary to introduce the surfactant into the reservoir and then to displace it (by a liquid free of surfactant) away from the injection well[81]. The gas in the foam may be carbon
dioxide[82] or natural gas[81].

When petroleum was displaced from an unconsolidated core (in this
instance a column of sand, grain size about 0.25 mm) by water alone, the
recovery was 43%. When 1% aqueous solution of sodium dioctyl sulfosuccinate was substituted for water, the displacement was 55% effective; this
was detergency, as in §190. However, when air foam of the above solution
was the displacing medium, 95% of the oil was recovered; this was an

effect specific for foam[83]. Only dense foams were suitable; 3 was the highest admissible expansion factor.

§193. Heat insulation.   Because the gas content of one bubble is separated from that of the neighboring bubble, convection effects in foams are weak; see §122. Consequently the heat conductance of foams is very low. This property is utilized on a large scale in solid foams but occasionally also liquid foams are employed as heat insulators. Thus, frost injury to cranberry or tomato plants can be prevented if a foam of saponin 5 + methyl cellulose 7.5 + water 1000 parts is spread over the plants as soon as the ambient temperature descends below the freezing point of water[84].

The heat-insulating properties are important also for fire-fighting foams. Several common foaming agents have been compared from this viewpoint by Perri[85]. Mixtures of hydrolyzed protein and a wetting agent, and sodium alkylarene sulfonates performed best; non-ionic surfactants of the type $R \cdot C_6 H_4 \cdot O(CH_2 CH_2 O)_n H$ were the next, and hydrolyzed protein alone and sodium alkyl sulfates occupied the last places.

§194. Foams in instrumentation and analysis.   It is well known that foam lamellae can be utilized for measuring the rate of gas flow. Usually the gas stream is separated into two currents. The main current continues in a wide tube, while a small aliquot of it is made to bubble through a soap solution and then to push one or several films along a narrow tube. The velocity of advance of these films is a measure of the gas flow in the mainstream as long as the diameters of the two tubes are given.

Gooderman[86,87] applied this technique to gas analysis. The gas mixture bubbles through a surfactant solution. The rate of sliding of the surfactant lamellae in a tube is measured. Then the gas passes through a reagent, which removes one of its constituents, and through another surfactant solution. The new rate of travel of foam films indicates the volume remaining, that is, also the volume removed. This sequence is repeated with another absorbent, and so on.

Volumes of a stationary gas can be determined by using foam films as "retaining walls." For instance, magnesum was precipitated as the carbonate $MgCO_3 \cdot (NH_4)_2 CO_3 \cdot 4H_2 O$ and the precipitate was decomposed with sulfuric acid in the presence of albumin. The carbon dioxide liberated was immobilized in foam, and the gas volume could be found from the volume of the former; and the magnesium amount in the initial sample was calculated from the gas volume. Analogously, ammonia was oxidized in an alkaline bromine solution to nitrogen. The solution contained saponin, nitrogen became the gas phase in a saponin foam, and the original amount of ammonia was calculated from the volume of this phase[88].

Probably the oldest utilization of foam for chemical analysis is the determination of total hardness of water, attributed to T. Clark (1847). It is based on the observation that solutions of sodium soaps do, and

solutions or suspensions of calcium soaps do not foam. The unknown
liquid is titrated with a sodium soap solution until stable foam can be
produced by shaking. The amount of soap needed to achieve this endpoint
depends on the amount of calcium present in the liquid.

In a "standard" method[89], a solution of 100 g Castile soap in 1 l. of
80% ethanol is diluted with aqueous ethanol or methanol until 6.40 cm^3
of the diluted liquid is needed to achieve permanent foam after shaking
with a solution of 0.004 g calcium carbonate in 50 cm^3 of water. "Permanent" means that "a lather remains unbroken for 5 min over the entire
surface of the water while the bottle (of 250 cc) lies on its side". Total
hardness is defined as parts of calcium carbonate in million parts of liquid;
thus the above solution has the total hardness of 80. If the total hardness of
a sample is x, it cannot be calculated simply by the equation $x/80 = X/6.40$,
if X is the volume (cm^3) of the above Castile soap solution used to titrate
50 cm^3 of the sample. This is so because approximately 0.7 cm^3 of the
soap solution is needed to achieve "permanent" foam on 50 cm^3 of water.
Consequently,

$$x = \frac{80X}{6.4-0.7} = 14X$$

or 1 cm^3 of the specified soap solution corresponds to 14 ppm of calcium
carbonate.

It may be simpler to titrate calcium carbonate or other calcium salt
solutions of various concentrations and to construct a calibration curve
with which then the result obtained with the unknown sample is compared. Presence of magnesium renders the titration data difficult to inter-
pret. This difficulty and the availability of other reagents for calcium
rendered the foam method obsolete.

The persistence of foam formed by shaking maize flour with water may
be three times as long with fresh maize as when the corn was spoiled by
long storage; thus foam stability can serve as an indicator of maize quality[90.]

The ratio of foaminess of urine to the value $\rho - 1$ (ρ being the specific
gravity) of the latter is particularly high when the urine contains small
amounts of bile degradation products or of proteins and may, in the future,
acquire some importance for diagnosis. When the urine has a high concentration of surface-inactive solids (as in diabetes), the above ratio is
abnormally low. The normal values of vt/V (see §53) lie between $2+200$
$(\rho-1)$ sec and $7+400(\rho-1)$ sec for 10 cm^3 samples; v is the volume of the
pneumatic foam and V is the gas volume injected in t sec^{91}.

§195.   A foam lamella can indicate pressure differences. The top of a
drum was closed by a soap film. When the drum (whose side walls contained openings) was rotated, the inside pressure decreased and the film
bulged inward. The pressure difference across the membrane was calculated
from the curvature[92].

Dewar[93] employed soap films for acoustical measurements. Various sounds (and air movements in general) gave rise to characteristic patterns in vertical lamellae.

The hydrostatic foam pressure (§115) was recommended for controlling such processes as boiling of milk or saturation of limed sugar juices with carbon dioxide[94,95]. When this pressure exceeds a predetermined value, it operates a device which depresses the heat supply (in the instance of boiling) or the rate of gas injection (in the instance of saturation). In an analogous arrangement, when the foam in the reaction vessel reaches an excessive height, it activates a float which enlarges the outlet tube for the reaction product and thus acts as a safety valve[96].

References

1. Plateau, J. *Mém. Acad. Roy. Sci. Belg.* 33 (1861), 5th series.
2. Reinold, A. W., and A. W. Rücker. *Trans. Roy. Soc. (London)* 177:627 (1886).
3. Drude, P. *Ann. Physik.* [3] 43:158 (1891).
4. Johonnot, E. S. *Phil. Mag.* [6] 11:746 (1906).
5. Perri, J. M., in J. J. Bikerman. *Foams.* New York: Reinhold. (1953), p. 189.
6. Fisher, G. M. U.S. 1 985 491 (1935).
7. Faintsimmer, R. Z., et al., U.S.S.R. 112 921 (1958). *Chem. Abstr.* 53:3679 (1959).
8. O'Brien, L. J. U.S. 3 379 260 (1968).
9. Bernard, G. G., and L. W. Holm. *Soc. Petrol. Engrs. J.* 10:9 (1970).
10. Albrecht, R. A., and S. S. Marsden. *Soc. Petrol. Engrs. J.* 10:51 (1970).
11. Bernard, G. G. U.S. 3 330 351 (1967).
12. Leach, R. O. U.S. 3 460 623 (1969).
13. Froning, H. R. U.S. 3 464 491 (1969). See also N. F. Ustinov, *Nefteprom. Delo* 1971, No. 11:38. *Chem. Abstr.* 76, No. 61511 (1972).
14. Vana, C. A. U.S. 2 481 457 (1949).
14a. W. R. Grace & Co., Britain. 1 162 039 (1969). *Chem. Abstr.* 76 No. 5178 (1972).
15. Loewenstein, L. French 849 855 (1939). *Chem. Abstr.* 35:7638 (1941).
16. Établissements Vermdrel, French 861 352 (1941). *Chem. Abstr.* 42:6052 (1948).
17. Thiegs, B. J. U.S. 2 916 855 (1959).
18. Zatsepin, V. G. *Tr. Vses. Nauchn.-Issled. Inst. Vet. Sanit.* 25:357 (1966). *Chem. Abstr.* 69, No. 26286 (1968).
19. Karreth, A. Britain. 379 892 (1931). *Chem. Abstr.* 27:3295 (1933).
20. Karreth, A. Britain. 368 423 (1929). *Chem. Abstr.* 27:3037 (1933).
21. Mattin, H. E., and L. G. M. Timson. U.S. 2 392 936 (1946). *Chem. Abstr.* 40:1954 (1946).
22. Strauch, C. B. U.S. 2 043 633 (1936).
23. Instytut Farmaceutyczny, Polish 43 305 (1960). *Chem. Abstr.* 58:5461 (1963).
24. Clark, W. G. U.S. 3 384 541 (1968).

25. Sanders, P. A. *Principles of Aerosol Technology.* New York: Van Nostrand Reinhold. (1970).
26. Davis, T. L. *The Chemistry of Powder and Explosives.* New York: Wiley. (1943), p. 120.
27. Aero Research Ltd., N. A. DeBruyne, and D. A. Hubbard, Britain. 549 496 (1942).
28. Marsh, R. D. Britain. 551 488 (1943).
29. Wilcox, F. S. U.S. 3 007 840 (1962).
29a. Radvan, B., and A. P. J. Gatward. *Tappi* 55:748 (1972).
30. Grubbe, R. J. U.S. 3 483 049 (1969).
30a. Lemmen, W. Ger. Offen. 2 030 304 (1971). *Chem. Abstr.* 76, No. 75841 (1972).
31. Crowe, C. W., and R. G. Duranleau. U.S. 3 436 262 (1969).
32. Eastman Kodak Co., German 951 546 (1956). *Chem. Abstr.* 53:17739 (1959).
33. Dunning, H. N., et al., *Petrol. Engr.* 31, No. 12:B28 (1959).
34. Lissant, K. J., and G. J. Samuelson. *Oil Gas J.* 58, No. 23:124 (1960).
35. Eakin, J. L., A. D. Hopkins, and H. N. Dunning. *Oil Gas J.* 58, No. 33:162 (1960).
36. Schmidt, R. G. *Petrol. Engr.* 32, No. 13:B51 (1960).
37. Dunning, H. N., J. B. Thielen, and D. L. Anderson. *Oil Gas J.* 59, No. 49: 167 (1961).
38. Eakin, J. L., and R. W. Taliaferro. *Oil Gas J.* 60, No. 94:131 (1962).
39. Billings, W. E., and R. M. Lasater. U.S. 3 273 643 (1966).
40. Anderson, D. L. U.S. 3 174 929 (1965).
41. Rogers, R. H., and J. T. Foley. U.S. 3 229 777 (1966).
42. Holman, W. E., and J. T. Patton. U.S. 3 251 417 (1966).
43. Luchinski, G. P. *Zh. Fiz. Khim.* 13:302 (1939). G. P. Luchinskii and M. I. Popova. *Zavodsk. Lab.* 8:552 (1939).
44. Pozin, M. E., I. P. Mukhlenov, and E. Ya. Tarat. *Foam Gas Purifiers, Heat Exchangers, and Absorbers.* Leningrad: Goskhimizdat. (1959).
45. Tubolkin, A. F., I. P. Mukhlenov, and E. S. Tumarkina. *Zh. Prikl. Khim.* 41:787 (1968).
45a. Sasvári, Gy., F. Hites, and T. Blickle. *Acta Chim.* (*Budapest*) 70:395 (1971).
46. Bransky, D. W., and F. F. Diwoky. *Refiner Natural Gasoline Mfr.* 19:75 (1940). U.S. 2 311 625 (1943).
47. Bretsznaider, S., et al., *Przemyzl Chem.* 35:564 (1956).
48. Žurakowski, S., et al., *Przemyzl Chem.* 49:139 (1970).
49. Helsby, F. W., and D. C. P. Birt. *J. Appl. Chem.* (*London*) 5:347 (1955).
50. Maminov, O. V., and S. I. Usmanova. *Tr. Kazansko. Khim.-Tekhnol. Inst.* No. 37:291 (1968). *Chem. Abstr.* 72, No. 123403 (1970).
51. Žurakowski, S., and R. Glaser. *Prezmyzl Chem.* 44:510 (1965).
52. Banczyk, L. *Chemik* 19:401 (1966). *Chem. Abstr.* 66, No. 67105 (1967).
53. Micafil A. G., Swiss 231 777 (1944). *Chem. Abstr.* 43:3186 (1949).
54. Rodionov, A. I., and V. M. Radikovskii. *Zh. Prikl. Khim.* 41:82 (1968).
55. Weissman, E. Y., and S. Calvert. *A.I.Ch.E.J.* 11:356 (1965).
56. Reich, I. M., and W. R. Johnston. U.S. 2 788 276 (1957).
57. Mukhlenov, I. P., and V. Ya. Demshin. *Zh. Prikl. Khim.* 28:922 (1955).

58. Zemskov, I. F., A. S. Stepanov, and A. V. Melkikh. *Zh. Prikl. Khim.* 35: 2467 (1962).
59. Katunin, V. Kh., and A. Ya. Shlyapnikova. *Gidroliz. i Lesokhim. Prom.* 12, No. 7:10 (1959). *Chem. Abstr.* 54:3942 (1960).
60. Price, F. H. *Trans. Inst. Mining Engrs. (London)* 105:244 (1946). *Chem. Abstr.* 40:6236 (1946).
61. Rozenberg, B. A., et al., *Zh. Prikl. Khim.* 37:908 (1964).
62. Erler, H., R. Stenzel, and G. Fischer. *Freiberger Forschungsh.* 85B:119 (1965). *Chem. Abstr.* 64:1705 (1966).
63. Balakirev, A. A., and V. K. Tikhomirov. *Kolloidn. Zh.* 30:191 (1968).
64. Harris, J. C. in *Nonionic Surfactants.* M. J. Schick, ed. New York: Dekker. (1967), p. 705.
65. Silverman, L. U.S. 3 338 665 (1967).
66. Nesmelov, V. V. et al., *Maslob.-Zhir. Prom.* 24, No. 6:20 (1958). *Chem. Abstr.* 52:19099 (1958).
67. Triandafilidi, I. G. *Khim. i Tekhnol. Topliv i Masel* 3, No. 11:27 (1959). *Chem. Abstr.* 53:3674 (1959).
68. Maminov, O. V., et al., *Izv. Vysshikh Uchebn. Zaved., Khim. i Khim. Tekhnol.* 1958, No. 5:149. *Chem. Abstr.* 53:9632 (1959).
69. Danyushevskaya, R. G., V. V. Nesmelov, and Z. N. Grechukhina. *Maslob. Zhir. Prom.* 29, No. 4:26 (1963). *Chem. Abstr.* 59:6240 (1963).
70. Nesmelov, V. V., et al., *Izv. Vysshikh Uchebn. Zaved., Khim. i Khim. Tekhnol.* 1958, No. 6:108. *Chem. Abstr.* 53:13569 (1959).
71. Lebedeva, N. M., V. V. Nesmelov, and R. Sh. Latypov. *Khim. i Khim. Tekhnol. Topliv. i Masel* 10, No. 7:32 (1965). *Chem. Abstr.* 63:9722 (1965).
72. Kuznetsova, I. M., and O. V. Maminov. *Izv. Vysshikh Uchebn. Zaved., Khim. i Khim. Tekhnol.* 10:589 (1967). *Chem. Abstr.* 67, No. 92324 (1967).
73. Bheda, M., and D. B. Wilson. *Chem. Engr. Prog. Symp. Ser.* 65, No. 97:274 (1969).
74. Bartkowiak, A., S. Lambiris, and M. G. Zabetakis. *Combustion Flame* 3:347 (1959). *Chem. Abstr.* 54:860 (1960).
75. Faber, L. D., and C. J. Carrell. U.S. 1 948 568 (1934).
76. Haefly, H. *Ind. Textile* 55:84 (1938).
77. Tarassuk, N. P., and E. N. Frankel. *J. Dairy Sci.* 38:438 (1955).
78. Davenport, J. E., J. G. Getsinger, and F. Carroll. *Ind. Eng. Chem., Process Des. Develop.* 4:84 (1965).
79. Patterson, G. G., J. R. Gahan, and W. C. Scott. *Ind. Eng. Chem. Process Des. Develop.* 6:393 (1967).
80. Bond, D. C., and O. C. Holbrook. U.S. 2 866 507 (1958).
81. Bond, D. C., and G. G. Bernard. U.S. 3 318 379 (1967).
82. Bernard, G. G., and L. W. Holm. U.S. 3 342 256 (1967).
83. Craig, F. F., and J. L. Lummus. U.S. 3 185 634 (1965).
84. Thiegs, B. J., and N. Wright. U.S. 2 875 555 (1959).
85. Perri, J. M., and C. Conway. *Ind. Eng. Chem.* 48:2021 (1956).
86. Gooderham, W. J. *Analyst* 72:520 (1947).
87. Gooderham, W. J. *Anal. Chim. Acta* 2:452 (1948).
88. Matsuura, N. *J. Chem. Soc. Japan, Pure Chem. Sect.* 71:168, 375, 458 (1950). *Chem. Abstr.* 45:4600, 6527 (1951).

89. *Standard Methods for the Examination of Water and Sewage.* New York: American Public Health Association. (1925), 6th ed., p. 28.
90. Zabrodskii, A. G., and A. F. Polozhishnik. *Kolloidn. Zh.* 15:238 (1953).
91. Bikerman, J. J. Instrument Society of America, Paper No. 54-13-4 (1954).
92. Satterly, J. *Am. J. Phys.* 19:448 (1951).
93. Dewar, J. *Proc. Roy. Inst. Gt. Brit.* 24:197 (1925).
94. F. Brukner and Arca-Regler A.-G., German 550 238 (1932). *Chem. Abstr.* 26:4113 (1932).
95. Schiebl, K. *Z. Zuckerind. Čechoslovak. Rep.* 55:476 (1931). *Chem. Abstr.* 26:1469 (1932).
96. Carter, R. P., and F. T. Parkinson. U.S. 3 154 577 (1964).

Indices

Author Index

Subject Index

Absorption in foam 311
Accumulation in foam 264, 268
    batch process 273
    continuous process 273
    multistage process 278
    and time of drainage 269, 287
Accumulation ratios 269
Acidity 105, 106, 110, 113, 126
Adhesives 310
Adsorption, amount 266
    equation 235, 241, 266
"Aerosol foams" 48, 308
Aerosols 312
Aging of foams 209, 251
Alcohols 79, 91, 102, 152
Alkane sulfonates 9, 85, 116
Alkyl benzene sulfonates 82, 87, 117,
    135, 177, 281, 299
    separation from sulfates 295
Alkyl sulfates 17, 22, 26, 28, 82, 85,
    113, 177, 188, 233, 283
    purification 293
Ammonium soaps 113
Animal fluids 128
Attachment of lamellae to solids 204
Attraction of bubbles 34

Bacteria 299
Beer 179, 252
    foam adhesion 254
    foam control by trace metals 253
Black films 25
    and composition 27, 28
    multiple 27
Boiling, nucleate 53
Boiling liquids, film persistence 66
    foam persistence 100, 156
Brittleness of films 236
Brush method 88
Bubble chain 201
Bubble column 48
Bubble diameter 189, 267, 271
    see also Bubble volume
Bubble formation, by agitation 48
    in chemical reactions 52
    at multiple orifices 38, 47, 81
    at single orifices 33
    at solid surfaces 52, 54
    and solubility 51, 52

Bubble pressure 40, 44, 47, 194
Bubble rise 2
Bubble shape 35, 46
    and diffusion 195, 197
    in foam 47, 62
    in gas emulsions 4
    under a film 17, 57
Bubble shrinkage 193
Bubble splitting 35, 49
Bubble volume, in boiling 53
    distribution 191, 197
    and drainage 172
    in foams 82, 85, 189, 198
    and frequency 36, 41, 45, 47, 48,
    53
    homogenizing 49
    maximum 55
    minimum 41
    and persistence 71, 94
    and pressure 4, 45
    and rate of bubbling 39, 41
    and temperature 37
    and turbulence 38, 39, 42, 44
    and wetting 36, 54
Bubbles, attraction 59
    contact with solids 56, 93
    contact of two 60, 75, 193
    at interfaces 57, 69
    persistent 127
    rearrangement 193, 197
    small 38
    under films 17, 57
Bubbling 42
    high speed 39, 44

Capillary attraction 59
Capillary pressure 3, 7, 10, 59, 193
    and bubble scission 35, 49
    and drainage 170
    and film thickness 194
    in foams 194, 199
    and nucleation 51
    and yield stress 206
Carry-over 247
Catenary 7
Catenoid 7
Cationic surfactants 22, 122, 177, 311
    accumulation in foam 280, 286,
    289